Agroecology and Agroecosystems

Agroecology and Agroecosystems

Edited by **Molly Ismay**

New York

Published by Callisto Reference,
106 Park Avenue, Suite 200,
New York, NY 10016, USA
www.callistoreference.com

Agroecology and Agroecosystems
Edited by Molly Ismay

International Standard Book Number: 978-1-63239-678-5 (Hardback)

Printed in China.

Contents

Preface

Agroecology studies the processes related to agriculture in an ecological system whereas agroecosystem is the fundamental unit of it. These systems are not limited to farming but also contain surroundings regions affected by these activities for example forest gardens. This book presents topics of key importance such as agricultural commercialization, climate change, small scale irrigation technologies, biotic and abiotic constraints, seed germination, etc. It also covers in detail some existent theories and innovative concepts revolving around these disciplines. Coherent flow of topics, student-friendly language and extensive use of examples make this book an invaluable source of knowledge. It is a beneficial text for ecologists, environmentalists, agriculturists, professionals and students alike.

This book has been the outcome of endless efforts put in by authors and researchers on various issues and topics within the field. The book is a comprehensive collection of significant researches that are addressed in a variety of chapters. It will surely enhance the knowledge of the field among readers across the globe.

It gives us an immense pleasure to thank our researchers and authors for their efforts to submit their piece of writing before the deadlines. Finally in the end, I would like to thank my family and colleagues who have been a great source of inspiration and support.

Editor

1

An Inventory of Flora in Um Dom Island (Khartoum State), Sudan

N. Mahmoud[1], M. S. Elhakeem[2], A. H. Abdallah[2*] and M. Kordofani[2]

[1]National Center for Research, Khartoum, Sudan.
[2]Department of Botany, Faculty of Science, Khartoum University, Sudan.

Authors' contributions

This work was carried out in collaboration between all authors. Authors MSE, NM, AHA and MK designed the study, wrote the protocol, collected samples and wrote the first draft of the manuscript. Author AHA managed the literature searches, analyses of the study and discuss the conclusion. All authors read and approved the final manuscript.

Editor(s):
(1) Maria Panitsa, Department of Environmental and Natural Resource Management, University of Patras, Greece.
(2) Abdel-Tawab H. Mossa, Environmental Toxicology Research Unit (ETRU), Pesticide Chemistry Department, National Research Centre, Egypt.
Reviewers:
(1) Chapungu Lazarus, Great Zimbabwe University, Zimbabwe.
(2) Anonymous, Chinhoyi University of Technology, Zimbabwe.
(3) Manuel Casares-Porcel, Granada University, Spain.
(4) Denise F. Dostatny, Plant Breeding and Acclimatization Institute, Błonie, Poland.
(5) Ana Maria Arambarri, La Plata National University, Argentina.
(6) Anonymous, University of Lome, Togo.

ABSTRACT

This study documents and updates the flora information of Um Dom Island, Khartoum State, Sudan. Herbarium specimens were prepared and deposited in the Herbarium of the Faculty of Science, University of Khartoum. Identified species were compared with specimens from the same Herbarium. This study showed that there are seventy four flowering plant species belonging to thirty three different families – there were no non-floral varieties identified. Three layers of flora have been identified in the area under study. The upper most, tree layer is dominated by *Acacia nilotica, Ziziphus spina-christi, Faidherbia albida*, and *Prosopis juliflora*. The second, shrub layer, is occupied by mainly *Calotropis procera*, *Tamarix nilotica,* and the third, a herb layer, is dominated by *Cyperus* sp. *Cenchrus prieurii.* Families with high diversity are Poaceae, Cyperaceae and

**Corresponding author: Email: amel762003@yahoo.com*

Amaranthaceae. This study is regarded as a departure point and reference for future studies, to assist and show the change in plant cover which ultimately may lead to land degradation and an arid environment.

Keywords: Floristic composition; flora of Um Dom Island; Blue Nile; Khartoum State area.

1. INTRODUCTION

Different parts of Khartoum State area have been studied by many authors in the past, including the vegetation of the semi desert north east of Khartoum by [1] and Omdurman desert by [2], in addition to some other simple contributions to the flora of Khartoum. There is a lack of comparative and related research into the floral composition of and around Khartoum State, which has resulted in a lack of supporting data for this study. Exceptions to this are [3] who studied the ecology and seed bank of some weeds of arable lands in Um Dom town, Khartoum State and [4] studied the flora of Tuti Island in Khartoum State.

Studies of regional flora of the Sudan include: the flowering plants of Northern and Central Sudan by [5]; the flora of Erkawit, Eastern Sudan by [6]; The flora of central Sudan by [7]; Important Trees of Northern Sudan by [8]; In addition to Ahti et al. [9] who studied the flora of the area around Wadi Halfa submerged by the Aswan Dam; also [10] made an illustrated guide to the plant of Erkawit, Eastern Sudan; [11] studied the flora of Jebel Marra, Western Sudan; and Braun et al. [12] produced an illustrated (drawings and photographs) account of the common weeds of central Sudan. In addition, a recent study of the Plants of Sudan and South Sudan is recorded byDarbyshire et al. [13].

Um Dom Island can be classified as one of the partly stabilized islands that were formed during the annual flooding 1951-1957. Um Dom Island is partially or wholly covered by the annual rise and fall of the Blue Nile rather than by rain. The influence of soil texture, soil stabilization and human interference are very clear. As the flood waters recedes numerous seedlings begin to appear. These belong mainly to various species of annuals plant which constitutes the major parts of vegetation, [14].

Previous studies of regional flora of Khartoum State have covered mainly different land areas around the capital city. However until now there has been no first-hand documentation and investigation into the floristic composition of Um Don Island, which is a relatively new topographical feature. This study is the first of its kind and as such it is a timely addition and contribution to changes the understanding of the flora of Sudan. This will help in further studies on floristic and the impact of seasonal flooding upon flora on Um Dom Island. The aim of this study is to document the present flora of Um Dom Island (January-December 2007) as well as to compare it with the regional floristic studies of Khartoum State.

2. MATERIALS AND METHODS

2.1 Description of the Study Area

2.1.1 Location of the study area

The study area Um Dom Island is located on the Blue Nile about 20 km south east Khartoum (32°-30̀ and 34° -45̀ east, 15° -30̀ and 16° -40̀ North). The approximate size of Um Dom area is about 6 km^2. Um Dom Island, the area under study, is located in the western part of the Um Dom area on the Blue Nile. The size of Um Dom town is 3.19 km^2, the settlement is located to the east of Um Dom area on the eastern bank of the Blue Nile, whilst the size of Um Dom Island is 2.95 km^2, in the western part of the area, with no permanent settlement on the island.

The waters of the Blue Nile come from a mountainous catchments area in the Ethiopian plateau and flows down towards Sudan at a relatively steep incline (1:10,000). The river has steep-sided channel that is 300-500 meters wide. The level of high water during the flood season (July, August and September) and that of low water (April) may vary by as much as 10 meters. At the time of flooding, the river is heavily laden with silt and sand [15] (Map 1).

2.1.2 Climate

The area of Um Dom Island according to [1] belongs to the dry tropics, characterized by dry mild winters (November–March) and hot rainy summers (April –October). The mean maximum temperature during winter in Khartoum State is 22C° and the minimum temperature is 5C° whilst during summer the maximum is 45C° and the minimum is 32C° (Sudan Metrological Authority). The weather in Um Dom Island in winter is cold and clear with a relative humidity 20% during the summer months the humidity is at 3%. The weather in Khartoum State during June when the south-west wind starts to blow, is often cloudy and the relative humidity as high as 40%. These conditions prevail until mid-September. The average annual rainfall of Khartoum during this period is 100-200 mm.

2.1.3 Geology

According to [16] in Sudan there are mainly four stable geological formations stretching across different geological eras. These are: 1. Sedimentary 2. Basal rock 3. Nubian rocks and 4. Island formation [17]. Most of the eastern part of the river Nile is covered by two types of Nubian formation: Nubian rocks, which are mainly formed by sandy rocks, and sedimentary soil, which lies between the Nubian rocks and the river bank. Sedimentary soil covers both sides of the Blue and the White Nile [18].

The topography of the area which lies to the eastern part of the Nile is characterized by a gentle gradient, north to south, towards the Blue Nile. The altitude of this area is 383m above sea level. The soil here is similar to areas within Khartoum State. The soil of Um Dom area is characterized by four types of soil; 1.clay soil, 2. colloidal, 3. silt and 4. dry semi desert soil. The clay soil is characterized by its high alkalinity and undergoes sever fissuring during the dry season, while the texture of colloidal soil is mainly sandy with high alkaline characteristics.

The silt soil is characterized by a high content of sedimentary soil which is deposited by the flooding of the river. Semi desert soil covers the area away from the river Nile in many directions.

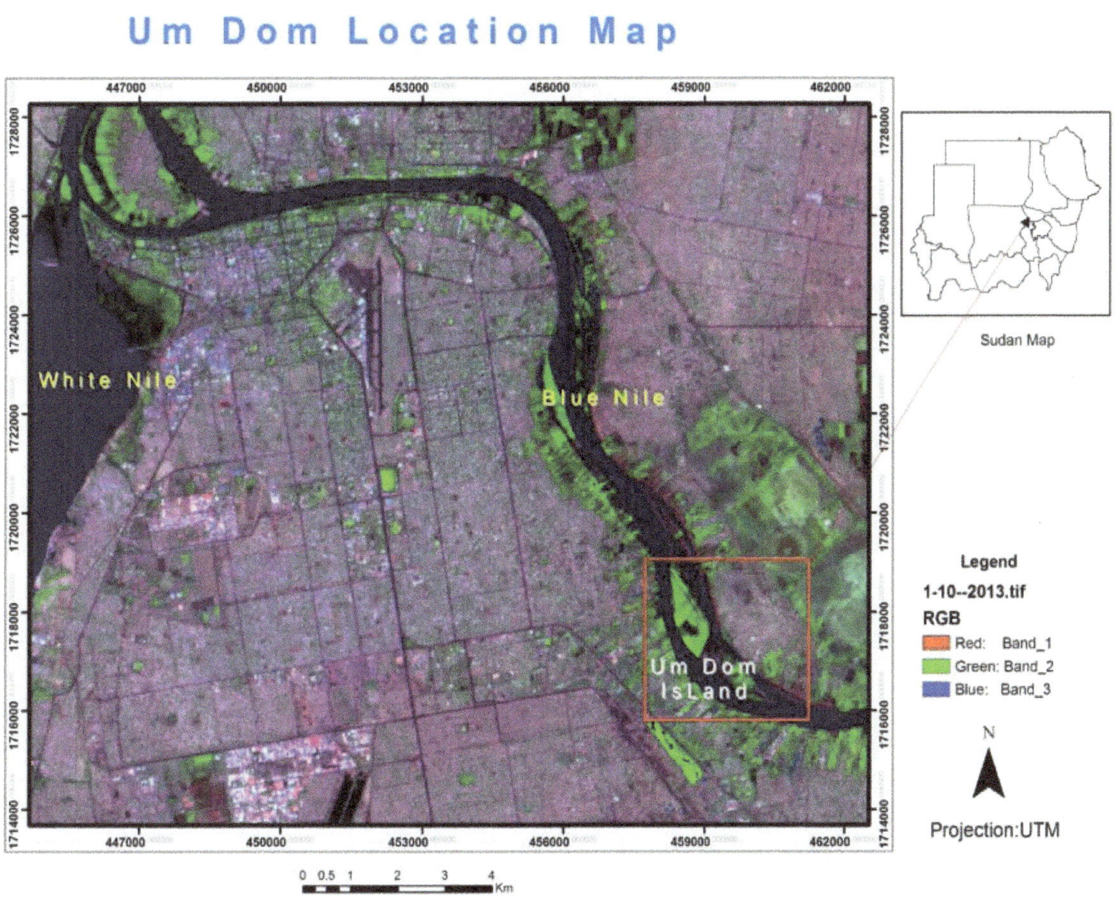

Map 1. Location of the study area
Source: Remote sensing authority. Sudan

2.1.4 Vegetation

According to [19] 72,000 square miles of semi-desert in Sudan belong to subdivision *Acacia tortilis –Maerua crassifolia.* The main reference for studying Khartoum State vegetation history is [19] who divided Khartoum State into two vegetation regions; flood plains and desert plain. The vegetation of Um Dom Island lies on the flood plain vegetation region.

Um Dom Island is covered with annual herbs and very dense trees (Plate 1). However, overgrazing has substantially affected the vegetation of the area [17], Dominant trees on the Island include: *Acacia nilotica*, *Ziziphus spina-christi*, *Faidherbia albida*, and *Prosopis juliflora* while the dominant herbs include: *Xanthium pungens*, and *Cenchrus prieurii.*

2.2 Data Collection and Species Identification

During a course of one year from January to December 2007 at three months intervals live plant specimens were collected from Um Dom Island and were mounted on herbarium sheets. Whole plant herbarium specimens were collected in the case of herbs, whereas twigs with leaves and flowers and/or fruits were collected from shrubs and trees and also mounted on Herbarium sheets. Identification of all materials was carried out using the available data relevant to African floras with a special attention to scientific publications on Sudan and the neighboring countries. Species were firstly verified using available taxonomical references: [5,20,6,21,22,23,24,11,25-31] and Braun et al. [12].

The collected and identified species were compared with identified herbarium specimens from the Faculty of Science Herbarium University of Khartoum. The nomenclature was updated according to Tropicos.org (Missouri Botanical Garden, USA) and the International Plant Names Index (IPNI) UK (Consulted: April 2015). In Tables 1 and 2, the species are treated in alphabetical order of scientific names (second column).

(1)

(2)

Plate 1. The study area Um Dom Island

Table 1. Flora of Um Dom Island

No.	Family	Scientific name	Vernacular name	Phenology
1	Malvaceae	*Abutilon figarianum* Webb	Hambouk/Gargadan	Herb
2	Mimosaceae	*Acacia nilotica* (L.) Willd. ex Delile	Sunt	Tree
3	Mimosaceae	*Acacia tortilis* Hayne subsp. *tortilis*	Samar	Tree
4	Amaranthaceae	*Aerva javanica* (Burm. f.) Juss.	Ras Elshaip	Herb
5	Amaranthaceae	*Alternanthera nodiflora* R.Br.	N.A	Herb
6	Amaranthaceae	*Amaranthus graecizans* L.	Lissan el Tair Saghir/Fissa elkalb	Herb
7	Amaranthaceae	*Amaranthus spinosus* L.	N.A	Herb
8	Lythraceae	*Ammannia multiflora* Roxb.var *parviflora*. (DC.) Koehne	N.A	Herb
9	Papaveraceae	*Argemone mexicana* L.	khshkhash	Herb

No.	Family	Scientific name	Vernacular name	Phenology
10	Lamiaceae	*Basilicum polystachion* Moench	Reihan	Herb
11	Elatinaceae	*Bergia suffruticosa* (Del.) Fenzl	N.A	Herb
12	Asclepidaceae	*Calotropis procera* (Aiton) R. Br.	Oshar	Shrub
13	Sapindaceae	*Cardiospermum halicacabum* L.	N.A	Herb
14	Poaceae	*Cenchrus prieurii* (Kunth)Maire.	Haskaneet	Herb
15	Euphorbiaceae	*Chrozophora plicata* (Vahl) A. Juss. ex Spreng	Al-gobera,al-argassi.	Herb
16	Cleomaceae	*Cleome gynandra* L.	Tamaleca	Herb
17	Convolvulaceae	*Convolvulus arvensis* L.	N.A	Herb
18	Tiliaceae	*Corchorus fascicularis* Lam.	N.A	Herb
19	Tiliaceae	*Corchorus depressus* (L.) Christens	Suteb/Um Shiteih	Herb
20	Fabaceae	*Crotalaria senegalensis* (Pers.) Bacle ex DC.	Sufera- saghira	Herb
21	Menispermaceae	*Cocculus pendulus* (J.R. Forst. & G. Forst.) Diels	N.A	Climber herb
22	Cucurbitaceae	*Cucumis melo* var.*agrestis* Naudin	Humeid/Agor	Climber herb
23	Cucurbitaceae	*Cucumis prophetarum* L.	Habashkol	herb
24	Convolvulaceae	*Cuscuta hyalina* Roth	Hamool/Silik	herb
25	Poaceae	*Cynodon dactylon* (L.) Pers.	Nageela	Herb (grass)
26	Cyperaceae	*Cyperus alopecuroides* Rottb.	N.A	Herb
27	Cyperaceae	*Cyperus bulbosus* Vahl.	N.A	Herb
28	Cyperaceae	*Cyperus digitatus* Roxb.	N.A	Herb
29	Cyperaceae	*Cyperus rotundus* L.	siada	Herb
30	Cyperaceae	*Cyperus vistatus* var *rigritanus*	Kelo-elatoud	herb
31	Poaceae	*Dactyloctenium aegyptium* (L.) Richt.	Um Assabi / Koreib	Herb
32	Solanaceae	*Datura stramonium* L.	Sakaran	Herb
33	Poaceae	*Dinebra retroflexa* (Vahl.)Panz.	N.A	Shrub
34	Poaceae	*Echinochloa colona* (L.) Link	Defra	Herb
35	Poaceae	*Eragrostis namaquensis* Nees ex Schrad.	Eish alfar	Herb
36	Euphorbiaceae	*Euphorbia aegyptiaca* Boiss.	Um Malbeina/Umleban	Herb
37	Euphorbiaceae	*Euphorbia hirta* L.	Um el labaina	Herb
38	Mimosaceae	*Faidherbia albida* (Delile) A. Chev	Haraz	Tree
39	Cyperaceae	*Fimbristylis biumbellata* Bubani	N.A	Herb
40	Boraginaceae	*Heliotropium supinum* L.	Danab el Agrab	Herb
41	Malvaceae	*Hibiscus panduriformis* Burm. f.	N.A	Shrub
42	Fabaceae	*Indigofera oblongifolia* Forsk.	Dahaseer	Shrub
43	Onagraceae	*Jussiaea erecta* L.	N.A	Herb
44	Asclepiadaceae	*Leptadenia heterophylla* (Del.) Decne.	Lwise	Twiner over shrub
45	Asclepiadaceae	*Leptadenia lancifolia* (Schumach.) Decne.	Lwise	Herb or Shrub
46	Cucurbitaceae	*Luffa cylindrica* (L.) M. Roem.	Leef	Herb
47	Convolvulaceae	*Merremia pentaphylla* (L.) Hallier f. basionimo *Convolvulus pentaphyllus* L.	N.A	herb
48	Caesalpinaceae	*Parkinsonia aculeate* L.	Seisaban	Shrub
49	Acanthaceae	*Peristrophe bicalyculata* Nees	El-nada, Hersha, Murkab	herb
50	Solanaceae	*Physalis angulata* L.	N.A	herb

No.	Family	Scientific name	Vernacular name	Phenology
51	Ploygonaceae	*Polygonum cespitosum* Blume	Safsaf	herb
52	Mimosaceae	*Prosopis juliflora* (Sw.) DC.	Mesquet	Tree
53	Fabaceae	*Rhynchosia minima* (L.) DC. var. *memnonia* (Delile) Meikle	Adan El far	Sub-Shrub
54	Euphorbiaceae	*Ricinus communis* L.	Khirwia	Sub Shrub
55	Poaceae	*Setaria acromelaena* (Hochst.) T. Durand & Schinz	Lossieg	Herbs
56	Poaceae	*Setaria verticillata* (L.) Beauv.	Lsseig/Ramida/umS ekkin	Herbs
57	Malvaceae	*Sida alba* L.	Um Hebiba / Um Shidayda	Herbs
58	Poaceae	*Sorghum arundinaceum* (Desv.) Stapf.	Adar	Herbs
59	Tamaricaceae	*Tamarix nilotica* (Ehrenb.) Bunge	Tarfa	Shrub or Small tree
60	Fabaceae	*Tephrosia apollinea* Klotzsch	Um ragiga	Herb
61	Aizoaceae	*Trianthema sedifolia* Viv.	Tarapa	Shrub
62	Saxifragaceae	*Vahilia dichotoma* (Murray) Kuntze	N.A	Herb
63	Asteraceae	*Xanthium brasilicum* Vell.	Rantouk	Herb
64	Asteraceae	*Xanthium pungens* Wallr.	Rantouk	Herb
65	Aizoaceae	*Zaleya pentandra* (L.) C.Jeffrey.	Al-raba'a	Herb
66	Rhamnaceae	*Ziziphus spina-christi* (L.) Desf.	Al-Sidir /Al-Nabag	Tree

Table 2. Cultivated species of Um Dom Island

No.	Family	Cultivated species	Vernacular name	Life form
67	Arecaceae	*Phoenix dactylifera* L.	Nakheel, Balah	Tree
68	Poaceae	*Sorghum vulgare* Pers.	Dura	Herb
69	Myrtaceae	*Psidium guajava* L.	Guava	Tree
70	Rutaceae	*Citrus limon* (L.) Osbeck	Limon	Tree
71	Fabaceae	*Phaseolus vulgaris* L.	Fasulia	Herb
72	Meliaceae	*Azadirachta indica* A. Juss.	Neem	Tree
73	Tiliaceae	*Corchorus olitorius* L.	Khudra/Mulokhia	Herb
74	Fabaceae	*Pisum sativum* L.	Bisila	herb

3. RESULTS AND DISCUSSION

The collected and presented flora shows that Um Dom Island is rich in flora, this may be attributed to its positioning on the Blue Nile and it being surrounded by water unlike Khartoum city which is located in the zone of semi desert with different type of *Acacia* sp. A total of 74 plants species belonging to 33 different families were recorded: sixty six were weedy and native flora to Um Dom Island Table 1. Eight species were cultivated species Table 2. Thirty families of which were Dicotyledons and three Monocotyledons.

Three layers of flora were identified in the area under study. The upper most, tree layer is dominated by *Acacia nilotica, Ziziphus spina-christi*. The middle shrub layer is consists mainly of *Calotropis procera*, *Tamarix nilotica* and the third, a herb layer, is dominated by *Cenchrus prieurii* and *Cyperus* spp. Families with high diversity are *Poaceae, Cyperaceae* and *Amaranthaceae* (Plate 2). Results show that Um Dom Island has a moderate vascular plant species diversity. This can be attributed to the location of the Island in the Blue Nile and the emergence of plant seedlings after annual flooding. A total of 89.2% of plant are spontaneous species and 10.8% are cultivated, for instance, *Tamarix nilotica* is an indigenous plant to the island while *Calotropis procera* has been introduced from the surrounding semi desert area.

Leptadenia heterophylla

Luffa cylindrica

Amaranthus spinosus

Alternanthera nodiflora

Pisum sativum

Leptadenia lancifolia

Merremia pentaphylla

Convolvulus arvensis

Calotropis procera

Phoenix dactylifera

Cucumis melo

Sorghum vulgare

Plate 2. Some of Um Dom Island flora

The geographical origin of the plants present on the island is diverse. For example there are seven different plant families identifies on Um Dom Island that originate in different parts of Sudan. Members of the family Poacae: *Echinochloa colona, Dactyloctenium aegyptium, Dinebra retroflexa* and *Cynodon dactylon* species, are widespread in different parts of Sudan, whereas the *Eragrostis naqamuensis* species is distributed in Equatoria State. The family Cyperaceae: species *Cyperus alopecuroides* is spread in Central and Southern Sudan and the *Cyperus bulbosus* species is distributed in Blue Nile State and Red Sea State. The family Asteraceae: the *Xanthium brasilicum* species is spread in Central Sudan. For the family Convolvulaceae: the *Merremia pentaphylla* and *Convolvulus arvensis* species have their origins in Northern and Central Sudan and the *Cuscuta hyaline* species originates from the Kordofan region. The family Solanaceae: species *Physalis angulata* is distributed in Kordofan. For the family Acanthaceae: the species *Peristrophe bicalyculata* is widespread throughout Sudan. Finally the family Arecaceae: the species *Phoenix dactylifera* is spread in Northern Sudan along the Nile river bank.

The middle part of Um Dom Island is given over to agricultural activities after the annual flooding period with different plants cultivated such as *Sorghum vulgare, Pisum sativum, Corchorus olitorius* and some fruit trees are planted like *Phoenix dactylifera* and *Psidium guajava*. Livestock, camels and goats, belonging to local people from the surrounding area graze on herbaceous like *Cynodon dactylon, Cyperus* sp. and *Sorghum vulgare* after harvesting have help in transferring species to and from the island along with water and wind transference seeds. Some species are not palatable such as *Abutilon figarianum, Aerva javanica, Argemone mexicana* and *Xanthium brasilicum* (Plate 1).

Previous research has studied the flora around Khartoum State. Eshak [3] studied the ecology and seed bank of some weeds of arable lands in Khartoum State. Eshak's study found that the floristic composition and diversity of Um Dom town occupied an intermediate position amongst the four sites studied throughout Khartoum State, and she also found that the species diversity of Um Dom town is 82 plant species. It is clear from her study that the arable land of Khartoum State is rich in weed flora and the species diversity correlates to the type of soil, agricultural practices and crops planted in Um Dom Town.

Abdelmageed [4] studied the flora of Tuti Island in Khartoum State, she found a total number of 130 species belonging to 107 genera and 40 families were collected through October 2004-2007.

4. CONCLUSION

This study reveals that Um Dom Island has a medium diversity of indigenous flora, some are spontaneous and some are introduced their geographical origins are widespread throughout Sudan. Um Dom Island is partly fixed or stable and is marked by a finer soil type. Um Dom Island can be reached from the eastern Blue Nile bank in winter without the need of a boat, local people are able to transfer their livestock from the town to the island during this period. During our study it was observed that human interference by overgrazing and cultivation are the most important factors that affect the vegetation of the area.

5. RECOMMENDATION

This work is regarded as the base for future studies to show the change in plant cover which ultimately leads to land degradation and arid environment.

ACKNOWLEDGEMENT

Really we are very grateful to all botany department members and to Dr. Dafaalla Ali Ibrahim in particular for their kind support.

COMPETING INTERESTS

Authors have declared that no competing interests exist.

REFERENCES

1. Halwagy R. The vegetation of the semi desert north east of Khartoum, Sudan. Oikos, Copenhagen. 1961;12(1):110.
2. Kassas MA. Land forms and plant cover in the Omdurman desert. Bull. Soc. Geogr.d, Egypt. 1955;29:43-58.
3. Eshak MM. The ecology and seed bank of some weeds of arable lands in Khartoum state. MSc. Thesis University of Khartoum; 2002.
4. Abdelmageed SA. Study on the flora of Tuti Island in Khartoum State. PhD. Thesis. Khartoum University; 2007.

5. Crowfoot GM. Flowering plants of northern and central Sudan. The Orphans Printing Press: Broad Street. London; 1928.
6. Andrews FW. The Flora of Erkwit. Trees and Shrubs. Forests Bull. No. 1, Khartoum; 1947.
7. Obeid M, Mahmoud A. The vegetation of Khartoum province. Sudan Notes and Records. 1968;50:134-159.
8. Sahni KC. Important trees of northern Sudan. United Nation and FAO: Khartoum; 1968.
9. Ahti T, Hamet-Ahti L, Petterson B. Flora of the inundated Wadi Halfa reach of the Nile, Sudanese Nubia, with notes on adjacent areas. Ann-Bot. Fenn. 1973;10:131-162.
10. Hassan HM. An illustrated guide to the plants of Erkawit, Khartoum University Press: Sudan; 1974.
11. Wickens GE. The flora of Jebel Mara and its geographical affinities. Her Majesty's a Stationery Office: London; 1976.
12. Braun M, Burgstaller H, Hamdoun AM, Walter H. Common weeds of central Sudan. GTZ: Eschbron; 1991.
13. Darbyshire I, Kordofani M, Farag I, Candiga R, Pickering H. The plant of Sudan and South Sudan Anannoted checklist. Kew: Royal Botanical Garden; 2014. ISBN: 978-1-84246-473-1.
14. Halwagy R. Studies on the succession of vegetation on some Islands and sand banks in the Nile near Khartoum, Sudan. Oikos, Copenhagen; 1963.
15. Allan WN, Smith RJ. Irrigation in the Sudan. In Tothil JD, editor. Agriculture in Sudan. Oxford University press. London; 1948;593-632.
16. Whiteman AJ. The geology of Sudan republic. Clarendon press: Oxford University Press. 1971;290.
17. Abu Sin MEI. The future of Sudan capital province. Khartoum University Press, Khartoum; 1991.
18. Andrew G. Geology of the Sudan. In: Tothil JD, (editor). Agriculture in the Sudan, London, Oxford University Press; 1948.
19. Harrison MN, Jackson JK. Ecological Classification of the vegetation of the Sudan. Forests Bull. No. 2, Khartoum; 1958.
20. Broun AF, Massey RE. Flora of the Sudan. Welling to House: London; 1929.
21. Andrews FW. The flowering plant of the (Anglo-Egyptian) Sudan. Vols. 1, 2, 3. Buncle & Co. Ltd., Arbroath; (1950, 1952, 1956).
22. Boulos L, El Hadidi N. Common weeds in Egypt. Dar-El-Maaref Press: Cairo; 1967.
23. Hutchinson J, Dalziel JM. Flora of West Tropical Africa – Vol. 1, 2, 3 in FN Hepper- Crown Agents of Overseas Government Administration, editor. London; 1954 – 1958, 1963, 1972.
24. Tackholm V. Students Flora of Egypt, eds. 1, 2 Cairo University Press, Cairo; 1974.
25. Hafliger E, Scholz H. Grass weeds 1,2. CIBA-GEIGY Ltd., Basle: Switzerland; 1980, 1981.
26. Whistler WA. Weed Handbook of Western Polynesia. GTZ: Eschbron; 1983.
27. Thirakul S. Manual of dendrology – Bahr El Ghazal and Central Regions, Sudan. Groupe Poulin, Theriault Ltee, Quebec; 1948.
28. Daoud HS. Flora of Kuwait. Dicotyledoneae. KPI, Ltd., London. 1985; 1.
29. Chaudhary SA. Grasses of Saudi Arabia. Ministry of Agriculture and Water: Riyadh; 1989.
30. El Amin HM. Trees and Shrubs of the Sudan. Ithaca Press: Exeter; 1990.
31. Turril WB, Milne-Redhead E, editors. Flora of tropical East Africa. Crown Agents of Overseas Government Administration: London; (1952-1991).

2

Selection Criteria for High Nitrogen Use Efficiency in Wheat (*Triticum aestivum* L.) Parents and their F_1 and F_2 Progenies

A. M. M. Al-Naggar[1*], R. Shabana[1], M. M. Abd El-Aleem[2] and Zainab El-Rashidy[2]

[1]Department of Agronomy, Faculty of Agriculture, Cairo University, Giza, Egypt.
[2]Wheat Research Department, FCRI, Agricultural Research Centre (ARC), Giza, Egypt.

Authors' contributions

This work was carried out in collaboration between all authors. Author AMMAN designed the study, wrote the protocol, produced the initial draft of manuscript and interpreted the data. Author ZER anchored the field study, gathered the initial data and performed preliminary data analysis. Authors RS and MMAEA managed the literature searches. All authors read and approved the final manuscript

Editor(s):
(1) Mirza Hasanuzzaman, Department of Agronomy, Faculty of Agriculture, Sher-e-Bangla Agricultural University, Bangladesh.
(2) Abdel-Tawab H. Mossa, Environmental Toxicology Research Unit (ETRU), Pesticide Chemistry Department, National Research Centre, Egypt.
Reviewers:
(1) Rezzoug Waffa, Ibn Khaldoun University, Algeria.
(2) Mirella Sari Gorla, University of Milano, Italy.

ABSTRACT

Developing nitrogen use efficient (NUE) genotypes of wheat could reduce crop N fertilizer requirement. However, selection for NUE alone can be difficult and less efficient than associating it with one or more of selection criteria. The aim of this study was to determine selection criteria for improving NUE in wheat genotypes and their F_1 and F_2 progenies under contrasting N environments. The results of this study indicated that high nitrogen uptake efficiency (NUPE), grain yield/plant (GYPP), grains/spike (GPS), spikes/plant (SPP), and grain protein content (GPC) of parents, high NUPE, GYPP, biological yield/plant (BYPP), GPC and GPS of F_1's and high NUPE, harvest index (HI), BYPP, SPP, GPC and GPS of F_2's have significant and strong correlation coefficients with high nitrogen use efficiency (NUE). It is observed that NUPE is commonly correlated with NUE in all studied genotypes (parents, F_1's and F_2's). Similarly, GPS is also strongly

*Corresponding author: Email: medhatalnaggar@gmail.com

correlated with NUE. Moreover, SPP is strongly correlated with NUE in parents and F_2's. A negative and significant correlation existed between NUPE and NUTE for parents under high-N (-0.82), F_1's under high-N (-0.73) and low-N (-0.36) and F_2's under high-N (-0.91) and low-N (-0.81). The traits NUPE, GYPP, SPP and GPS showed high heritability in narrow sense and could therefore be offered to wheat breeders for use in breeding programs as selection criteria for improving nitrogen use efficiency.

Keywords: Correlations; bread wheat; NUE; secondary traits; heritability; N-uptake.

1. INTRODUCTION

Low soil nitrogen is among the major abiotic stresses threatening cereal production and limiting food security and economic growth [1]. The incidence of low-N stress may increase, due partly to global climate changes, declines in soil organic matter and reduction in soil fertility and water holding capacity [2]. One approach to reducing the impact of N deficiency on wheat production may be to select cultivars that are superior in N-use efficiency, either due to enhanced uptake capacity or because of more efficient use of absorbed N in grain production [3].

Nitrogen use efficiency has been described in various ways, but these definitions generally describe two types of efficiency, either uptake efficiency or utilization efficiency. Nitrogen uptake efficiency (NUPE) has been defined as total plant nitrogen content per unit content per unit N available in the soil. Nitrogen utilization efficiency (NUTE) was described by Moll et al. [4] as the grain production per unit N concentration in the plant. They recommended the development of genotypes with both high uptake and high utilization efficiencies.

Various studies worldwide have identified genetic associations between grain yield and NUE components under contrasting N conditions. In general, these studies indicated that NUPE accounts for more of the genetic variation in NUE at low-N than at high N supply, e.g. amongst 10 spring wheat cultivars in Mexico [5] and 20 winter wheat cultivars in France [6]. However, Dhugga and Waines [7] comparing 12 spring wheats in California, found that NUPE was the most important component of NUE at all N levels.

Genetic gains in NUE with breeding under low N supply have been related mainly to improvements in NUPE in spring wheat in Mexico [5] and Finland [8] and to NUTE in winter wheat in France [9] and the UK [10]. In these studies, modern cultivars normally had higher yields than old cultivars under low N input conditions [5,9,10]. Under high N supply, several studies worldwide concluded that wheat breeding did not result in consistent improvements in NUPE but in improved NUTE associated with higher harvest index, e.g. in Mexico [11], Argentina [12], France [9] and in various countries [13,14]. In contrast, studies in the UK [10], Mexico [5] and Finland [8] found that increases in NUE were explained approximately equally by NUPE and NUTE.

Numerous studies of cultivars and segregating populations have shown inverse relationships between grain yield and grain N concentration (e.g. Kibite and Evans [15] and Triboi et al., [16]. Assuming a constant N, harvest index (the proportion of above-ground N (AGN) at harvest in the grain; NHI), grain yield will be positively associated with AGN and/ or negatively associated with grain nitrogen content (GNC). Under low N supply, genetic gains in yield were positively associated with AGN and NHI and negatively associated with GNC [5,9,10]. Whereas under high N supply, genetic gains in yield were positively associated with AGN [5,9,10], positively [5,9] or neutrally [10] associated with NHI, and negatively [5,9] or neutrally [10] associated with GNC. Essentially, under both low and high N supply, wheat crops with higher NUE compared to current cultivars will require an increase in NUPE to raise AGN and/or an increase in NUTE. The latter may be achieved by raising NHI and/or lowering GNC.

There have to date been relatively few attempts to use 'direct selection' breeding to improve NUE in wheat. CIMMYT in Mexico adopted a strategy to select for grain yield in medium-to-high fertility conditions, since at this fertility level both NUPE and NUTE contribute to the observed variation in NUE, resulting in lines which were more N-efficient [5]. More recently, it has been suggested that this method of selection may not be as efficient as selecting lines under alternating high-low N selection regimes commencing with high N in the F_2 [17].

A selection criterion for a given trait should be of strong association with this trait and of high heritability. Therefore, the objectives of the present study were (i) to evaluate interrelationships among traits for determining the strongest ones with N-use efficiency in wheat parents and their F_1 and F_2 progenies and (2) to estimate heritability and expected genetic advance from selection for such criteria in order to identify those of strong correlation with NUE coupled with high heritability and high selection gain under low-N to be used as selection criteria for NUE.

2. MATERIALS AND METHODS

This study was carried out at Giza Research Station of the Agricultural Research Center(ARC), Giza Egypt (30° 02'N latitude and 31° 13'E longitude with an altitude of 22.50 meters above sea level), in 2005/2006 season and at Noubarya Research Station of the ARC, Noubarya, Egypt (30°66'N latitude and 30°06' E longitude with an altitude of 15.00 meters above sea level), in 2006/2007, 2007/2008 and 2008/2009 seasons.

Six bread wheat genotypes (*Triticum aestivum* L.) were chosen for their divergence in nitrogen use efficiency to be used as parents of diallel crosses, based on previous field screening carried out by Wheat Res. Dept., Field Crops Res. Inst., ARC, Egypt. Three of them were promising breeding lines of high yield under low-N (L25, L26 and L27) and three were commercial local cultivars of low yield under low-N (Gemmeiza 7; Gem7, Gemmeiza 9; Gem9 and Giza 168; Gz168).

In season 2005/2006, a half diallel of crosses involving the six parents (without reciprocals) was done at Giza Agric. Res. Stat., Agric. Res. Center, to obtain the F_1 seeds of 15 crosses. In summer 2006, a part of F_1 seeds was sown in greenhouse of Wheat Res. Dept. under controlled conditions to obtain the F_2 seeds. In season 2007/2008, the half diallel of crosses was again done to increase quantity of F_1 seeds and in summer 2007 the F_1seeds were again sown in the greenhouse under controlled conditions to obtain more seeds of 15 F_2 crosses. In the seasons 2007/2008, 2008/2009, parents (6), F_1's (15) and F_2's (15) were sown on 17th of November each season in the field of Noubarya Res. Stat., under two levels of nitrogen fertilizer; the low level was without fertilization (LN) and the high level was 75 kg Nitrogen/ feddan (HN); this is the recommended level of Ministry of Agriculture. This level of nitrogen fertilizer (168 kg Urea/fed) was added in two equal doses, the first dose was added just before the sowing irrigation and the second dose just before the second irrigation (21 days after irrigation). In this experiment, a split plot design in lattice (6x6) arrangement was used with three replications. The two levels of nitrogen were allotted to the main plots and the genotypes to the sup plots. Each parent or F_1 was sown in two rows and each F_2 was sown in four rows; each row was three meter long; spaces between rows were 30 cm and 10 cm between plants, and the plot size was 1.8 m^2 for parent or F_1 and 3.6 m^2 for F_2. All other agricultural practices were done according to the recommendation of Ministry of Agriculture for growing wheat in Noubarya region.

Available soil nitrogen in 30 cm depth was analyzed immediately prior to sowing and N application at the laboratories of Water and Environment Unit, ARC, Egypt in the two seasons. Soil nitrogen was found to be 55 and 57 kg N/ fed in the seasons 2007/2008, 2008/2009, respectively. Available soil nitrogen after adding nitrogen fertilizer was therefore 55 and 130 kg N/fed in the first season and 57 and 132 kg N/fed in the second season for the two treatments, i.e. LN and HN, respectively. The available nitrogen to each plant (including soil and added N) was calculated for each environment to be 0.79, 1.85 g/plant in 2007/2008 season and 0.81 and 1.89 kg/fed in 2008/2009 season, with an average across the two seasons of 0.80 and 1.87 g/plant for the two environments LN and HN, respectively. The soil analysis of the experimental soil at Noubarya Research Station, as an average of the two growing seasons, indicated that the soil is sandy loam (67.86% sand, 7.00% silt and 25.14% clay), the pH is 8.93, the EC is 0.55 dSm^{-1}, the soluble cations in meq l^{-1} are Ca^{2+} (5.30), K^+ (0.70), Na^+ (0.31), Mg^{2+} (2.60) and the soluble anions in meq l^{-1} are CO_3^{2-} (0.00), HCO_3^- (2.10), Cl^- (5.30) and SO_3^{2-} (1.51).

A random sample of 10 plants of each genotype of parents and F_1's and 30 plants of F_2's was used to collect data for 14 traits: Days to 50% heading (DTH) as number of days from sowing date to the date at which 50% of main spike awns/ plot have completely emerged from the flag leaves, days to maturity (DTM) measured as number of days from sowing date to the date at which 50% of main peduncles/ plot have turned to yellow color (physiological maturity), plant height (PH) measured as plant length from the soil surface to the tip of the spikes, excluding awns, number of spikes/plant (SPP) as number of fertile spikes per plant, number of grains/ spike (GPS), 100 grain weight (100 GW) measured as weight of 100 grains taken from each guarded plant, grain yield/ plant (GYPP) measured as weight of the grains of each individual plant, biological yield/ plant (BYPP) measured as weight of the grains and stem of each individual plant and harvest index (HI%) according formula: HI= 100 (GYPP/ BYPP). At physiological maturity stage, five random guarded plants were removed from each plot by cutting at the soil surface. The plants were bulked as one sample per plot. They were separated into straws (including leaves, stems and spike residues) and grains. Samples were oven dried at 70℃ to a constant weight and each part was weighed separately. Samples were ground in powder and nitrogen of straws (N straw) and grains (N_g) was determined using Kjeldahl procedure according to A.O.A.C. [18]. Total plant nitrogen (N_t) was calculated as follows: $N_t = N_g + N_{straw}$. Data were collected for: nitrogen use efficiency (NUE) g/g= (GYPP / N_s), nitrogen uptake efficiency (NUPE)% =100 (N_t / N_s), nitrogen utilization efficiency (NUTE) (g/g)= (GYPP/N_t), nitrogen harvest index (NHI%)= 100(N_g/ N_t), and grain protein content (GPC) measured as follows: GPC%= N_g x 5.70 according to AACC [19], where GYPP is grain yield/ plant in gram, N_t is total nitrogen in the whole plant (grains and straw), N_s is available nitrogen in the soil for each plant, and N_g is grain nitrogen content. Nitrogen efficiency parameters were estimated according to Moll et al. [4].

The analysis of variance (ANOVA) of the split plot design was performed on the basis of individual plot observation using the MIXED procedure of SAS ® [20]. Moreover, each environment (HN and LN) was analyzed separately as lattice design for the purpose of determining genetic parameters using Genestat10th addition windows software. Least significant differences (LSD) values were calculated to test the significance of differences between means according to Steel et al. [21]. Simple correlation coefficients were calculated between pairs of studied traits under stress and non-stress conditions in parents, F_1 and F_2 generations according to Singh and Narayanan [22]. The genetic parameters and ratios were calculated according to methods developed by Jinks and Hayman [23], Jinks [24] and Hayman [25,26] and described by Sharma [27]. Narrow-sense heritability (h^2_n) in F_1's was estimated using the following equation: h^2_n = [1/4D / (1/4D + 1/4H_1– 1/4F + Ê]. Narrow-sense heritability (h^2_n) in F_2 was estimated using the following equation: h^2_n = [1/4D / (1/4D + 1/16H_1– 1/8F + Ê)], where Ê = expected environmental component of variance. D = variance due to additive effects of the genes. F = mean of the covariance of additive and dominance effects across all arrays. H_1= variance component due to dominance deviation. The expected genetic advance (GA) from direct selection as a percentage of the mean (x) was calculated according to Singh and Narayanan [22] based on 1% selection intensity as follows: GA = 100[(k.h^2_n δ_{ph})/x] Where: k = 2.64 (selection differential for 1% selection intensity), and δ_{ph}= square root of the dominator of the narrow sense heritability.

3. RESULTS AND DISCUSSION

3.1 Analysis of Variance

Combined analysis of variance across years (Y) of the split plot design in lattice (6x6) arrangement across 2008/2009 and 2009/2010 seasons for the studied 36 wheat genotypes (6 parents , 15 F_1's and 15 F_2's) under two levels of nitrogen was performed (data not presented). Mean squares due to years were highly significant for nine studied traits, namely days to heading (DTH), plant height (PH), spikes/plant (SPP), grains/ spike (GPS), 100 grain weight (100 GW), grain yield/ plant (GYPP), biological yield/ plant (BYPP, nitrogen utilization efficiency (NUTE) and nitrogen harvest index (NHI), indicating significant effect of climatic conditions on most studied traits,

Results also exhibit that mean squares due to nitrogen levels (N) were highly significant for all studied traits, indicating that the N level has an obvious effect on all studied traits of studied wheat genotypes. Mean squares due to genotypes (G) were highly significant for all studied traits, indicating that wheat genotypes used in this study were significantly (P≤ 0.01) different for all studied traits. Moreover, mean squares due to genotypes x nitrogen levels, *i.e.* G x N were significant (P ≤ 0.01 or 0.05) for all studied traits, indicating that genotypes ranks differently from one nitrogen level to another and that selection can be done under a specific soil nitrogen environment as proposed by Al-Naggar et al. [28-37]. The significant G×N interaction for NUE was also a good evidence for varying responses of these wheat genotypes at various

N levels [38,39]. The interactions G x Y and G x Y x N were also significant ($P \leq 0.01$ or 0.05) for all studied traits, indicating that genotypes ranks differ from one combination of Y x N to another.

Combined analysis of variance of a lattice design for all studied traits under each environment (high N and low N) across two seasons was also performed (data not presented). Mean squares due to genotypes, parents, F_1's and F_2's under the two levels of nitrogen were highly significant for all studied traits. Significant differences among parents of diallel crosses in all studied traits are pre-requisite for performing the diallel analysis for estimating the inheritance of studied traits under different N- application rates. Mean squares due to parents *vs.* F_1's and F_1's *vs* F_2's were highly significant for all studied traits under the two levels of nitrogen, except for nitrogen uptake efficiency (NUPE) which were not significant for F_1's *vs* F_2's, indicating the presence of significant heterosis for all studied traits and the presence of inbreeding effects for most studied traits. Mean squares due to the interaction G x Y were highly significant for all studied traits under the two levels of nitrogen, except for NUPE and GPC under low N which were not significant. Mean squares due to the interaction P x Y under high level of nitrogen were significant or highly significant for 10 studied traits and non significant for DTH, BYPP, NUE and GPC. Mean squares due to the interactions F_1's x Y and F_2's x Y under high-N were significant or highly significant for all studied traits, except NUPE for F_1's x Y and 100GW for F_2's x Y, which were not significant. Mean squares due to the interactions F_1's x Y and F_2's x Y were significant or highly significant for all studied traits under low N, except for 100GW, GYPP, NUE, NUPE, NUTE and GPC for F_1's x Y and NUE and NUPE for F_2's x Y. Mean squares due to the interactions P's *vs* F_1's x Y and F_1's *vs* F_2's x Y under the two levels of nitrogen were significant and highly significant for all studied traits, except GPC and NHI for F_1's *vs.* F_2's x Y and NHI for P's *vs* F_1's x Y. The significance of the interactions P's *vs* F_1's x Y and F_1's *vs* F_2's x Y indicates that heterosis and inbreeding effects differ from season to season in most studied traits.

3.2 Mean Performance

A comparative summary of means of all studied traits across all 36 genotypes (6 parents, 15 F_1's and 15 F_2's) subjected to two levels of nitrogen conditions and across two years is presented in Table (1). In general, low N caused a significant reduction in 7 out of 8 studied traits, namely GYPP, SPP, 100 GW, GPS and HI. Mean grain yield/plant (GYPP) was significantly decreased due to low-N by an average of 18.96, 21.17, and 15.40% for parents, F_1's and F_2's, respectively. Reduction in grain yield of wheat due to low soil nitrogen was reported by several investigators. A positive relationship between N application levels and the grain yield has already been shown in many studies [39,40]. Significant reduction in grain yield as a result of low-N was associated with significant reductions in all yield components traits, *i.e.* SPP, 100GW and GPS. These reductions were relatively high in magnitude for number of spikes/ plant (SPP) for parents (23.65%), F_1's (23.99%) and F_2's (43.52%). This indicates that SPP is the most determining component of grain yield / plant of wheat under low-N stress. The importance of this trait (number of spikes or fertile tillers per plant) in wheat for grain productivity under abiotic stress conditions was previously reported by several investigators; e.g. Al-Naggar et al. [41]. Al-Bakry et al. [42] observed that increasing nitrogen application increased the number of fertile tillers per unit area. Geleto et al. [43] reported that grain yield is closely related to the number of spikes per unit area. Fertilized plots produced more spikes than control. Such response can be attributed to the adequate nitrogen availability which might facilitate the tillering ability of plants, resulting in a greater spike population. Ayoub et al. [44] also reported that spike population increased with increase in nitrogen level.

Moreover, low nitrogen caused a significant reduction in biological yield / plant (BYPP) by 12.49, 12.27 and 11.24%, grain protein content (GPC) by 25.06, 29.18, and 23.31% and harvest index (HI) by 6.57, 8.97 and 3.69% for parents, F_1's and F_2's, respectively. It was observed that low- N caused slight but significant earliness of DTM by 3.13, 5.33 and 4.52% (4.17, 7.16 and 6.07 days) and DTH by 0.70, 4.50 and 5.55 days for parents, F_1's and F_2's, respectively.

On the contrary, low–N caused increases in the averages of nitrogen use efficiency (NUE) by 89.56, 97.60 and 84.28% for parents, F_1's and F_2's, respectively. Significant increase in NUE due to low- N stress was associated with significant increases in averages of NUPE and NUTE. In agreement with these results, Ortiz-Monasterio et al. [5] also reported significant influence of N application rate on NUTE in which the highest efficiencies were measured at the lowest application rate. Sinebo et al. [45] also found similar results in which they reported NUTE values of 29.3- 43.9 and 31.8-48.3 kg kg-1 N, respectively.

In the present study, magnitude of increase in NUPE (65.87, 83.08 and 78.70%) was much higher than that in NUTE (14.39, 3.93 and 20.21%) for parents, F_1's and F_2's, respectively. This indicates that NUPE is the most determinant component of NUE. Al-Naggar et al. [33-37] showed that N efficiency diminishes as N fertilizer rates increase. Gorny et al. [46] reported that NUPE, NUTE and NHI tended to increase under low-N. According to Le Gouis et al. [47], in wheat, Al-Naggar et al. [33,37] in maize and Al-Naggar et al. [28,29] in grain sorghum , N uptake in biomass was the most important factor in NUE determination regardless of N level. Abeledo et al. [48] reported that both N conversion and N capture have played a role in the improvement of NUE. On the contrary, Gaju et al. [49] found that NUPE effect explained only a small amount of phenotypic variance in NUE amongst cultivars, but NUTE affected it up to 61% and 77% under high –N low-N, respectively. The differences in results might be attributed to the different genetic background of genotypes used in different studies. In the present study, it is also observed that low- N stress caused a significant (but slight) increase in plant height of F_1's (6.80%) and F_2's (4.80%).

Table 1. Means of studied wheat traits under low–N (0 Kg N/fed) and high–N (75 Kg N/fed) and relative reduction compared to high–N combined across parents, F_1's and F_2's across two seasons

Traits	Parameter	Parents		F_1 crosses		F_2 crosses	
		High-N	Low-N	High-N	Low-N	High-N	Low-N
DTH	Average	88.64	87.94	89.61	85.11	89.04	83.49
	Reduction%	---	0.78	---	4.95**	---	6.13**
DTM	Average	132.33	128.17	133.33	126.17	132.76	126.69
	Reduction%	---	3.13*	---	5.33**	---	4.52**
PH(cm)	Average	82.74	81.21	89.54	83.96	102.95	99.05
	Reduction%	---	1.74	---	6.22**	---	3.78**
GPS	Average	80.23	69.81	79.95	71.76	74.48	64.78
	Reduction%	---	13.47**	---	9.80**	---	12.47**
100 GW(g)	Average	4.66	4.05	4.33	3.84	3.37	2.61
	Reduction%	---	12.96**	---	10.51**	---	21.72**
SPP	Average	11.88	9.11	12.13	9.14	12.95	7.31
	Reduction%	---	18.96**	---	23.99**	---	43.52**
GYPP(g)	Average	27.53	22.41	29.12	22.83	25.65	21.54
	Reduction%	---	18.96**	----	21.17**	---	15.40**
BYPP(g)	Average	63.14	54.98	64.87	56.78	59.29	52.59
	Reduction%	---	12.94**	---	12.27**	---	11.24**
HI (%)	Average	43.67	40.73	45.11	40.51	43.50	41.37
	Reduction%	----	6.57**	---	8.97**	---	3.96
NUE(g/g)	Average	14.72	28.03	15.57	28.53	13.72	26.91
	Reduction%	----	-.89.56**	----	-97.60**	----	-84.28**
NUPE (%)	Average	16.00	26.87	16.14	29.43	17.13	29.97
	Reduction%	-----	-65.87**	-----	-83.08**	---	-78.70**
NUTE	Average	0.94	1.07	0.98	1.00	0.84	0.98
(g/g)	Reduction%	----	-14.39**	----	-3.93	---	-20.21**
GPC (%)	Average	14.35	11.33	14.61	12.64	14.86	12.91
	Reduction%	------	21.3**	-------	14.71**	------	13.12**
NHI (%)	Average	56.10	55.94	56.21	58.08	56.75	57.73
	Reduction%	-----	0.28	-----	-3.33	------	-1.73

*N= nitrogen, * and** indicate significance at 0.05 and 0.01 probability levels, respectively reduction% = 100 [(HN-LN)/HN]*

3.3 Interrelationships

Estimates of simple correlation coefficients between each of grain yield /plant (GYPP), nitrogen uptake efficiency (NUPE) and nitrogen utilization efficiency (NUTE) and other studied traits across the two seasons under high-N and low-N were calculated across all parents, F_1's and F_2's and presented in Table (2). Grain yield/ plant showed perfect positive association with NUE (r=1.00) under both high-N and low-N for all groups of genotypes, *i.e.,* parents, F_1's and F_2's, that is why the estimates of correlation coefficients between GYPP and other traits are very close to those between NUE and the same traits.

The strongest correlations (r ≤ 0.90) was observed between grain protein content (GPC) which is an expression of grain nitrogen content (N_g) and each of NUPE and NUE (GYPP) for all genotypes under both environments, *i.e.* high-N and low-N.

Concerning the components of grain yield, significant and high-in-magnitude correlations were observed between GYPP and each of GPC (N_g) (0.92), HI (0.82), GPS (0.71),100GW (0.63) and SPP (0.43) across parents, NUPE (0.80), HI (0.67) and NUTE (0.62) across F_1s and HI (0.82) and SPP (0.32) across F_2's under high-N and GPS (0.86), GPC (0.95) HI (0.55), SPP (0.57) and 100GW (0.36) across parents, GPS (0.60), BYPP (0.74), HI (0.71), GPC (0.44) and SPP (0.36) across F_1s and SPP (0.29), GPS (0.24), NUPE (0.39), HI (0.55) and BYPP (0.38) across F'_2s under low-N conditions. These results indicate the importance of SPP, GPS, GPC and NUPE for getting high grain yield / plant under both high and low N conditions. Sinclair and Jamieson [50] reported that, it is logical that the number of grains produced is not only important for determining the yield, but is also related to the NUE. However, the results imply that breeding efforts improving yield by increasing the number of grains should also result in higher NUE as well according to Moll et al. [5].

The results indicate also that under low-N, it is possible to obtain wheat genotypes (parents, F'_1s and F'_2s) that have high grain yield of high grain protein contents. This conclusion was previously reported by some investigators. A further improvement of grain protein content under N-limited conditions without substantial depressions in yielding capacity seems to be possible in bread wheat as suggested by Laperche et al. [51]. In other studies, Ayoub *et al.* [44] and Ortiz-Monasterio et al. [5] reported negative associations of grain yield with protein content. Sinebo et al. [45] found significant negative correlations of NUTE with grain protein content.

Concerning the components of NUE of parents, results (Table 2) showed that NUE had positive and significant correlations with NUPE (0.83 and 0,88) and negative and significant correlation coefficients with NUTE (-0.62 and -0.46) under low-N and high-N, respectively. So, parents data indicate that NUE is based mainly on NUPE, especially under low-N. Data of F_1's showed that NUE is depending only on NUTE under high-N (r=0.61) and on NUPE under low-N (0.75), suggesting that NUE of F_1 hybrids depends more on NUPE under high-N and NUTE under low-N. Regarding F_2 populations, data showed that NUE had significant and positive, but low-in-magnitude correlations with NUPE under low-N (r=0.39) and high–N (r=0.24) and not correlated with NUTE. This indicates that F_2 hybrids express their NUE due to NUTE under both low and high N conditions, with higher degree under low-N.

In general, across all genotypes, NUE depends mainly on NUPE, except F_1's under high-N which depends on NUTE. Thus under low-N, NUE increases by increase of NUPE for parents, F_1's and F_2's. The same conclusion was valid under high-N, for parents and F_2's, but for F_1's, NUE increases by increasing NUTE. Numerous studies indicated that NUPE accounts for more of the genetic variation in NUE at low N than at high N supply, e.g. amongst 10 spring wheat cultivars in Mexico [5] and 20 winter wheat cultivars in France [47]. Moreover, Dhugga and Waines [52] comparing 12 spring wheats in California, found that NUPE was the most important component of NUE at all N levels. Genetic gains in NUE with breeding under low N supply have been related mainly to improvements in NUPE in spring wheat in Mexico [5] and Finland [8] and to NUTE in winter wheat in France [9] and the UK [10].

In general, grain yield or nitrogen use efficiency (NUE) were significantly correlated with nitrogen uptake efficiency (NUPE) in most cases; this correlation was more pronounced in parents (0.88 and 0.83) than in F_1 crosses (0.80 and 0.75) and F_2 crosses (0.24 and 0.39) under high-N and low-N, respectively. Harvest index (HI) showed significant and positive correlation coefficients in most cases with GYPP and NUE for parents, F_1's and F_2's of medium to high magnitude under low-N and high-N.

Since NUPE and HI showed strong correlations with NUE or GYPP in this study and nitrogen translocation efficiency (NTRE) (grain N/total plant N) proved be strongly correlated with HI in other studies [33], grain yield or NUE of parents, F_1's and F_2's in the present study could therefore be considered of strong association with N-uptake and N-translocation efficiencies. Similar conclusion was reported by Al Naggar et al. [35-37] in maize and Gorny et al. [46] in wheat.

Table 2. Correlation coefficients among studied traits across parents, F_1's and F_2's under high and low-N

Genotype	High N			Low N		
	GYPP	NUPE	NUTE	GYPP	NUPE	NUTE
Parents						
DTH	0.71**	0.96**	-0.93**	0.51**	0.15	0.66**
DTM	0.14	0.10	-0.05	-0.21	-0.33*	0.12
PH	0.23	0.22	-0.25	-0.75**	-0.48**	-0.60**
SPP	0.43**	0.79**	-0.97**	0.57**	0.82**	-0.22
GPS	0.71**	0.85**	-0.73**	0.86**	0.91**	0.16
100GW	0.63**	0.84**	-0.79**	0.36*	0.28*	0.21
GYPP	---	0.88**	-0.45**	---	0.83**	0.54**
BYPP	0.07	0.85**	-0.73**	0.38*	0.40**	0.50**
HI%	0.82**	0.11	0.27	0.55**	0.58**	0.17
NUE	1.00**	0.88**	-0.46**	1.00**	0.83**	0.54**
NUPE	0.88**	----	-0.82**	0.83**	---	-0.02
NUTE	-0.45**	-0.82**	---	0.54**	-0.02	---
GPC	0.92**	0.99**	-.0.76**	0.95**	0.99**	0.40*
NHI	0.32	-0.11	0.58**	-0.47**	-0.41*	0.55**
F_1 crosses						
DTH	0.12	-0.17	0.21*	0.70**	0.57**	0.50**
DTM	0.18	0.18	0.35**	0.69**	0.60**	0.48**
PH	0.05	0.16	-0.05	0.52**	0.45**	0.57**
SPP	0.35**	0.61**	-0.46**	0.36**	0.44**	-0.45**
GPS	0.30**	0.07	0.17	0.60**	0.28*	0.06
100GW	0.14	0.41**	-0.24*	-0.22*	-0.07	-0.64**
GYPP	--	0.08	0.62**	----	0.75**	0.06
BYPP	0.42**	0.76**	-0.30**	0.74**	0.69**	-0.11
HI%	0.67**	-0.53**	0.86**	0.71**	0.43**	0.51**
NUE	1.00**	0.08	0.61**	1.00**	0.75**	0.06
NUPE	0.80**	----	-0.73**	0.75**	----	-0.36**
NUTE	0.62**	-0.73**	---	0.36**	-0.36**	---
GPC	0.41*	0.37*	-0.77**	0.44**	0.84**	-0.56**
NHI	-0.37*	-0.61	0.23	-0.68**	-0.50**	0.10
F_2 crosses						
DTH	-0.33**	-0.55**	0.40**	0.20	0.23*	0.14
DTM	-0.11	-0.62**	0.56**	0.41**	0.49**	-0.20
PH	0.39**	0.29**	-0.08	-0.05	-0.15	0.14
SPP	0.32**	0.47**	-0.48**	0.29**	0.34**	-0.10
GPS	-0.04	0.56**	-0.54**	0.24*	0.43**	-0.07
100GW	0.15	0.43**	-0.39**	0.18	0.50**	-0.57**
GYPP	---	0.24*	0.11	----	0.39**	-0.08
BYPP	0.24*	0.59**	-0.57**	0.38**	0.39**	-0.19
HI%	0.82**	-0.12	0.41**	0.55**	-0.05	0.16
NUE	1.00**	0.24*	0.11	1.00**	0.39**	0.39**
NUPE	0.11	---	-0.91**	0.39**	---	-0.81**
NUTE	-0.91**	-0.91**	---	-0.08	-0.81**	----
GPC	0.27*	0.36*	-0.24	0.26*	0.89**	-0.69**
NHI	-0.11	-0.60**	0.61**	0.02	0.89**	0.31**

** and** indicate significant at 0.05 and 0.01 probability levels, respectively*

It is observed that a negative and significant correlation existed between NUPE and NUTE for parents under high-N (-0.82), F_1's under high-N (-0.73) and low-N (-0.36) and F_2's under high-N (-0.91) and low-N (-0.81). This observation was reported by many previous investigators (e.g. Gorny et al., [46]). The two processes (N-uptake and N-utilization) appear to be governed by different genetic factors [53-55]. For instance results of extensive molecular studies, on wheat and maize [51,56-58] revealed that different sets of genes (QTL regions) controlled various components of the two major measures of N efficiency, and that expression of the genes (loci) was considerably dependent upon soil N status. Hence, the appearance of the above mentioned negative relationship between NUPE and NUTE in the examined hybrids and their parents may be a genetic quandary. Most of modern parents of wheat were developed under enhanced N fertilization with strong selection for yielding capacity and grain quality. Thus, such a strategy may have favored high-yielding genotypes with increased NUTE, but not those with a high NUPE. This implies that the proposal of Martre et al. [59] to search for genotypes able to translocation of larger amounts of N grains without reducing plant biomass and photosynthetic capacity under N shortage has become particularly challenging for wheat breeders'.

4. HERITABILITY

4. 1 In F_1 Generation

Broad-sense heritability (h^2_b) in F_1's for all studied traits (except GPC) in this experiment was of high magnitude and ranged from 61.7% (SPP) to 99.9% (GPS) under high-N and from 41.0% (NHI) to 100.0% (GPS) under low-N environments (Table 3), indicating that environment had a small effect on the phenotype of F_1's for most studied traits. Grain protein content (GPC) trait showed very small h^2_b value under both high and low N environments (8.52 and 20.00%, respectively), indicating a large effect of environment on this trait.

Narrow-sense heritability (h^2_n) in F_1's (Table 3) was generally of medium magnitude and ranged from 8.50% (NUTE) to 56.89% (GPS) under high-N and from 9.00% (NHI) to 73.55% (GPS) under low-N. It is observed that GPC trait recorded 0.0% h^2_n under both high- and low- N; the reason may be because its D value was negative; that is why we considered it zero.

The big difference between broad- and narrow-sense heritability estimated from F_1's in this experiment could be attributed to the high estimates of dominance as compared to additive component. It is observed that narrow-sense heritability (h^2_n) in F_1's of the present study was generally higher in magnitude under low-N than under high-N for all studied traits, except for plant height. This increase in h^2_n under low-N compared to high-N was more pronounced in SPP, GPS, GYPP and NUE traits; *i.e.* the most important components of grain yield and consequently nitrogen use efficiency in wheat.

Expected genetic advance (GA) from selection (based on 1% selection intensity) across two years for F_1's ranged from 0.12% for NHI to 23.9% for HI under high-N and from 1.02% for NHI to 23.26% for GPS under low-N. In general, the values of GA were higher under low-N than under high-N (for 11 out of 14 studied traits). The traits SPP, GPS, GYPP, BYPP, NUE and NUPE showed much higher estimates of GA (23.25, 23.26, 19.70, 10.44, 19.62 and 22.43%, respectively) under low-N than those under high-N (2.50, 15.87, 5.11, 5.44, 9.44 and 11.25%, respectively).

4. 2 In F_2 Generation

Broad-sense heritability (h^2_b) in F_2's for all studied traits (except GPC under both environments and NHI under low N) in this experiment was of high magnitude and ranged from 40.12% (SPP) to 100.00% (GPS) under high-N and from 63.71% (100 GW) to 100.00% (GPS) under low-N environments (Table 4), indicating that environment had a small effect on the phenotype of F_2's for most studied traits. In F_2 generation, grain protein content (GPC) under both high and low N environments (13.00 and 20.00%, respectively) and NHI trait under low-N (14.30%) showed very small h^2_b value, indicating a large effect of environment on these traits.

Narrow-sense heritability (h^2_n) in F_2's (Table 4) was generally of low to high magnitude and ranged from 2.11% (NHI) to 82.90% (GPS) under high-N and from 6.80% (NHI) to 93.00% (GPS) under low-N. It is observed that in F_2's, GPC trait recorded 0.0% h^2_n under both high- and low- N; the reason may be because its D value was negative; that is why we considered it zero. The big difference between broad- and narrow- sense heritability estimated from F_2's in this experiment could be attributed to the high estimates of dominance as compared to additive component.

Table 3. Estimates of heritability in broad (h^2_b) and narrow (h^2_n) sense and genetic advance from selection (GA) for studied traits under high and low-N in F_1 populations of 15 diallel crosses across two seasons

Parameter	DTH	DTM	PH	SPP	GPS	100GW	GYPP
			High N				
h^2_b%	91.15	85.88	94.50	61.70	99.90	81.00	91.27
h^2_n%	28.80	15.85	39.71	9.20	56.89	21.40	20.41
GA%	2.13	0.80	4.66	2.50	15.87	9.24	5.11
			Low N				
h^2_b%	80.27	87.78	96.00	54.42	100.00	57.50	91.00
h^2_n%	42.52	21.80	33.11	66.32	73.55	27.43	67.14
GA%	2.27	1.30	4.38	23.25	23.26	9.90	19.70
Parameter	**BYPP**	**HI**	**NUE**	**NUPE**	**NUTE**	**GPC**	**NHI**
			High N				
h^2_b%	96.32	87.30	91.20	96.00	96.60	8.52	63.85
h^2_n%	29.10	9.60	39.50	40.70	8.50	0.00	1.90
GA%	5.54	23.90	9.44	11.25	4.00	0.00	0.12
			Low N				
h^2_b%	97.88	93.50	90.75	96.60	87.30	20.00	41.00
h^2_n%	40.20	15.87	66.90	46.36	11.11	0.00	9.00
GA%	10.44	5.42	19.62	22.43	5.63	-5.70	1.02

Table 4. Estimates of heritability in broad (h^2_b) and narrow (h^2_n) sense and genetic advance from selection (GA) for studied traits under high and low-N in F_2 populations of 15 diallel crosses across two seasons

Variance components	DTH	DTM	PH	SPP	GPS	100GW	GYPP
			High-N				
h^2_b%	74.40	61.00	97.41	40.12	100.00	65.83	81.76
h^2_n%	56.00	46.75	16.45	16.10	82.90	30.60	42.70
GA%	3.00	1.40	2.60	3.20	20.40	12.80	7.70
			Low-N				
h^2_b%	82.59	76.59	98.40	70.00	100.00	63.71	88.65
h^2_n%	79.10	37.80	17.72	43.76	93.00	25.56	85.10
GA%	5.38	1.70	3.20	22.00	28.12	12.31	23.14
Variance components	**BYPP**	**HI**	**NUE**	**NUPE**	**NUTE**	**GPC**	**NHI**
			High-N				
h^2_b%	90.37	82.00	81.64	95.03	91.83	20.00	59.53
h^2_n%	65.40	19.30	42.64	33.37	20.41	0.00	2.11
GA%	7.00	3.60	7.70	9.52	6.80	0.00	0.12
			Low-N				
h^2_b%	96.00	78.20	88.60	88.10	67.57	13.00	14.30
h^2_n%	55.44	33.38	85.00	72.80	27.00	0.00	6.80
GA%	13.00	7.50	23.10	27.43	6.80	0.00	6.70

It is observed that narrow-sense heritability (h^2_n) in F_2's of the present study was generally higher in magnitude under low-N than under high-N for all studied traits, except for DTM, 100 GW and BYPP. This increase in h^2_n under low-N compared to high-N was more pronounced in SPP, GPS, GYPP, NUE and NUPE traits; *i.e.* the most important components of grain yield and nitrogen use efficiency in wheat.

Our results of heritability in F_1's and F_2's are in agreement with some researchers, e.g. Shabana et al. [60] , Blum [61,62] and Al-Naggar et al. [28-37], who support the idea that heritability is higher under stressed than non-stressed environment. On the contrary, other investigators reported that heritability is higher under good (non-stressed) environment [63-65].

Expected genetic advance (GA) from selection (based on 1% selection intensity) across two years ranged from 0.12% for NHI to 20.40% for GPS under high-N and from 1.70% for DTM to 28.12% for GPS under low-N (Table 4). In general, the values of GA were higher under low-N than under high-N (for 12 out of 14 studied traits). The traits SPP, GPS, GYPP, BYPP, NUE, NUPE and NHI showed much higher estimates of GA (22.00, 28.12, 23.14,13.00, 23.10, 27.43 and 6.70 %, respectively) under low-N than those under high-N (3.20, 20.40, 7.70, 7.00, 7.70, 9.52 and 0.12%, respectively).

These results in F_1 and F_2 generations indicated that to improve grain yield and nitrogen use efficiency traits in the present germplasm, it is better to practice selection for these traits under low-N conditions to obtain higher values of selection gain.

5. ALTERNATIVE SCREENING CRITERIA FOR NUE

The selection criterion for a given abiotic stress tolerance character, such as low-N tolerance, should be of strong association with tolerance to low-N, *i.e.* with nitrogen use efficiency and of high heritability in narrow-sense under low-N conditions. The results in Tables (2, 3 and 4) concluded that NUPE, GYPP, GPS, and SPP of parents, NUPE, GYPP, BYPP, DTH and GPS of F_1's and NUPE, HI, BYPP, SPP and GPS of F_2's had strong associations with high nitrogen use efficiency.

The importance of number of spikes (fertile tillers per plant) in wheat for grain productivity under abiotic stress conditions was previously reported by several investigators [41]. CIMMYT breeders found that maize grain yield under low-N was closely related to some secondary traits such as improved N-uptake efficiency, high plant nitrate content, high-specific leaf-N content and late leaf senescence [3,65]. These results for NUPE and SPP are in consistency with those reported by Al-Naggar et al. [30,31,33-37].

Numerous studies indicated that NUPE accounts for more of the genetic variation in NUE at low N than at high N supply, e.g. amongst 10 spring wheat cultivars in Mexico [5] and 20 winter wheat cultivars in France [47]. Moreover Dhugga and Waines [52] comparing 12 spring wheats in California, found that NUPE was the most important component of NUE at all N levels. Genetic gains in NUE with breeding under low N supply have been related mainly to improvements in NUPE in spring wheat in Mexico [5] and Finland [8], but to NUTE in winter wheat in France [9] and the UK [10]. Differences in results could be attributed to genotypes used in different studies; that might differ in genes controlling NUPE and genes controlling NUTE.

Harvest index (HI) and nitrogen harvest index (NHI) provide a means for determining the allocation of important nutrients within the plant tissues to grain during reproductive growth. Since NUPE and HI showed strong correlations with NUE or GYPP in this study and nitrogen translocation efficiency (NTRE) (grain N/total plant N) proved be strongly correlated with HI in other studies [33] , grain yield or NUE of parents, F_1's and F_2's in the present study could therefore be considered of strong association with N-uptake and N-translocation efficiencies. Similar conclusion was reported by Al-Naggar et al. [33] in maize and Gorny et al. [46] in wheat.

In the present study, a negative and significant correlation existed between NUPE and NUTE for parents under high-N (-0.82), F_1's under high-N (-0.73) and low-N (-0.36) and F_2's under high-N (-0.91) and low-N (-0.81). Both processes, N uptake and N-utilization, appear to be governed by different genetic factors [53-55]. For instance results of extensive molecular studies , on wheat and maize [51,56-58] revealed that different sets of genes (QTL regions) controlled various components of the two major measures of N efficiency, and that expression of the genes (loci) was considerably dependent upon soil N status. Hence, the appearance of the above mentioned negative relationship between NUPE and NUTE in the examined hybrids and their parents may be a genetic quandary. Most of modern parents of wheat were developed under enhanced N fertilization with strong selection for yielding capacity and grain quality. Thus, such a strategy may have favored high-yielding genotypes with increased NUTE, but not those with a high NUPE. This implies that the proposal of Martre et al. [59] to search for genotypes able to translocation of larger amounts of N grains without reducing plant biomass and photosynthetic capacity under N shortage has become particularly challenging for wheat breeders.

Selection based on improved performance under low N based on NUE alone has often been considered efficient, but the use of secondary traits can increase selection efficiency [33]. Plant breeders have advocated the judicious incorporation of secondary traits within breeding programs [61,62]. Results of the present study suggest that to maximize the genetic gain from selection, for improving NUE in wheat and consequently, grain yield under low-N, future research should focus on the incorporation of secondary traits such as NUPE, GYPP, SPP and GPS in the selection programs along with NUE trait.

6. CONCLUSION

The results of this study indicated that GYPP, GPS, SPP and NUPE of parents, GYPP, NUPE, BYPP, DTH and GPS of F_1's and HI, NUPE, BYPP, SPP and GPS of F_2's are considered selection criteria for high nitrogen use efficiency (NUE). Nitrogen uptake efficiency (NUPE) is a common selection criterion for NUE in all studied genotypes, *i.e.* parents, F_1's and F_2's. Grains/spike (GPS) as a selection criterion for grain yield/plant (GYPP) is also selection criterion for NUE, since GYPP is perfectly correlated with NUE. Moreover, spikes/plant (SPP) is strongly correlated with both GYPP and NUE in parents and F_2's. All these selection criteria (NUPE, GYPP, SPP and GPS) have high heritability in narrow sense and could be offered to wheat breeders for use in selection programs for getting high selection gain when improving nitrogen use efficiency. Results suggest that future research should focus on the incorporation of such secondary traits (NUPE, GYPP, SPP and GPS) in the selection programs along with NUE trait in order to maximize the genetic gain from selection for improving wheat NUE.

COMPETING INTERESTS

Authors have declared that no competing interests exist.

REFERENCES

1. Banziger M, Diallo AO. Progress in developing drought and N stress tolerant maize cultivars for Eastern and Southern Africa. Proceeding of the 7th Eastern and Southern Africa regional maize conference, Nairobi, Kenya. 2001;189-194.
2. Banziger M, Edmeades GO, Lafitte HR. Physiological mechanisms contributing to the increased N stress tolerance of tropical maize selected for drought tolerance. Field Crops Res. 2002;75:223-233.
3. Lafitte HR, Edmeades GO. Improvement for tolerance to low soil nitrogen in tropical maize. I. Selection criteria. Field Crops Res. 1994;39:114.
4. Moll RH, Kamprath EJ, Jackson WA. Analysis and interpretation of factors which contribute to efficiency to nitrogen utilization. Agron. J. 1982;75:562–564.
5. Ortiz Monasterio JI, Sayre KD, Rajaram S, Mc Mahom M. Genetic progress in wheat yield and nitrogen use efficiency under four nitrogen rates. Crop Sci. 1997;37:898–904.
6. Le Gouis J, Beghin D, Heumez E, Pluchard P. Genetic differences for nitrogen uptake and nitrogen utilization efficiencies in winter wheat. Eur. J. Agron. 2000;12:163–173.
7. Dhugga KS, Waines JG. Analysis of nitrogen accumulation and use in bread and durum wheat. Crop Sci. 1989;29: 1232–1239.
8. Muurinen S, Slafer GA, Peltonen Sainio P. Breeding effects on nitrogen use efficiency of spring cereals under northern conditions. Crop Sci. 2006;46:561–568.
9. Brancourt Hulmel M, Doussinault G, Lecomte C, Berard P, Le Buanec B, Trottet M. Genetic improvement of agronomic traits of winter wheat cultivars released in France from 1946 to 1992. Crop Sci. 2003; 43:37–45.
10. Foulkes MJ, Sylvester Bradley R, Scott RK. Evidence for differences between winter wheat cultivars in acquisition of soil mineral nitrogen and uptake and utilization of applied fertiliser nitrogen. J. Agric. Sci. (Camb.) 1998;130:29–44.
11. Fischer RA, Wall PC. Wheat breeding in Mexico and yield increases. J. Aust. Inst. Agric. Sci. 1976;42:139–148.
12. Calderini DF, Dreccer MF, Slafer GA. Genetic improvement in wheat yield and associated traits: A re-examination of previous trends and latest trends. Plant Breed. 1995;114:108–112.
13. Paccaud FX, Fossanti A, Cao HS. Breeding for quality and yield in winter wheat: Consequences for nitrogen uptake and nitrogen partitioning efficiency. Z. Pflanzenzucht. 1985;94:89–100.
14. Feil B. Breeding progress in small grain cereals, a comparison of old and modern cultivars. Plant Breed. 1992;108:1–11.
15. Kibite S, Evans LE. Causes of negative correlations between grain yield and grain

protein concentration in common wheat. Euphytica. 1984;33:801–810.
16. Triboi E, Martre P, Girousse C, Ravel C, Triboi Blondel AM. Unravelling environmental and genetic relationships between grain yield and nitrogen concentration for wheat. Eur. J. Agron. 2006;25:108–118.
17. Van Ginkel M, Ortiz Monasterio I, Trethowan R, Hernandez E. Methodology for selecting segregating populations for improved nitrogen use efficiency in bread wheat. Euphytica. 2001;119:223–230.
18. AOAC. Official methods of association of analytical chemists. 15th ed. Washington D.C, USA. 1990;290.
19. AACC. American Association Cereal Chemists. Approved methods of the american association cereal chemists. American Association of Cereal Chemists, Inc, St. Paul, Minnesota; 2000.
20. Littell RC, Milliken GA, Stroup WW, Wolfinger RD. SAS system for mixed models. SAS Inst, Cary, NC; 1996.
21. Steel RGD, Torrie JH, Dickey D. Principles and procedure of statistics. A Biometrical approach 3rd Ed. McGraw Hill Book Co. Inc, New York. 1997;352-358.
22. Singh P, Narayanan SS. Biometrical techniques in plant breeding. Kalayani Publishers, New Delhi, India; 2000.
23. Jinks JL, Hayman I. The analysis of diallel crosses. Maize Genetics Cooperation Newsletter. 1953;27:48-54.
24. Jinks JL. The analyses of continuous variation in diallel cross of Nicotiana rustica varieties. Genetics. 1954;39:767-88.
25. Hayman BL. The theory and analysis of diallel crosses. Genetics. 1954a;39: 789-809.
26. Hayman BL. The analysis of variance of diallel tables. Biometrics. 1954b;10: 235-244.
27. Sharma RJ. Statistical and biometrical techniques in plant breeding. New Delhi. Second Edition. 2003;432.
28. Al Naggar AMM, El Kadi DA, Abo Zaid Zeinab SA. Genetic parameters of grain sorghum traits contributing to low – N tolerance. Egypt. J. Plant Breed. 2006; 10(2):79-102.
29. Al Naggar AMM, El Kadi DA, Abo Zaid Zeinab SA. Inheritance of nitrogen use efficiency traits in grain sorghum under low- and high-N. Egypt. J. Plant Breed. 2007;11(3):181-206.
30. Al-Naggar AMM, Shabana R, Mahmoud AA, Shaboon SAM. Genetic improvement of maize for low-soil nitrogen tolerance via S_1 recurrent selection. Egypt. J. Plant Breed. 2008;12(2):255-277.
31. Al Naggar AMM, Atta MMM, Amein MM. Maize genotypic differences in nitrogen use efficiency under low soil-N conditions. Egypt. J. of Appl. Sci. 2009;24(3B): 528-546.
32. Al Naggar AMM, Shabana R, Al Khalil T H. Tolerance of 28 maize hybrids and populations to low-nitrogen. Egypt. J. Plant Breed. 2010;14(2):103-114.
33. Al Naggar AMM, Shabana R, Al Khalil TH. Alternative screening criteria for selecting nitrogen use efficient genotypes of maize. Egypt. J. Plant Breed. 2011;15(1):27-40.
34. Al Naggar AMM, Shabana R, Atta MMM, Al Khalil TH. Genetic parameters controlling some maize adaptive traits to elevated plant densities combined with reduced N-rates. World Research Journal of Agronomy. 2014;3(2):70-82.
35. Al Naggar AMM, Shabana R, Atta MMM, Al Khalil TH. Matching the optimum plant density and adequate N-rate with high-density Tolerant genotype for maxmizing maize (*Zea mays* L.) crop yield. Journal of Agriculture and Ecology Research. 2015a; 2(4):237-253.
36. Al Naggar AMM, Shabana R, Atta MMM, Al Khalil TH. Maize response to elevated plant density combined with lowered N-fertilizer rate is genotype-dependent. The Crop Journal. 2015b;(3):96-109.
37. Al Naggar AMM, Shabana R, Atta MMM, Al Khalil TH. Response of genetic parameters of low-N tolerance adaptive traits to decreasing soil-N rate in maize (*Zea mays* L.). Applied Science Reports. 2015c;9(2):110-122.
38. Earl CD, Ausubel FM. The genetic engineering of nitrogen fixation. Nutritional Review. 1983;41:1-6.
39. Austin RB, Bingham J, Blackwell RD, Evans LT, Ford MA, Morgan CL, Taylor M. Genetic improvements in winter wheat yields since 1900 and associated physiological changes. Journal of Agricultural Science. 1980;94:675-689.
40. Desai RM, Bahatia CR. Nitrogen uptake and nitrogen harvest index in durum wheat

cultivars varying in their grain protein concentration. Euphytica. 1978;27: 561-566.

41. Al Naggar AMM, Atta MMM, Sobieh SES Al Azab KHF. Predicted genetic parameters from F_1 and F_2 diallel analyses and actual progress from selection for drought tolerance in wheat. Egypt. J. Plant Breed. 2013;17(4):33–58.
42. Al Bakry MRI, Al Naggar AMM, Moustafa HAM. Improvement of grain yield of a glaucous wheat mutant line via backcrossing. Egypt. J. Plant Breed. 2008; 12(2):123-131.
43. Geleto T, Tanner DG, Mamo T, Gebeyehu G. Response of rain fed bread and durum wheat to source level and timing of nitrogen fertilizer on two Ethiopian vertisole S. I. yield and yield components. Comm. in Soil Sci. and Plant Analysis. 1995;26: 1773-1794.
44. Ayoub M, Guertin S, Smith DL. Nitrogen fertilizer rate and timing effect on bread wheat protein in eastern Canada. Crop Sci. 1995;174:337-349.
45. Sinebo W, Gretzmacher R, Edelbauer A. Genotypic variation for nitrogen use efficiency in Ethiopian barley. Field Crops Res. 2003;85:43–60.
46. Gorny AG, Banaszak Z, Lugowska B, Ratajczak D. Inheritance of the efficiency of nitrogen uptake and utilization in winter wheat (*Triticum aestivum* L.) under diverse nutrition levels. Euphytica. 2011; 77:191–206.
47. Le Gouis J, Beghin D, Heumez E, Pluchard P. Diallel analysis of winter wheat at two nitrogen levels. Crop Sci. 2002;42: 1129–1134.
48. Abeledo LG, Calderini DF, Slafer GA. Nitrogen economy in old and modern malting barleys. In Field Crops Research. 2008;106:171–178.
    DOI: 10.1016/j.fcr.2007.11.006
49. Gaju O, Allard V, Martre P, Snape JW, Heumez E, Gouis JLE, Moreau D, Bogard M, Griffiths S, Orford S, Hubbart S, Foulkes MJ. Identification of traits to improve the nitrogen-use efficiency of wheat genotypes. In Field Crops Research. 2011;123:139–152.
    DOI: 10.1016/j.fcr.2011.05.010
50. Sinclair TR, Jamieson PD. Grain number, wheat yield and bottling beer: An analysis. Field Crop Res. 2006;98:60–67.
51. Laperche A, Brancourt Hulmel M, Heumez E, Gardet O, Hanocq E, Deviennearret F, Le Gouis J. Using genotype x nitrogen interaction variables to evaluate the QTL involved in wheat tolerance to nitrogen constraints. Theor. Appl. Genet. 2007;15: 399–415.
52. Hirel B, Le Gouis J, Ney B, Gallais A. The challenge of improving nitrogenuse efficiency in crops plants: Towards a more central role of genetic vari-ability and quantitative genetics within integrated approaches. J. Exp. Bot. 2007;58; 2369–2387.
53. Gallais A, Hirel B. An approach of the genetics of nitrogen use efficiency in maize. Journal of Experimental Botany. 2004;55:295–306.
54. Good AG, Shrawat AK, Muench DG. Can less yield more? Is reducing nutrient input into the environment compatible with maintaining crop production? Trends Plant Sci. 2004;9:597–605.
55. Bertin P, Gallais A. Physiological and genetic basis of nitrogen use efficiency in maize. II. QTL detection and coincidences. Maydica. 2001;46:53–68.
56. Quarrie SA, Steed A, Calestani C. A high-density genetic map of hexaploid wheat (*Triticum aestivum* L.) from the cross Chinese Spring 3SQ1 and its use to compare QTLs from grain yield across a range of environments. Theoretical and Applied Genetics. 2005;110:865–880.
57. Habash DZ, Bernard S, Schondelmaier J, Weyen J, Quarrie SA. The genetics of nitrogen use in hexaploid wheat: N utilisation, development and yield. Theor. Appl. Genet. 2007;114:403–419.
58. Martre P, Porter JR, Jamieson PD, Triboi E. Modeling grain nitrogen accumulation and protein composition to understand the sink/source regulations of nitrogen utilization in wheat. Plant Physiol. 2003; 133:1959–1967.
59. Shabana R, Bailey T, Fery KJ. Production traits of oats selected under low; medium and high productivity. Crop Sci. 1980;20: 739-744.
60. Blum A. Breeding crop varieties for stress environments. Crit. Rev. Plant Sci. 1988a; 2:199-238.
61. Blum A. Plant breeding for stress environments. CRC press Inc, Florida, USA. 1988b;78-84.

62. Rosielle AA, Hamblin J. Theoretical aspescts of selection for yield in stress and non-stress environments. Crop Sci. 1981; 21:43-46.
DOI:10.2135/cropsci1981.0011183X002100010013x
63. Atlin GN, Frey KJ. Selecting oat lines for yield in low productivity environments Crop Sci. 1990;30:556-561.
64. Banziger M, Lafitte HR. Efficiency of secondary traits for improving maize for low-nitrogen target environments. Crop Sci. 1997;37:1110-1117.
65. Worku M. Genetic and Crop-Physiological Basis of Nitrogen Efficiency in Tropical Maize. Ph.D. Thesis. Fac. Agric. Hannover Univ. Germany. 2005;122.

3

Diagnosis of Soil Fertility and Nutritional Status and Characterization of Olive Orchard in Central Morocco

Karima Bouhafa[1,2*], Lhoussaine Moughli[2], Amal Hadiddou[1], Khadija Bouchoufi[1] and Abdelkader Sdouq[1]

[1]Regional Agricultural Research Center, P.O.Box 578, Meknes, Morocco.
[2]Department of Soil Science, Hassan II Institute of Agronomy and Veterinary Medicine, P.O.Box 6202, Rabats-Instituts, Rabat, Morocco.

Authors' contributions

This work was carried out in collaboration between all authors. Author KB designed the study, completed the surveys and collection of soil and leaf samples, performed the statistical analysis, and wrote the first draft of the manuscript. Author LM supervised the study and managed the literature searches. Authors AH and AS has contributed to the investigation and collection of soil and leaf samples. Author K. Bouchoufi managed the laboratory analyzes. All authors read and approved the final manuscript.

Editor(s):
(1) Małgorzata Pietrowska-Borek, Department of Biochemistry and Biotechnology, Poznań University of Life Sciences, Poland.
Reviewers:
(1) Raul Leonel Grijalva Contreras, Instituto Nacional de Investigaciones Forestales Agrícolas y Pecuarias, Mexico.
(2) Walid Fediala Abd El-Gleel Mosa, Alexandria University, Alexandria, Egypt.

ABSTRACT

This study was conducted in 2009/2010 with 58 olive orchards randomly selected at the Sais region. The main objectives of this work are: i) Characterization of olive orchard ii) Determining the soil fertility levels and the assessment of nutritional status of these olive orchards. Composite soil samples from two depths (0-30 cm and 30-60 cm) and composite leaf samples were collected during the December/January period. Surveys were conducted on these orchards to collect data for characterization and analysis of different fertilization modes adopted in relation to olive yield. More

Corresponding author: Email: bouhafakarima@gmail.com

than half of the surveyed groves (57%) are conducted in rain fed conditions. The surveys have revealed shortcomings as regards fertilization practices of olive cultivation. Indeed, approximately 48.3% of the farmers surveyed provide no fertilizer for their olive trees. The laboratory analyzes of soil and leaf samples collected in these orchards have confirmed this. The study showed no correlation between the olive yield and soil fertility parameters and between the olive yield and the olive leaf parameters. The relationship between the olive nutrition parameters and the soil fertility parameters, at the 0-30 cm horizon, showed that olive leaf parameters were not linked to those of soil fertility. However, much of the leaf nitrogen content (about half) has been explained by the soil nitrate content in irrigated orchards. The determination of the olive tree fertilization standards is a necessary step for a rational management of fertilizers for the olive trees. These standards should be developed through field trials.

Keywords: Olea europaea; nitrogen; phosphorus; potassium; soil; leaves; surveys.

1. INTRODUCTION

The general situation of the olive orchards productivity throughout the world seems to be unsatisfactory. Indeed, about 70% of the olive orchards are traditional and marginal with a medium to very low productivity due, in a significant degree, to the lack of appropriate orchard management [1]. Under-application of inputs may lead to unsatisfactory yields, under the potentiality of the orchard [2]. The new intensive orchards (about 30% of the total) present a suitable productivity but are often associated with higher environmental impacts [3,4]. Olive growing sustainability in the Mediterranean region is highly dependent on sustainability of soil resource management practices [1]. However, there is almost no information on the overall effect of soil status on olive yield [5].

In Morocco, the olive tree is the main fruit species cultivated, with about 840 000 ha until 2011 [6], and a large proportion of its total area located in rainfed areas. Olive growing is an intensive agricultural activity which generates almost 100 000 permanent jobs [6]. The olive sector contributes up to 5% of upstream agricultural GDP and accounts for 15% of agricultural foodstuff exports. The annual production for 2014 was 100 000 tones of table olives and 120 000 tones of olive oil. Almost 75% of the olives produced in Morocco go for olive oil whereas table olive represents 25% of olive production [6]. Olive yields are still below the olive sector potential. Indeed, the weather, in the drought case, and weak technical management are the main factors generating the low productivity registered for olives. Fertilization is one of the techniques non mastered by the majority of Moroccan growers and which consequently leads to low yields of the olive tree.

This work done at Sais region in Central Morocco, has as main objectives:

- Characterization of olive orchard at Sais region in central Morocco.
- Determining the soil fertility levels and the assessment of nutritional status of these olive orchards.

2. MATERIALS AND METHODS

2.1 Surveys

This study was conducted at Sais region, through surveys, soil testing and leaf analyzes.

58 olive orchards, distributed on the territory of the study area (Moulay Driss, Mhaya, Ain Jemaa, Dkhissa, Agourai, Ras Jerry, Oued Jdida, Ain Kerma, Ain Orma, Ain Chkef, Sfassif, Laksir, Sebt Jehjouh, Ait Ouallal) were chosen at random. A plug survey was developed and served as a support for data collection in the field. These relate to the culture system, planting density, tree age, variety, soil type, and the technical route of culture including mineral and organic fertilization adopted by farmers. These data were used for the characterization of the sampled groves and analysis of different modes of fertilization adopted in relation to the olive yield.

2.2 Soil Testing

At these selected orchards, composite soil samples were collected in the diagonal method, at two depths 0-30 cm and 30-60 cm, in the December/January period. They were the subject of analyzes for their characterization (limestone, organic matter, pH) and the determination of their fertility levels (nitrogen, phosphorus and potassium). The analytical methods used are: extraction by chromotropic acid [7] for nitrates,

Olsen [8] for phosphorus, Walkley and Black [9] for organic matter, Drouineau [10] for active limestone extraction and ammonium acetate [11] for potassium.

2.3 Leaves Analyzes

Composite leaf samples were also collected from the same orchards and in the same period of soil sampling. The nitrogen and phosphorus mineralization was created by concentrated sulfuric acid and the potassium with perchloric acid. The leaf nitrogen contents were determined by Kjeldahl distillation, those of phosphorus by colorimetry and those of potassium by photometrically [12].

2.4 Statistical Analysis

Correlation and regression analyzes between some parameters were performed using the software Genstat.

3. RESULTS AND DISCUSSION

3.1 Characterization of Olive Orchards

3.1.1 Age

The sample is characterized by different age groups ranging from a minimum of 4 years to 70 years. Young orchards whose age does not exceed 10 years represent only 5% of all surveyed orchards while older orchards (> 40 years) have a part of 35% of the studied orchards. The orchards belonging to the age group between 11 and 40 years account for more than half (60%) of all surveyed groves (Fig. 1).

3.1.2 Driving mode

More than half of the surveyed groves (57%) are conducted in rainfed conditions (Fig. 2). At irrigated orchards, the gravity irrigation system dominates with a part of 24% of all surveyed orchards. As for drip irrigation, it is only present in 19% of studied farms.

3.1.3 Density

Most plantations (84.5%) have a planting density between 100 and 350 trees.ha^{-1} (Fig. 3). Extensive densities (<100 trees.ha^{-1}) represent only 5% of respondents orchards. The rest (10%) is represented by orchards with the density exceeds 350 trees.ha^{-1}.

3.1.4 Intercropping

The percentage of surveyed orchards where the olive tree was associated with intercropping was important (48.3%). The species used by farmers are mainly cereals (barley, wheat, and oats); legumes (Lentil, bean, peas, chickpeas, and beans); the vegetable crops (tomato, potato, onion, turnip, zucchini, cauliflower, pepper); fodder (alfalfa).

3.1.5 Fertilization

48.3% of surveyed growers use no mineral fertilizer for their olive trees. The main fertilizers used by farmers are ammonium nitrate, sulphate of ammonia, diammonium phosphate, phosphate monoamonium, potassium nitrate, urea and some fertilizer formulas such as 14/28/14, the 10/25/20/S and the 0/16/30/S. 41.4% of farmers do not use manure. The type of manure used in these orchards is dominated by sheep/beef mixture (Table 1).

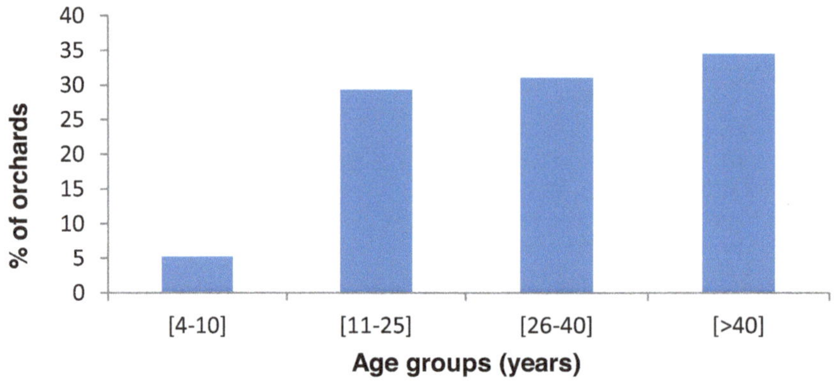

Fig. 1. Olive orchards distribution according to their ages

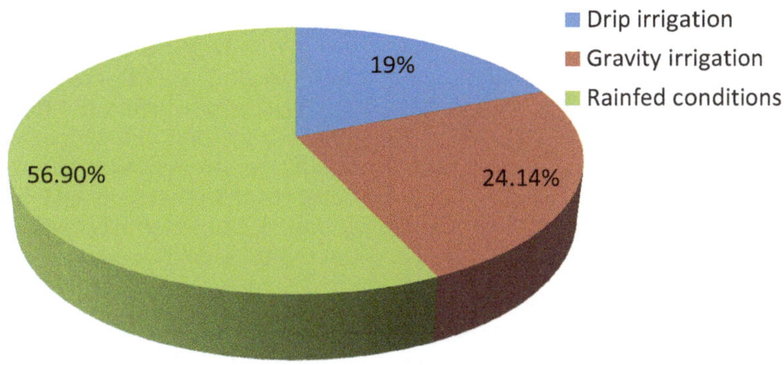

Fig. 2. Olive orchards distribution according to the driving mode

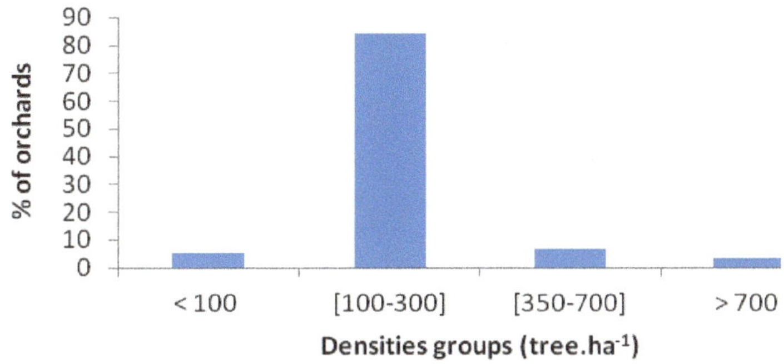

Fig. 3. Olive orchards distribution according to their planting densities

Table 1. Use of manure by olive growers

Manure	% of orchards
Sheep manure	12.1
Cattle manure	1.7
Sheep manure + Cattle manure	44.8
No	41.4

3.1.6 Variety

The dominant variety in olive orchards studied was the Moroccan Picholine with a percentage of 98.3% of the total surveyed orchards.

3.1.7 Yield

The yields reported by growers surveyed vary widely from one farm to another. They vary between 0 and 14.3 $t.ha^{-1}$. Olive trees under drip irrigation recorded an average yield of 5.1 $t.ha^{-1}$, those led by gravity irrigation recorded an average yield of 1.9 $t.ha^{-1}$. For orchards in rainfed conditions, the average yield is 1.8 $t.ha^{-1}$.

3.2 Assessment of Soil Fertility of Olive Orchards Sampled

Based on analyzes of samples taken from soil, the majority of these are basic, calcareous, and largely poor to moderately filled in organic matter.

3.2.1 Organic matter content

The organic matter content was highly variable from one floor to another. It was between 0.1 and 4.5% for the soil layer 0-30 cm, and between 0.04 and 2.7% for the soil layer 30-60 cm. The surface horizon is usually provided by organic matter (mean = 2.1%) than the 30-60 cm horizon (mean = 1.3%). The olive trees are reported to grow quite well on soils containing more than 1% of organic matter [13], even if a threshold of 1.5% is considered low in other conditions [14].

3.2.2 Active limestone content

The average content of active limestone was 12.5% and 10.9%, respectively, for the soil layers 0-30 cm and 30-60 cm. It varied between 0 and

18% for the 0-30 cm horizon and 0 to 19.8% for the 30-60 cm horizon. Excellent yield and vegetative growth can be observed, in both soils with small amount of limestone and where this amount reaches 50%, with a limit of 76% [15].

3.2.3 pH

The soil pH ranged from 6.3 to 8.8 for the 0-30 cm horizon and 6.4 to 8.9 for the 30-60 cm horizon. Alkaline soils predominate in the olive orchards studied. Indeed, in the two soil layers 0-30 and 30-60 cm, 96.6% and 93.1% of soils, respectively, are basic soils according to standards set by [16] (Fig. 4). It is well known that the olive-tree tolerates a wide margin of soil pH, but the neutral, slightly alkaline values to alkaline ones, between 7 and 8.5, assure its best development [17].

3.2.4 Soil fertility

For a given site, soil fertility levels have declined in the lower layers of the soil in comparison to the topsoil [5]. In our study, the soil nitrate levels ranged between 1.8 and 71.4 $mg.kg^{-1}$ and between 1.5 and 40 $mg.kg^{-1}$, respectively, for the soil layers 0-30 cm and 30-60 cm. For phosphorus, soils showed levels ranging from 1.3 to 59.3 $mg.kg^{-1}$ for the 0-30 cm layer and from 1.4 to 41.7 $mg.kg^{-1}$ for the 30-60 cm layer. The soil potassium ranged between 43.8 and 1456.5 $mg.kg^{-1}$ for the 0-30 cm horizon, and between 34.4 and 997.7 $mg.kg^{-1}$ for 30-60 cm horizon. The soil P and K levels are higher than the values found in other studies in Tunisia, Syria and Spain. Indeed, in a soil fertility survey of olive trees in Sidi Bouzid (northwest Tunisia), [18] reported values between 6 and 19.9 $mg.kg^{-1}$ to P (average of 11.8 $mg.kg^{-1}$) and from 90 to 270 $mg.kg^{-1}$ for K (average of 133 $mg.kg^{-1}$) in the 0-30 cm soil layer. In Syria, [5] reported in the topsoil (0-20 cm), the P values ranging from 2.4 to 23.2 $mg.kg^{-1}$ and K values between 14 and 386 $mg.kg^{-1}$. [19] reported, in Andalusia, soil potassium contents between 61 and 171 $mg.kg^{-1}$) in the 0-30 cm layer. Pa against, [20] found, in the Andalusian olive, soil potassium contents between 105 and 565 $mg.kg^{-1}$ and soil phosphorus contents between 60 and 162 $mg.kg^{-1}$ in the 0-30 cm layer.

3.3 Determination of the Nutritional Status of the Olive Orchards

The olive leaf analysis revealed low leaf nitrogen, phosphorus and potassium contents which have varied, respectively from 0.224 to 0.602%; from 0.036 to 0.255 and from 0.338 to 1.078%. [18] found values between 0.058% and 0.14% for P and between 0.43% and 0.93% for K. [5] reported values ranging from 0.95 to 1.56% for nitrogen, from 0.07 to 0.14% for phosphorus and from 0.12 to 1.07 for potassium. The leaf mineral contents were in most cases lower than the deficiency levels reported in the literature. Indeed, all of the studied orchards require nitrogen inputs, according to the fertilization standards defined by [21] and [22] (Table 2). 91% of orchards need potassium fertilization. As for phosphorus, it was less of a problem compared with nitrogen and potassium as about a third of sampled soils need a supply of phosphorus.

Table 2. Percentage orchards should be fertilized based on leaf analyzes

Mineral element	% of orchards
N	100 %
P	36.2 %
K	91.4 %

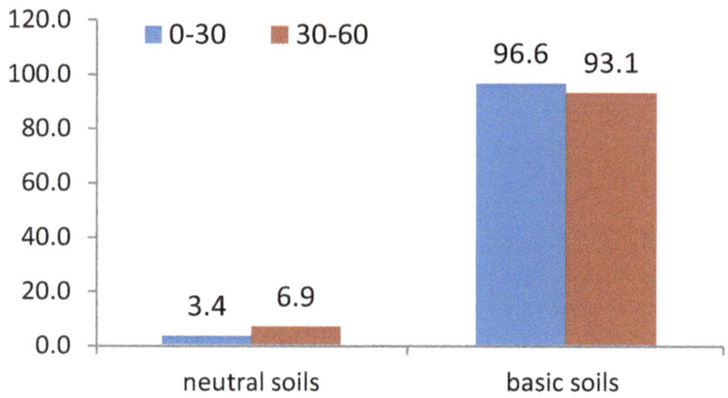

Fig. 4. pH of soil sampled groves

3.4 Relations between the Parameters Studied

The data statistical analysis showed no correlation between the olive yield and soil nitrate or phosphorus or exchangeable potassium contents (Table 3). [5] in their study in Syria, found that the most important factor was the amount of potassium available in the root zone explaining the two-thirds of the yield variability (R^2 = 68%) followed by total nitrogen in the root zone that was ranked second with 58% of the explained variability and mineral nitrogen in the root zone, which explained 44% of variability (significant correlation).

The olive yield is not correlated with the leaf nitrogen or phosphorus or potassium contents (Table 4). [5] found the same results for nitrogen and phosphorus when they noted a significant correlation between the olive yield and the leaf potassium contents with 26% of the explained variability. A strong negative correlation was observed by [23] between the olive yield and leaf phosphorus and potassium content.

We also studied the relationship between the olive leaf nitrogen, phosphorus and potassium contents, at the 58 sampled orchards, and soil nitrate, available phosphorus and exchangeable potassium contents, at the 0-30 cm horizon. Figs. 5, 6 and 7 present regressions curves produced between these parameters for each driving mode orchards adopted by farmers (rainfed conditions, gravity irrigation and drip irrigation). These have shown that nutrition parameters of olive leaves are not linked to those of soil fertility. With the exception of nitrogen where much (about half) of the leaf content has been explained by the soil nitrate content but only in irrigated orchards (Fig. 5). Note that all these regressions are not statistically significant. [18] also showed no relationship between the olive leaf phosphorus and potassium content and the soil phosphorus and potassium content, respectively.

Table 3. Regressions between the olive yield and the soil nitrates, phosphorus and potassium contents, according to the three driving modes of olive groves

Yield	Rainfed conditions		Gravity irrigation		Drip irrigation	
	Model	**R^2**	**Model**	**R^2**	**Model**	**R^2**
Soil nitrates	$y = -0,0005x^2 + 0,0257x + 1,5725$	$R^2 = 0,0067$	$y = 2,96e\text{-}0,036x$	$R^2 = 0,3033$	$y = -0,0007x^2 - 0,0638x + 6,6668$	$R^2 = 0,076$
Soil phosphorus	$y = 0,0002x^2 - 0,0365x + 2,3283$	$R^2 = 0,0255$	$y = -0,0032x^2 + 0,1696x + 0,3605$	$R^2 = 0,1802$	$y = -0,017x^2 + 0,9779x - 3,967$	$R^2 = 0,4261$
Soil potassium	$y = 1E\text{-}06x^2 - 0,0013x + 2,0524$	$R^2 = 0,0062$	$y = -1E\text{-}05x^2 + 0,0087x + 0,4994$	$R^2 = 0,1166$	$y = 0,0002x^2 - 0,094x + 17,129$	$R^2 = 0,1734$

Table 4. Regression between the olive yield and the leaf mineral contents, according to the three driving modes of olive groves

Yield	Rainfed conditions		Gravity irrigation		Drip irrigation	
	Model	**R^2**	**Model**	**R^2**	**Model**	**R^2**
% Leaf nitrogen	$y = -8,8455x^2 + 8,6368x - 0,1061$	$R^2 = 0,0045$	$y = 184,75x^2 - 118,7x + 20,193$	$R^2 = 0,3515$	$y = -166,6x^2 + 125,2x - 17,805$	$R^2 = 0,0299$
% Leaf phosphorus	$y = 193,49x^2 - 38,882x + 3,4603$	$R^2 = 0,0327$	$y = 300,44x^2 - 68,193x + 4,9764$	$R^2 = 0,6275$	$y = -720,24x^2 + 253,39x - 13,875$	$R^2 = 0,1631$
% Leaf potassium	$y = -12,073x^2 + 16,425x - 3,3019$	$R^2 = 0,0299$	$y = 4,2404e\text{-}1,676x$	$R^2 = 0,3184$	$y = -185,75x^2 + 201,21x - 46,614$	$R^2 = 0,2352$

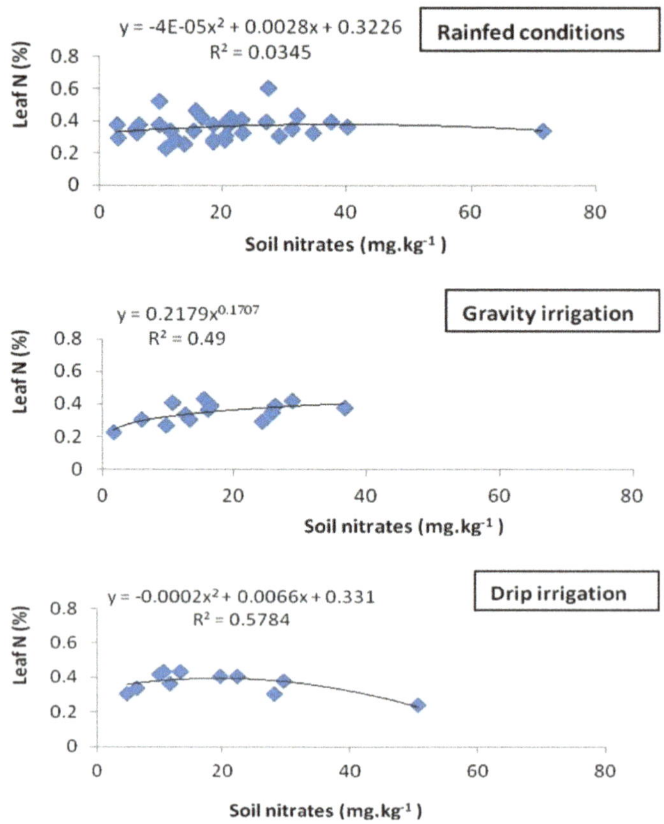

Fig. 5. Evolution of the leaf nitrogen content according to the soil nitrate content (0-30 cm) for the three driving modes of the studied orchards

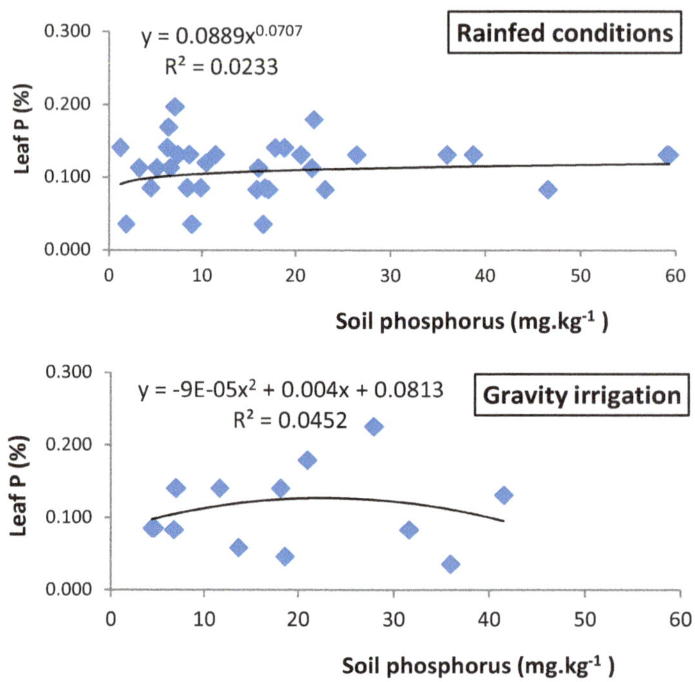

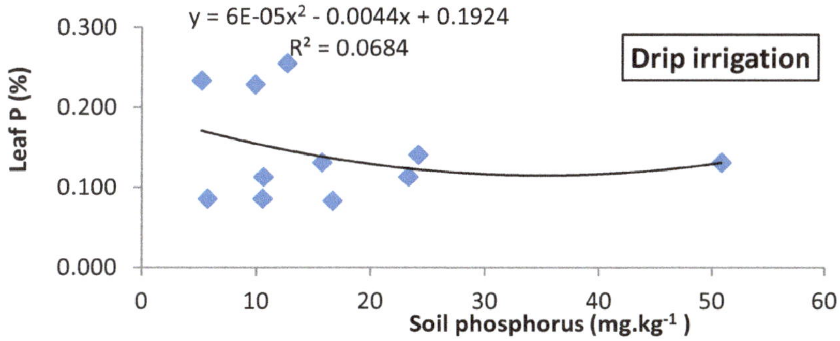

Fig. 6. Evolution of the leaf phosphorus content depending on the soil available phosphorus content (0-30 cm) for the three driving modes of the studied orchards

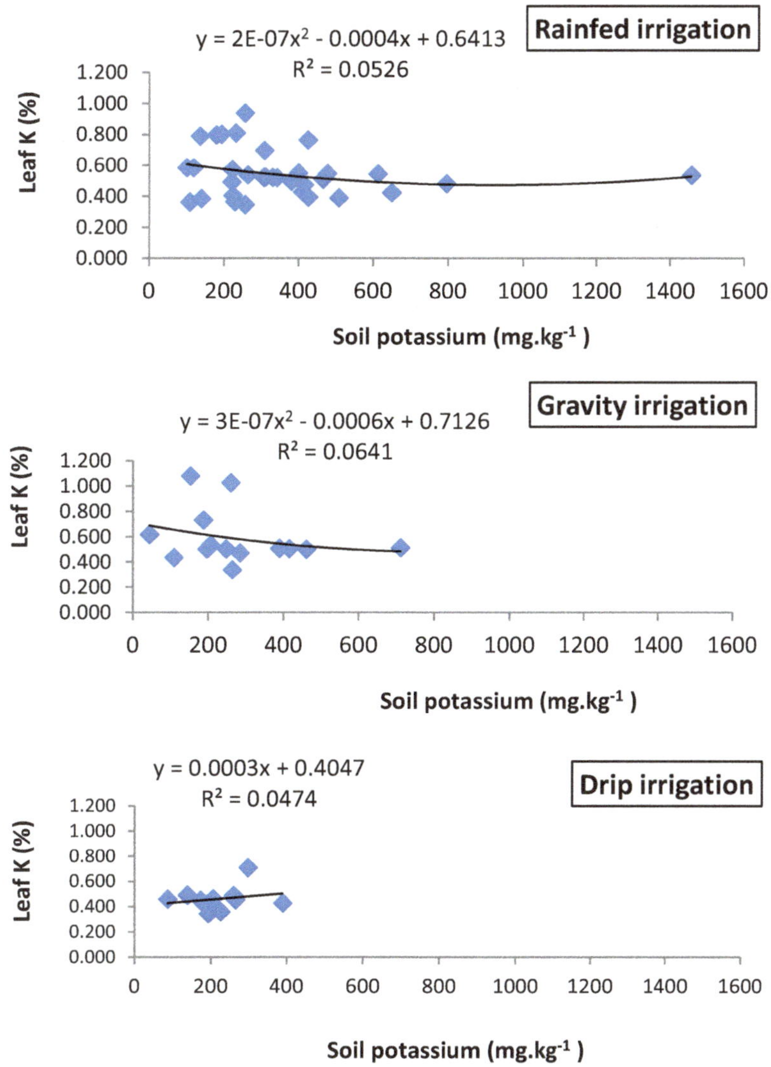

Fig. 7. Evolution of the leaf potassium content depending on the soil exchangeable potassium content (0-30 cm) for the three driving modes of the studied orchards

4. CONCLUSION

- The surveys have revealed shortcomings as regards fertilization practices of olive cultivation at Sais region. Indeed, approximately 48.3% of the farmers surveyed provide no fertilizer for their olive trees.
- The laboratory analyzes of soil and leaves samples collected in these orchards have confirmed these failures: Much of olive orchards require additions in fertilizers.
- The study of the correlation between yield and some soil parameters showed no correlation between the olive yield and soil fertility parameters.
- Also, the olive yield was not correlated with the leaf nitrogen or phosphorus or potassium contents.
- The relationship between the olive nutrition parameters and the soil fertility parameters, at the 0-30 cm horizon in the olive groves, showed that olive leaf parameters were not linked to those of soil fertility. However, much of the leaf nitrogen content (about half) has been explained by the soil nitrate content but only in irrigated orchards.

This diagnostic work done at Sais region has identified problems concerning, especially, fertilization of olive trees. The determination of the olive tree fertilization standards is a necessary step for a rational management of fertilizers for the olive trees. These standards will be developed through field trials.

ACKNOWLEDGEMENT

Our thanks go to Ms Sahar Bennani for her assistance in statistical analysis.

COMPETING INTERESTS

Authors have declared that no competing interests exist.

REFERENCES

1. Gargouri K, Sarbeji M, Barone E. Assessement of soil fertility variation in an olive orchard and its influence on olive tree nutrition. Second international seminar. Biotechnology and quality of olive tree products around the mediterranean basin. Marsala-Mazara del Vallo, Italy. 5-10 November; 2006.
2. Bouma J. Precision agriculture: introduction to the spatial and temporal variability of environmental quality. In: Lake JV, Bock GR, Goode JA. (Eds.), Precision agriculture: Spatial and temporal variability of environmental quality. Ciba Foundation Symposium, 210. Wiley, Wageningen, the Netherlands. 1997;5-17.
3. Michelakis N. Olive orchard management: advances and problems. Proc. 4th International Symposium on Olive Growing. Eds. C. Vitagliano and G.P. Martelli. Acta Hort. 2002;586:239-245.
4. Touzani A. Tour d'horizon de l'oléiculture mondiale et programme d'intervention du COI. Olivea. 1998; French.
5. Tubeileh A, Turkelboom F, Al-Ibrahem A, Thomas R and Tubeileh K.S. Modelling the Effects of Soil Conditions on Olive Productivity in Mediterranean Hilly Areas. International Journal of Agronomy. 2014;(2014)12. Article ID 672123.
6. General description of olive growing in Morocco. Available:http://www.internationaloliveoil.org
7. Sims, James R, Jackson, Grant D. Rapid analysis of soil nitrate with chromotropic acid. Soil Science Society of America; 1971.
8. Olsen SR, Cole CV, Watanabe FS, Dean LA. Estimation of available phosphorus in soils by extraction with bicarbonate. U.S. Dep. Agric. Circ. 939, USA; 1954.
9. Allison LE. Organic Carbon. Dans CA, Black ed. Methods of soil analysis. Part II, chap. 90. American Society of Agronomy. Inc., Madison, Wis. 1965;1372-1376.
10. Drouineau G. Rapid determination of the active limestone soil. Reportation new data on the nature of the limestone fractions. Ann. Agron. French. 1942;12:441-450.
11. Chapman HD. Cation-exchange capacity. Dans C. A. Black, ed. Methods of soil analysis. Part II, Chap. 57 and 58. American Society of Agronomy. Inc. Madison, Wis. 1965;891-903.
12. El Gharous M, El Amrani M, El Mjahed K. Manuel des Méthodes d'Analyses de Sol et de la Plante. Aridoculture, Institut National de la Recherche Agronomique. Morocco. 1995; French.
13. Soyergin S, Moltay I, Genç C, Fidan AE, Sutçu AR. Nutrient status of olives grown in the Marmara region. Acta Hort. 2002; 586:375-379.

14. Freeman M, Carlson RM. Mineral nutrient availability. In: Olive production manual, ferguson et al. eds. Publication 3353. University of California. 1994;69-75.
15. Llamas JF. Basis of fertilization in olive cultivation and the olive tree's vegetative cycle and nutritional needs. Int. Course on the fertilization and intensive cultivation of the olive. Cordoba. 1983;1-25.
16. Chafai Elalaoui A. Guide engrais fertilisation. Edition: Ecole Nationale d'Agriculture. Meknès, Maroc; 2003. French.
17. Martinez Raya A. Suitable land for olive cultivation. Inter. Course on the Ferti. and Intens. Cult. Of the Olive, Spain. 1983;26-29.
18. Gargouri K, Mhiri A. Relationship between soil fertility and phosphorus and potassium olive plant nutrition. Options Mediterranean's: Serie A. 2002;50:199-204.
19. Restrepo-Diaz H, Benlloch M, Navarro C and Fernández-Escobar R. Potassium fertilization of rainfed olive orchards. Scientia Horticulturae. 2008;16(4):399-403.
20. Ferreira Llamas J. Tests and standardized trials used for olive tree fertilization. In international course on the fertilization and intensive cultivation of the olive. UNDP/FAO/Spanish national institute for agricultural research, Cordoba, Spain, 1984;30-52.
21. Marin L, Fernández-Escobar R. Optimization of nitrogen fertilization in olive orchards. In: Val J, Montanés L, Monge E. (Eds), Proceeding of the third International symposium on mineral nutrition of deciduous fruit trees, Zaragoza, Spain. 1997;411-414.
22. Fernandez-Escobar R, Moreno R, Garcia-Creus M. Seasonal changes of mineral nutrients in olive leaves during the alternate-bearing cycle. Scientia Horticulturae. 1999;82:25-45.
23. Jasrotia A, Sing RP, Bhutani VP, Sing IM. Influence of tree characters and nutrients status of olive trees on their productivity. Acta Horticulturae. 1999;474:313-315.

4

The Utilization of Vermicomposting Outputs in Ecology Soilless Culture of Lettuce

M. A. Abul-Soud[1*], M. S. A. Emam[1], A. M. H. Hawash[1], M. H. Mohammed[1] and Z. Y. Maharik[1]

[1]Central Laboratory for Agricultural Climate, Agricultural Research Centre, Giza, Egypt.

Authors' contributions

This work was carried out in collaboration between all authors. All authors read and approved the final manuscript.

Editor(s):
(1) Abdel-Tawab H. Mossa, Environmental Toxicology Research Unit (ETRU), Pesticide Chemistry Department, National Research Centre, Egypt.
Reviewers:
(1) Anonymous, Institute for Tropical and Sub-tropical Crops, South Africa.
(2) Rezzoug Waffa, Ibn Khaldoun University, Algeria.
(3) Anonymous, SDM College of Engineering and Technology Dharwad, India.
(4) Anonymous, National Research Centre, Cairo, Egypt.

ABSTRACT

The increase demands for food security, the expected future of climate change impacts, environmental concerns, water shortage and the need for recycling different organic wastes and mitigate their CO_2 emission were the driving forces for developing the use of vermicomposting outputs and ecology soilless culture. Two experiments were carried out during autumn seasons of 2011, 2012 and 2013 at Central Laboratory for Agricultural Climate (CLAC), Giza, Egypt under net house. First study investigated the ability of vermicomposting outputs as substrate and organic nutrient solution. Vermicompost mixed with perlite, vermiculite and sand (30: 70% *v/v*) compared to peat moss + perlite (50: 50% *v/v*) combined with different sources of nutrient solutions (vermi-tea, vermi-liquid and chemical solution) on the yield of lettuce. Study the effect of vermicompost mixed with perlite in different proportions of 1:1, 1:2 and 1:3 (v/v) compared to peat moss: perlite combined with different sources of nutrient solution (vermi-tea, compost-tea and chemical) on lettuce yield take a place in second study.Iceberg lettuce type *cv.* Robinson F1 hybrid was cultivated in both two experiments in split plot design with three replicates.
Data revealed that using vermicompost as a substrate combined with different substrates recorded

**Corresponding author: Email: abul_soud1@yahoo.com*

the highest values of the physical properties of head lettuce during the two successive seasons compared to peat moss + perlite. Vermicompost + sand followed by vermicompost + perlite recorded the highest results of the physical properties of head lettuce. Otherwise, obtained results of the second experiment illustrated that increasing the vermicompost rate decreased the physical properties of head lettuce. The proportion of vermicompost + perlite (1: 3 *v/v*) gave the highest records of the physical properties of head lettuce compared to the control. The vermi-liquid and vermi-tea gave positive effects on the physical properties of head lettuce in the first and second study respectively. The chemical nutrient solution recorded the highest N (%) content in both experiments.

The study supported the use of vermicompost and vermi-liquid instead of peat moss and chemical nutrient solution in ecology soilless production of lettuce plants. The vermicomposting could be used for recycling the organic urban wastes to produce input materials of soilless, green roof and ecology culture besides mitigating greenhouse gases (GHG's) emission.

Keywords: Ecology soilless culture; vermicompost; vermi-liquid; substrate culture organic urban wastes; manure; lettuce.

1. INTRODUCTION

With the drastic increase in population globally, there is no doubt that food supply will have to be increased in order to meet the demand. All modern agricultural techniques (soilless culture, vermicomposting and etc.) are playing a vital role as maximizing yield of crops has become one of the priorities of the farmers these days.

The needs for optimizing the soilless culture inputs and maximize the production with the concern of the environmental impacts led to the development of the ecology soilless culture system via alternate the peat moss and chemical nutrient solution by vermicomposting outputs. Peat moss is the most wide use substrate in horticulture activities (seedlings production and soilless culture) for its desirable physical and chemical properties and the high production output but this substrate is un-regenerated natural resource. While the environmental and ecological concerns in the recent years led to minimize the use of peat because its harvest is destroying endangered wetland ecosystems worldwide [1]. At the same time, the need to produce local substrate instead of importing it drive many researchers to develop different substrate to play the role of peat moss. Several studies revealed that peat can be substituted by various compost types without any negative effects on a variety of crops raised in these substrates [2,3].

On the other hand, the commercial soilless culture progress slowly while it's expected to grow so fast through the next years according to the increase demands for food security, the expected future of climate change impacts and water shortage will be the driving forces to pay more attention on soilless culture. Soilless culture depends greatly on chemical nutrient solution regardless of the type system (hydroponic, aeroponic and substrate).

Via the vermicomposting, multi products could be offer such as vermicompost, vermicompost-tea, vermi-liquid (liquid collected during vermicomposting process) and earthworm biomass. Vermicomposting also had a friendly environmental impact through mitigating GHG's emission of different organic wastes instead of burial or incineration. Vermicomposting (Worm composting) is defined as a process in which earthworms play a major role with microbes in the conversion of organic solid waste into more stabilized dark, earth-smelling soil conditioner and nutrient-rich compost that is rich in major and micronutrients. During vermicomposting, organic matter is stabilized by the enhanced decomposition (humification) in presence of earthworms [4], but by a non-thermophilic process [5]. Different organic wastes can be used in vermicompost production by different species of earthworms which include horse waste [6]; urban solid waste [7]; city leaf litter and food wastes [8,9]; paper waste and residues of plant decomposition.

Several studies assessed the effect of vermicompost amendments in potting substrates on seedling emergence and growth of a wide range of marketable fruits cultivated in greenhouses [10,11], as well as on growth, yields [11-12]. As the price of peat is increasing each year, farmers have gradually decreased the use of peat and changed to other substrates such as coco peat but the cost and handling of coco peat still not enough profit. Using local resources for soilless substrate can decrease costs.

Vermicomposting process produces leachate due to the activities of micro-organism and draining of leachate is important to prevent saturation of the vermicomposting unit. Leachate thus derived from vermicomposting are regarded as beneficial and can be used as liquid fertilizer due to high concentration of plant nutrients [13,14].

The term vermi-wash (Vermi-liquid) was coined by [15]. Leachate is generated along with vermicomposting process commonly referred to as vermicomposting leachate or worm-bed leachate, [16] Extract from vermicompost is known as vermicompost extract, [17]. The preparations of these vermicomposting derived liquids are different. Vermicomposting derived liquids contain valuable nutrients that promote plant growth. Substrates that have been used in these liquids production are mainly animal and agricultural waste.

Merrill and McKeon [18] reported that interest in organic teas for use in agriculture and horticulture has grown rapidly during the last decade. Many scientists suggest that certain liquid extractions of manures or composts (herein called "Organic Teas"), at various stages of decay, can supply plants with at least four major benefits [19-21]: A source of plant nutrients; a source of beneficial organic compounds, an ability to suppress certain plant diseases; a way to build soil structure when applied as a drench.

The main objectives of the current study were to assess ecology soilless culture under Egyptian condition, to investigate the use of vermicomposting outputs in lettuce production and to offer alternative sources for peat moss and chemical nutrient solution.

2. MATERIALS AND METHODS

Two experiments were conducted in the experimental station at the Central Laboratory for Agricultural Climate (CLAC), Agriculture Research Center (ARC), Egypt, during the autumn seasons of 2011 and 2012for the first experiment while the second take place in 2012 and 2013 under double span net house (18 x 60 x 4.5 m).

2.1 Plant Material

Lettuce (Iceberg type) Robinson F1 hybrid seeds were sown on 2nd week of October in cultivated seasons in polystyrene trays. After the fifth true leaf stage, lettuce seedlings were transplanted to the substrate culture. Seven seedlings of lettuce were planted in horizontal polyethylene bags (1 m length - 30 L volume) in open system. The bags were placed in double rows. The final plant spacing was 20 cm in the row, 15 cm between the rows.

2.2 The Vermicomposting Process

The Epigiec earthworms *Lumbriscus rubellus* (Red Worm), *Eisenia fetida* (Tiger Worm), *Perionyx excavatus* (Indian Blue) and *Eudrilus eugeniae* (African Night Crawler) were used in the vermicomposting beds system. Epigiec earthworms (fife Kg) were placed in each bed system. Worm diameter: 0.5 – 5 mm and worm length: 10 – 120 mm. Bed system of vermicomposting was used in this investigation for producing the vermicompost and vermi-liquid. Eight Beds were established under black net house by digging the soil and mulched with black polyethylene plastic sheet 0.5 ml to perform a bed with length 2.5 m, width 1.2 m and depth 50 cm. A slope 1.5 % had been done by using water balance to collect the vermin-liquid through water bucket. Mixing the different raw materials: horse manure (H. M) + vegetable and fruit wastes (V. F. W) + shredded paper (Sh. P) in the rate of 2: 2: 1 (*v/v*) respectively was done by using turning machine and pre-composting of different raw materials for 7 to 10 days before feed it to worms to avoid the thermophilic stage (increase temperature above 35℃ cause the death of earthworms in vermicompost systems). After pre-composting done, the final mix soaked in water for 0.5 to 1 hour to make sure there was no anymore dry parts, then put it in lines along the bed with the soaked water. The composition of the different organic wastes presented in Table 1. The feeding of earthworm done every two days and every 21 days the earthworms were fasting for 7 days to give them the opportunities to re-eat the cast and to avoid non composted wastes. Moisture content was in the range of 60 – 70%.

2.3 The First Study Treatments

Two factors were studied, first: Three substrate mixtures were used by substituting a commercial peat medium with vermicompost (VC) as follows: vermicompost + perlite, vermicompost + sand, vermicompost + vermiculite in proportion of (30%: 70% *v/v*) compared to peat moss + perlite (control) (50%: 50% *v/v*). Second, three sources of nutrient solutions vermicompost-tea, vermi-liquid and chemical nutrient solution (control).

The experimental design was a split plot with3 replicates. Each experimental plot contained 7 plants. The sources of fertigation were assigned as main plots and vermicompost mixes as subplots.

2.4 The Second Study Treatments

Three different rates of vermicompost (V) mixed with perlite (P) as follows: 1:1, 1:2 and 1:3 (v/v) compared to peat moss: Perlite (control) combined with different sources of nutrient solution vermi-tea, compost-tea and chemical (control).

The experimental design was a split plot with3 replicates. Each experimental plot contained 7 plants. The sources of nutrient solutions were assigned as main plots and vermicompost rates as subplots.

2.5 System Materials

Horizontal polyethylene bags (90 x 20 cm with 30 L volume) were filled with different substrate mixes and placed in two rows in open substrate system.

The compost-tea and vermicompost-tea were prepared by soaking 10 kg of both in water tank (50 L) for 24 hours (active extract) to get the concentrated extractions that were going to be used as nutrient solutions. Filtration was made before using the compost-tea and vermicompost-tea to get the clear solution for fertilizing the lettuce and to prevent the dust included in to block the dripper [22]. On the other hand, the vermi-liquid was used directly after filtered. Different nutrient solutions pumped via submersible pump (110 watt). Water tanks 120 L were used in open system of soilless culture. Plants were irrigated by using drippers of 4 l/hr. capacity. The fertigation was programmed to work 8 times / day and the duration of irrigation time depended upon the season. The Electrical conductivity (EC) of the different nutrient solutions were adjusted by using EC meter to the required level (1.5 $mmhos^{-1}$). The chemical compositions of vermicompost-tea, vermi-liquid, compost-tea and chemical nutrient solution were illustrated in Table 2.

Substrate physical properties were estimated according to [23,24] as followed:

The bulk density (*B.D*) was simply measured as dry weight/volume (g/cm^3 or kg/l)

Total pore space (T.P.S) was percentage pore space and the proportion and amount of water and air that was present in pore space

Total pore space
= *(1 – bulk density/ true density) x 100*

Table 1. The chemical composition (%) of the different agricultural wastes

Raw material	C/N ratio	Macro elements %				
		N	P	k	Ca	Mg
H. M	26.41	1.29	0.48	2.39	1.45	1.52
V. F. W	62.60	0.34	0.19	0.64	0.81	0.43
Sh. P	166.81	0.016	0.01	0.00	0.20	0.01
The mix	78.18	0.78	0.31	0.73	0.81	0.59

H. M = horse manure. V. F. W = vegetable and fruit wastes. Sh. P = shredded paper

Table 2. The chemical composition of different sources of nutrient solutions used in the study

Nutrient source	Macro nutrients (ppm)				
	N	P	K	Ca	Mg
Vermi-tea	103	12	258	111	46
Vermi-liquid	148	32	345	98	61
Compost-tea	130	15	220	105	27
Chemical	120	30	180	120	30
	Micro nutrients (ppm)				
	Fe	Mn	Zn	B	Cu
Vermi-Tea	4.25	1.50	0.26	0.31	0.20
Vermi-liquid	11.05	2.37	0.31	0.32	0.16
Compost-tea	5.65	0.84	0.22	0.60	0.34
Chemical	3.00	0.75	0.25	0.25	0.15

Water hold capacity % (W.H.C) is the amount of water present after the substrate in a container has been saturated and allowed to drain.

Water hold capacity % = ((FW- DW)/ VB) x 100

FW (fresh weight) = weight of substrate after stop draining

DW (dry weight) = dry weight of substrate after 24 hours at a temperature 80 – 90℃.

Air porosity % (A.P) was the proportion of the volume of substrate *(VB)* that contains air after it has been saturated with water and allowed to drain. Collect the volume of water leached plus the volume of air present after the substrate in a container allowed to drain.

Air porosity % = T.P.S – W.H.C.

The pH of the potting mixtures were determined using a double distilled water suspension of each potting mixture in the ratio of 1:10 (w:v) [25] that had been agitated mechanically for 2 h and filtered through Whatman no.1 filter paper. The same solution was measured for electrical conductivity with a conductance meter that had been standardized with 0.01 and 0.1M KCl.

Samples of three plants of each experimental plot were taken to determine growth parameters after harvesting (55 days from the transplanting date) as follows head weight (gm), head volume (cm^3), Density (g/cm^3), the inner diameter such as the head length (cm) and head width (cm).Total dry weight was determined after oven-drying the leaf samples at 70°C for 48 hours.

For mineral analysis of leaves (N, P and K %), Three plant leaf samples of each plot were dried at 70℃ in an air forced oven for 48 h. Dried leaves were digested in H_2SO_4 according to the method described by [26] and N, P and K contents were estimated in the acid digested solution by colorimetric method (ammonium molybdate) using spectrophotometer and flame photometer [27]. Total nitrogen was determined by Kjeldahl method according to the procedure described by [28]. Phosphorus content was determined using spectrophotometer according to [29]. Potassium content was determined photo-metrically using Flame photometer as described by [27].

Samples of different organic materials and their bulk mix were analyzed before vermicomposting and after according to [27-29].

The calculations of sequestrate carbon dioxide (CO_2) and save the nutrients in the soil were calculated as follows:

Sequestrate CO_2/ton
= C % (raw material) x 10

Nutrient save /ton
= Nutrient % (after composting) x 10

Analysis of data was done by computer, using SAS program for statistical analysis. The differences among means for all traits were tested for significance at 5% level according to Waller and Duncan (1969).

Crop management practices were in accordance with standard recommendations for commercial growers.

3. RESULTS AND DISCUSSION

3.1 The Environmental Impact Assessment of Vermicomposting

The results in Table 5 showed that the vermicomposting process increased the total N, P, K, Ca and Mg % of the vermicompost as compared to the bulk raw materials while C/N ratio decreased as a result of N fixation, concentrated the nutrients and bulk reduction. In addition, the nutrient saved (Kg / tone) via using vermicomposting process from non-significant organic sources such as kitchen wastes and shredded newspapers gave good evidences on recycling the urban organic wastes and the application of the output as Fig. 1 presented. Needless to say that the most important point of utilizing vermicomposting was mitigating the GHG's emission from the different organic wastes used instead of incineration or buried through converted to vermicompost that could be utilize in ecology soilless culture of different vegetables led to more mitigation of CO_2 emission. However, the determined calculation measured the organic carbon of organic wastes used in this study that treated by vermicomposting and produced in different forms was estimated by 605.3 Kg per each tone while after vermicomposting process decreased to 187.1 Kg / tone.

However, in recent years, researchers have become progressively interested in using another related biological process for stabilizing organic wastes, which does not include a thermophilic stage, but involves the use of earthworms for breaking down and stabilizing the organic wastes[4, 5]Never the less, vermicomposting is defined as a low cost technology system for processing organic wastes [30]. More information proved that earthworms act as mechanical blenders besides fragmenting the organic matter which modify its physical and chemical status by gradually reducing the ratio of C: N and increasing the surface area exposed to microorganisms inducing much more favorable media for microbial activity and further decomposition [31].

3.2 The Physical and Chemical Properties of Different Substrate Mixtures

The physical and chemical properties of different substrates mixtures for both two experiments are illustrated in Tables 4 and 5 respectively.

The obtained data that shown in Table 4 indicated that the highest values of total porosity, air porosity and E.C were recorded by the mixture of vermicompost: perlite (30 : 70 % *v/v*) while vermicompost: vermiculite (30 : 70 % *v/v*) gave the highest W.H.C and pH.

The mixture of vermicompost: sand (30: 70% *v/v*) recorded the highest B.D and the lowest T.P.S, W,H.C and A.P. Dealing with heavy substrate mixture may be undesirable for small plants like lettuce because of the difficulty of handling it through preparing the substrate and establish the system. Otherwise, on commercial scale, the handling with light substrate mixture (The lowest bulk density) that vermicompost: perlite introduced is preferable.

Due to mentioned results, the need to apply different proportions of perlite with vermicompost was investigated as Table 5 illustrated. The results demonstrated that there was a significant decrease in the bulk density, total porosity, water holding capacity, E.C and pH, with decreasing vermicompost mixture proportion, while decreasing vermicompost mixture proportion up to 1 : 3 perlite led to increase the air porosity as a result of increasing the perlite that characterized by high air porosity as Table 5 presented. The mixture proportion 1: 1 of vermicompost: perlite presented the highest bulk density, total porosity, and water holding capacity, E.C and pH while the highest air porosity and lowest total porosity and water holding capacity were recorded by proportion 1:3. The cost of different mixture proportion plays a vital role in deciding the suitable substrate mixture beside the physical and chemical properties of substrate. The vegetable crops varied strongly in the desirable physical and chemical properties for achieving

Table 3. The chemical composition (%) of the bulk raw material before and after vermicomposting

Vermicomposting	C/N ratio	Organic C (%)	N (%)	P (%)	K (%)	Ca (%)	Mg (%)
Before	78.18	60.53	0.78	0.31	0.73	0.81	0.59
After	12.8	18.71	1.46	0.59	1.06	1.08	0.89

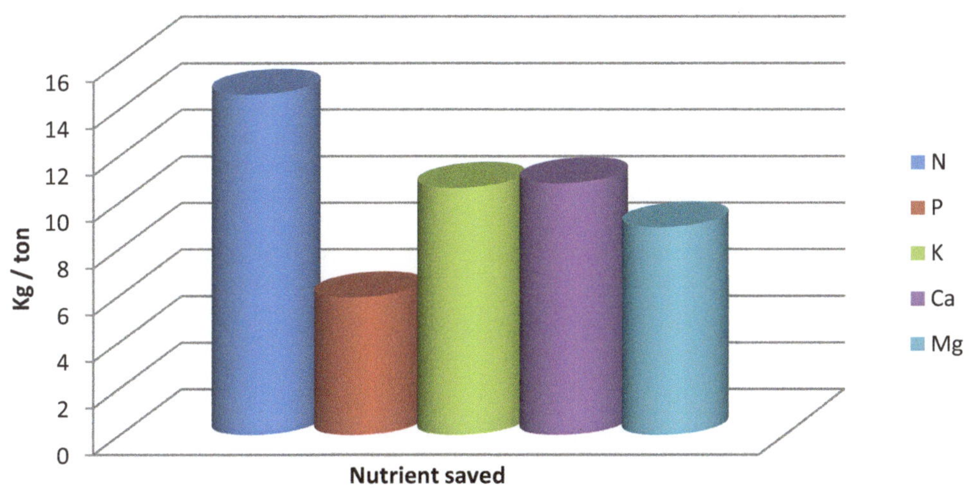

Fig. 1. The nutrient saved of raw organic wastes bulk (Kg/ton) via vermicomposting

the optimum growth and high yield also many factors affected on selecting the suitable commercial substrate such as the availability of energy, the water quality, the used system (open or close)and etc. The suitable commercial substrate doesn't mean the highest or the lowest value of some or all physical and chemical properties but mean mainly the profitable substrate of some crop under specific system and environmental conditions.

3.3 First Experiment

3.3.1 Vegetative growth

Table 6 presented the effect of vermi-output as nutrient solution (N.S) and substrate mixtures on the physical properties of lettuce. The obtained data indicated that the use of vermi-liquid as a nutrient solution gave the highest results of head fresh weight, volume, density, inner length and width of head lettuce while vermi-tea recorded the highest value of dry matter content. On the other hand, the lowest records of physical properties of lettuce were observed from the chemical nutrient solution. Vermicompost-liquid and tea could be used as source for nutrient solution in lettuce production.

Regarding to the impact of vermicompost as substrate, vermicompost had positive significant effect on the vegetative growth of lettuce comparing to control substrate. Vermicompost + sand gave the highest significant results of fresh weight, volume and inner width of head lettuce, while the highest values of density, dry matter content and inner length of head lettuce were gave by vermicompost + perlite compared to control substrate that had the lowest results as presented in Table 6. Accordingly, vermicompost should have a great potential in the horticultural and agricultural industries as media for plant growth. There are only few research studies that have examined the responses of plants to the use or substitution of vermicompost to soil or greenhouse container media [32,33]. Most of these studies confirmed that vermicompost have beneficial effects on plant growth. Vermicompost has considerable potential for substituting peat in horticultural potting substrates.

The data in Table 6 illustrated that the interaction between vermi-liquid combined with vermicompost + sand followed by vermi-liquid combined with vermicompost + perlite recorded the highest results of the physical properties of lettuce. On the contrary, the lowest values were given by chemical nutrient solution combined with peat + perlite (control).

3.3.2 Chemical properties of lettuce

The data in Table 7 illustrate that applying chemical nutrient solution gave the highest contents of N, P and K (%) of head lettuce while vermi-tea treatment as nutrient solution recorded the lowest values of N, P and K of head lettuce.

On the other hand, the effect of substrate mix varied on N, P and K contents of lettuce. The highest N content of head lettuce was recorded by vermicompost + sand (30:70% *v/v*) while Vermicompost + perlite gave the highest contents of P and K. On the other hand, the lowest contents of N and K were presented by control treatment (peat + perlite). Vermicompost + vermiculite gave the lowest content of P as shown in Table 7.

The interaction effect of different nutrient solution sources and substrate mixtures indicated that chemical nutrient solution combined with vermicompost + perlite substrate recorded the highest significant results of P and K while the highest N value given by chemical combined with vermicompost + sand. The most important point in this study was the content of N of head lettuce regarding to the undesired characteristics of high N content in leafy vegetables because of negative hazards of high free N on human nutrition and health [34,35].

The obtained results illustrated that, vermicompost mixed with perlite could introduce good alternative of peat moss referring to the lowest contents of N and highest values of P and K of head lettuce.

Needless to say this, vermi-liquid presented good ability to utilize as nutrient solution instead of chemical nutrient solution but the availability of vermi-liquid limit its application.

Mixing vermicompost with peat and perlite resulted in nutrient content decreased, due to the dilution, and kept nutrients within acceptable, or optimal, ranges for growing tomato transplants [36]. Differences in growth responses were attributed to differences in nutrient content of potting mixes. Although the present study was focused more on effects of vermicompost on plant growth rather than on causes leading to these effects, the results indicated that availability of nutrients is an important factor influencing plant growth [37]. But changes in physical and biological properties of the substrate could also be responsible for observed differences [38].

3.4 Second Experiment

3.4.1 Physical properties of lettuce

Increasing the vermicompost rate mixed with perlite had a negative impact on the physical properties of lettuce (reduce the lettuce head weight) as revealed data of Table 8 indicated. Increasing the rate of perlite in substrate mixtures led to increase the fresh weight, volume, density and dry weight of head lettuce. The mixture of perlite + vermicompost (3: 1 *v/v*) recorded the highest results of fresh weight, volume, density and dry weight of head lettuce. Otherwise, the lowest values were given by perlite + vermicompost (1: 1 *v/v*). These results could be explained due to the high content of nutrients and organic matter content in substrate caused root burning and salinity disorders [3,10] Vermicompost has considerable potential in horticultural potting substrates in low rate mixture of the substrate.

Regarding to the effect of different organic tea on the physical properties of lettuce, the physical properties of head lettuce varied in their response to the nutrient solution source. While chemical nutrient solution recorded the highest value of fresh weight of head lettuce, compost-tea gave the highest result of volume. On the other hand, the highest records of density, dry weight and dry matter presented by vermi-tea (Table 8).

Table 8 illustrated effect of different organic-tea and perlite mixtures on the physical properties of lettuce. The interaction of perlite + vermicompost (3: 1 *v/v*) combined with vermi-tea had the highest data of fresh weight, density, dry weight and dry matter of head lettuce while the highest volume were recorded by perlite + vermicompost (3: 1 *v/v*) combined with compost-tea. The lowest results of physical properties of head lettuce presented by perlite + vermicompost (1: 1 *v/v*) combined with compost-tea.

3.4.2 Chemical properties of lettuce

The revealed data of Table 9 show that the use of chemical nutrient solution recorded the highest values of N, P and K content of head lettuce while vermi-tea gave the lowest N content of head lettuce. On the other hand, compost-tea had the lowest P and K contents of head lettuce. These results regarded mainly for the nutrients balance in the chemical nutrient solution that meet the lettuce nutrient requirements in the contrary for the composition of vermi-tea and compost-tea. The lowest N content of head lettuce is desirable character and considered an advantage for using Vermi-tea.

Table 4. The physical and chemical property of different substrates mixes of first study

Substrate	Physical				Chemical	
	B.D Kg/l	*T.P.S %*	*W.H.C %*	*A.P %*	E.C mmhos^{-1}	pH
Control	0.14 D	65.2 C	52.8 C	12.5 A	0.45 D	7.60B
VC:Sa	1.58 A	44.0 D	39.5 D	4.5 C	0.61 C	7.66AB
VC:Ver	0.91 B	71.5 B	65.5 A	6.0 B	0.90 B	7.83 A
VC : Pr	0.43 C	74.0 A	62.5 B	11.5 AB	1.21 A	7.42B

** Bulk density (B.D), Total pore space (T.P.S), water holding capacity (W.H.C), Air porosity (A.P)*
***Control peat moss + perlite (50 %: 50 % v/v), VC: Sa vermicompost: sand (30: 70 % v/v), VC: Ver vermicompost: vermiculite (30 : 70 % v/v), VC : Pr vermicompost: perlite (30: 70 % v/v)*
***Capital letters indicate the significant difference of each factor (P<0.05)*

Table 5. The physical and chemical property of different substrates mixes of second study

Substrate	Physical				Chemical	
	B.D Kg/l	*T.P.S %*	*W.H.C %*	*A.P %*	E.C mmhos^{-1}	pH
Control	0.140 D	65.2 BC	52.8 B	12.5 C	0.45 D	7.60 A
V: P 1:1	0.495 A	73.0 A	62.0 A	11.0 D	1.55 A	7.45 AB
V: P 1:2	0.383 B	66.7 B	51.3 B	15.5 B	1.47 B	7.40 B
V: P 1:3	0.284 C	64.0 C	46.0 C	18.0 A	1.17 C	7.30 B

**Bulk density (B.D), Total pore space (T.P.S), water holding capacity (W.H.C), Air porosity (A.P)*
***Control peat moss + perlite (50%: 50% v/v), V: P vermicompost: Perlite 1: 1, 2 and 3 v/v);*
***Capital letters indicate the significant difference of each factor (P<0.05)*

Table 6. Effect of different nutrient solution sources and substrate mixtures on the physical properties of lettuce

Nutrient solution	First season 2011/2012					Second season 2012/2013				
	Substrate					Substrate				
	Control	VC : Pr	VC:Sa	VC:Ver	Mean(B)	Control	VC : Pr	VC:Sa.	VC:Ver	Mean(B)
Fresh weight (g/plant)										
Vermi-tea	399 gh	704 c	724 c	611 e	609.5 B	411 gh	725 c	735 c	629 e	625.4 B
Vermi-liq	419 g	740 b	804 a	644 d	651.7 A	431 g	762 b	828 a	663 d	671.2 A
Chemical	373 h	662 d	696 c	571 f	575.5 C	384 h	681 d	716 c	588 f	592.7 C
Mean (A)	397.0 D	702.0 B	741.3 A	608.7 C		409.0 D	723.1 B	760.1 A	626.9 C	
Volume (cm^3)										
Vermi-tea	510.0 i	592.7 g	830 c	768.3 d	675.2 B	520.0 h	607.5 f	855.0 b	768.2 d	687.6 B
Vermi-liq	556.3 h	632.4 f	910 a	739.7 de	709.6 A	556.2 g	637.3 e	920.0 a	743.0 d	714.2 A
Chemical	506.6 i	645.6 f	866 b	713.4 e	682.9 B	480.0 i	595.0 f	825.0 c	636.6 e	634.2 C
Mean (A)	524.3 D	623.6 C	868.7 A	740.5 B		518.7 D	613.3 C	866.7 A	634.2 C	
Density (g/cm^3)										
Vermi-tea	0.78 ef	1.19 a	0.87 cd	0.80 def	0.91 A	0.78 fg	1.19 b	0.88 de	0.85 e	0.93 B
Vermi-liq	0.75 f	1.17 a	0.88 c	0.87 cd	0.92 A	0.85 e	1.26 a	0.93 d	0.84 ef	0.97 A
Chemical	0.74 f	1.03 b	0.80 def	0.85 cde	0.84 B	0.74 g	1.03 c	0.83 ef	0.85 e	0.86 C
Mean (A)	0.76 C	1.13 A	0.85 B	0.84 B		0.79 D	1.16 A	0.88 B	0.85 C	
Dry weight (g/plant)										
Vermi-tea	38.9 j	84.5 a	73.8 e	53.2 g	62.6 A	39.1 h	78.9 c	74.2 d	53.5f	61.4 B
Vermi-liq	39.5 i	81.8 b	74.4 d	53.1 g	62.2 B	39.7 h	84.8 a	75.0 d	53.4 f	63.3 A
Chemical	38.8 j	78.4 c	69.5 f	51.7 h	59.6 C	39.0 h	80.8 e	69.9 e	52.0 g	60.4 C
Mean (A)	39.1 D	81.6 A	72.5 B	52.7 C		39.3 D	81.5 A	73.1 B	52.9 C	
Inner head length (cm)										
Vermi-tea	11.4 c	15.1 b	14.8 b	15.0 b	14.1 B	12.0 c	16.0 b	15.7 b	15.9 b	14.9 B
Vermi-liq	12.5 e	16.6 a	16.3 a	16.4 a	15.5 A	13.2 e	17.6 a	17.3 a	17.4 a	16.4 A
Chemical	11.9 f	15.0 b	14.1 d	14.3 d	13.8 C	12.6 f	15.9 b	15.0 d	15.1 d	14.6 C
Mean (A)	11.9 C	15.6 A	15.1 B	15.2 B		12.6 C	16.5 A	15.9 B	16.1 B	
Inner head width (cm)										
Vermi-tea	10.6 g	13.7 c	14.3 b	12.3 e	12.7B	11.2 i	14.5 c	15.2 b	13.0 f	13.5 B
Vermi-liq	11.8 f	14.4 b	14.9 a	13.5 c	13.6 A	12.5 g	15.3 b	15.8 a	14.3 cd	14.5 A
Chemical	10.3 h	13.0 d	13.5 c	11.6 f	12.1 C	10.8 j	13.6 e	14.2 d	12.2 h	12.7 C
Mean (A)	10.9 D	13.7 B	14.2 A	12.5 C		11.5 D	14.5 B	15.0 A	13.2 C	

Control peat moss + perlite (50 %: 50 % v/v), VC:Sa vermicompost: sand (30 : 70 % v/v), VC:Ver vermicompost: vermiculite (30 : 70 % v/v), VC : Pr vermicompost: perlite (30 : 70 % v/v)

** Similar letters indicate non-significant at 0.05 levels.** Capital letters indicate the significant difference of each factor ($P<0.05$)*

**** Small letters indicate the significant difference of interaction ($P<0.05$*

Table 7. Effect of different nutrient solution sources and substrate mixtures on the chemical properties of lettuce

Nutrient solution	First season 2011/2012					Second season 2012/2013				
	Substrate					Substrate				
	Control	VC : Pr	VC:Sa	VC:Ver	Mean(B)	Control	VC : Pr	VC:Sa.	VC:Ver	Mean(B)
						N (%)				
Vermi-tea	2.11 k	2.57 i	2.93 g	2.78 h	2.60 C	2.17 i	2.56 h	3.17 de	2.89 c	2.70 C
Vermi-liq	2.51 j	3.11 f	3.45 c	3.41 d	3.12 B	2.71 g	3.17 e	3.39 b	3.31 b	3.15 B
Chemical	3.26 e	3.56 b	3.67 a	3.67 a	3.54 A	3.34 b	3.41 b	3.12 e	3.54 a	3.35 A
Mean (A)	2.63 D	3.08 C	3.35 A	3.29 B		2.74 C	3.05 B	3.23 A	3.25 A	
						P (%)				
Vermi-tea	0.43 g	0.53 e	0.56 d	0.38 h	0.48 C	0.54 e	0.54 e	0.55 e	0.45 f	0.52 C
Vermi-liq	0.58 d	0.74 b	0.68 c	0.47 f	0.62 B	0.57 e	0.77 a	0.67 c	0.56 e	0.64 B
Chemical	0.72 b	0.78 a	0.57 d	0.69 c	0.69 A	0.73 b	0.78 a	0.63 d	0.71 b	0.71 A
Mean (A)	0.58 C	0.68 A	0.60 B	0.51 D		0.61 B	0.70 A	0.62 B	0.57 C	
						K (%)				
Vermi-tea	0.64 i	0.76 h	0.89 g	0.88 g	0.79 C	0.58 h	0.79 g	0.84 f	0.84 f	0.76 C
Vermi-liq	0.87 g	1.21 c	1.09 e	1.13 d	1.08 B	0.86 f	1.31 ab	1.15 d	1.11 e	1.11 B
Chemical	1.11 d	1.34 a	1.19 f	1.26 b	1.23 A	1.12 de	1.33 a	1.28 bc	1.27 c	1.25 A
Mean (A)	0.87 C	1.10 A	1.06 B	1.09 A		0.85 C	1.14 A	1.09 B	1.07 B	

Control peat moss + perlite (50 %: 50 % v/v), VC:Sa vermicompost: sand (30 : 70 % v/v), VC:Ver vermicompost: vermiculite (30 : 70 % v/v), VC : Pr vermicompost: perlite (30 : 70 % v/v)

** Similar letters indicate non-significant at 0.05 levels.** Capital letters indicate the significant difference of each factor ($P<0.05$)*

**** Small letters indicate the significant difference of interaction ($P<0.05$)*

Table 8. Effect of different organic-tea and perlite mixtures on the physical properties of lettuce

Nutrient solution	First season 2012/2013					Second season 2013/2014				
	Substrate					Substrate				
	Control	V: P 1:1	V: P 1:2	V: P 1:3	Mean(B)	Control	V: P 1:1	V: P 1:2	V: P. 1:3	Mean(B)
Fresh weight (g/plant)										
Vermi-tea	574 d	403 f	636 b	692 a	576.3B	577.4 f	405.4 i	639.8 d	707.5 a	582.5 B
Compost-tea	526 e	342 g	536 e	543 e	486.8C	529.2 h	344.1 j	539.2 g	546.3 g	489.6 C
Chemical	607 c	393 f	643 b	690 a	583.3A	620.0 e	404.0 i	653.9 c	689.7 b	591.9 A
Mean (A)	569 C	379 D	605 B	641 A		575.5 C	384.5 D	611.0 B	647.8 A	
Volume (cm^3)										
Vermi-tea	526.6 e	523.4 e	557.9 d	581.5 d	547.4 C	558.2 e	554.8 e	602.6 d	612.2 d	581.9 C
Compost-tea	665.8 b	462.2 g	661.8 b	733.8 a	630.9A	710.7 b	489.9 g	723.0 b	784.2 a	676.9 A
Chemical	619.4 c	503.9 f	618.3 c	616.1c	589.4 B	656.6 c	534.1 f	668.2 c	653.1 c	627.9
Mean (A)	603.9 B	496.5 C	612.6 B	643.8 A		641.8 C	526.3 D	664.6 B	683.2 A	
Density (g/cm^3)										
Vermi-tea	1.09 abc	0.77 d	1.14 ab	1.19 a	1.05 A	1.27 ab	0.85 e	1.26 bc	1.35 a	1.17 A
Compost-tea	0.79 d	0.74 d	0.81 d	0.74 d	0.77 C	0.84 e	0.78 e	0.86 e	0.88 e	0.84 C
Chemical	0.98 c	0.78 d	1.04 bc	1.12 abc	0.98 B	1.03 d	0.83 e	1.10 cd	1.22 bc	1.05 B
Mean (A)	0.95 A	0.76 B	1.00 A	1.02 A		1.05 B	0.82 C	1.06 B	1.15 A	
Dry weight (g/plant)										
Vermi-tea	46.0 d	34.9 f	62.4 b	71.3 a	53.6 A	48.7 e	36.9 g	64.2 b	75.5 a	56.3 A
Compost-tea	38.5 f	27.9 g	40.6 e	38.8 ef	36.5 C	40.8 fg	29.6 h	43.0 f	41.1 fg	38.6 C
Chemical	51.5 c	35.6 f	58.5 b	54.0 c	49.9 B	54.6 d	37.7 g	56.8 d	57.2 c	51.6 B
Mean (A)	45.3 B	32.8 C	53.8 A	54.7 A		48.1 B	34.7 D	54.7 B	48.1 A	
Dry matter content (%)										
Vermi-tea	8.01 cdef	8.65 bc	9.82 a	10.30 a	9.19 A	8.5 bcd	8.8 bc	9.1 b	10.9 a	9.5 A
Compost-tea	7.32 ef	8.17 bcde	7.57 def	7.14 f	7.54 C	7.8 d	8.3 cd	7.7 d	7.8 d	7.9 C
Chemical	8.49 bcd	9.05 b	9.09 b	7.83 cdef	8.61 B	8.9 bc	9.2 b	9.6 b	8.5 bcd	9.1 B
Mean (A)	7.94 B	8.62 A	8.82 A	8.42 AB		8.4 B	8.8 AB	9.1 A	9.1 A	

Control peat moss + perlite (50 %: 50 % v/v), V: P vermicompost: Perlite 1: 1, 2 and 3 v/v)

** Similar letters indicate non-significant at 0.05 levels. ** Capital letters indicate the significant difference of each factor ($P<0.05$)*

**** Small letters indicate the significant difference of interaction ($P<0.05$)*

Table 9. Effect of different organic-tea and perlite mixtures on the chemical properties of lettuce

Nutrient solution	First season 2012/2013					Second season 2013/2014				
	Substrate					Substrate				
	Control	V: P 1:1	V: P 1:2	V: P 1:3	Mean(B)	Control	V: P 1:1	V: P 1:2	V: P. 1:3	Mean(B)
					N (%)					
Vermi-tea	2.13 g	3.36 cd	3.08 e	2.74 f	2.83B	2.20 g	3.45 c	3.12 e	2.76 f	2.88 B
Compost-tea	2.15 g	3.17 e	3.04 e	2.65 f	2.75 B	2.23 g	3.34 d	3.19 e	2.84 f	2.90 B
Chemical	3.18 de	3.89 a	3.62 b	3.46 bc	3.54 A	3.16 e	3.94 a	3.61 b	3.31 c	3.51 A
Mean (A)	2.49 D	3.47 A	3.25 B	2.95 C		2.53 D	3.58 A	3.31 B	2.97 C	
					P (%)					
Vermi-tea	0.38 g	0.73 bc	0.59 de	0.47 fg	0.54 C	0.41 f	0.77 b	0.64 d	0.53 e	0.59 B
Compost-tea	0.49 ef	0.78 b	0.65 cd	0.51 ef	0.61 B	0.46 ef	0.76 bc	0.66 cd	0.50 ef	0.60 C
Chemical	0.58 de	0.89 a	0.75 bc	0.65 cd	0.72 A	0.65 d	0.91 a	0.73 bcd	0.65 d	0.74 A
Mean (A)	0.48 D	0.80 A	0.66 B	0.54 C		0.51 C	0.81 A	0.68 B	0.56 C	
					K (%)					
Vermi-tea	0.71 g	1.31 ab	1.14 cd	0.95 ef	1.03 B	0.68 f	1.34 ab	1.21 cd	0.87 e	1.03 B
Compost-tea	0.58 g	1.15 cd	1.02 de	0.86 f	0.90 C	0.61 f	1.24 bc	1.09 d	0.81 e	0.94 C
Chemical	1.07 cde	1.40 a	1.28 ab	1.19 bc	1.24 A	1.20 cd	1.45 a	1.24 bc	1.23 bc	1.28 A
Mean (A)	0.79 D	1.29 A	1.15 B	1.00 C		0.83 D	1.34 A	1.18 B	0.97 C	

Control peat moss + perlite (50 %: 50 % v/v), V: P vermicompost: Perlite 1: 1, 2 and 3 v/v)

** Similar letters indicate non-significant at 0.05 levels. ** Capital letters indicate the significant difference of each factor (P<0.05)*

**** Small letters indicate the significant difference of interaction (P<0.05)*

Increasing perlite proportion from 1 up to 3 mixed with vermicompost led to decrease the mineral content (N, P and K) of head lettuce compared to the peat: perlite substrate that gave the lowest contents of N, P and K as shown in Table 9. The highest content of N, P and K were illustrated by substrate treatment perlite: vermicompost 1: 1 (*v/v*).

Table 9 presented the interaction effect of different organic-tea and perlite mixtures on the chemical properties of lettuce. The results show that the chemical nutrient solution combined with perlite: vermicompost 1: 1 gave the highest data of N, P and K contents of head lettuce while the lowest N and P were recorded by vermi-tea combined with peat: perlite.

Nutrients in vermicompost are present in readily available forms for plant uptake such as nitrates, exchangeable phosphorus, potassium, calcium, and magnesium [33.39]. Studies show greatest plant growth responses when vermicompost constituted a relatively small proportion (10–20%) of the total volume of the substrate mixture, with higher proportions of vermicompost in the mixture not always improving plant growth [11].

4. CONCLUSION

The recommended treatments of this study are Vermicompost + sand combined with vermi-liquid and perlite + vermicompost (3: 1 *v/v*) combined with vermi-tea. More feasibility study need to investigate the economic value added of using vermicomposting outputs in ecology soilless culture as a result of the limited use of vermicomposting technology under Egyptian conditions where the application of vermicomposting concern on research scale not on commercial scale yet. Vermicomposting technology plus soilless culture could contribute strongly in food security.

ACKNOWLEDGEMENTS

The authors acknowledge the financial support provided by "Integrated environmental management of urban organic wastes using vermicomposting and green roof (VCGR) project" No. 1145, funded by Science and Technology Development Fund, Egypt.

COMPETING INTERESTS

Authors have declared that no competing interests exist.

REFERENCES

1. Robertson RA. Peat, horticulture and environment. Biodivers. Conserv. 1993; 2:541-547.
2. Eklind Y, Raemert B, Wivstad M. Evaluation of growing media containing farmyard manure compost, household waste compost or chicken manure for the propagation of lettuce (*Lactuca sativa L.*) transplants. Biol. Agric. Hortic. 2001;19: 157–181.
3. Hashemimajd K, Kalbasi M, Golchin A, Shariatmadari H. Comparison of vermicompost and composts as potting media for growth of tomatoes. Journal of Plant Nutrition. 2004;27:1107-1123.
4. Atiyeh RM, Edwards CA, Subler S, Metzger JD. Pig manure vermicompost as a component of a horticultural bedding plant medium: Effects on physicochemical properties and plant growth. Bioresource Technology. 2001;78:11-20.
5. Elvira C, Sampedro L, Benitez E, Nogales R. Vermicomposting of sludges from paper mill and dairy industries with Eisenia Andrei: A pilot-scale study. Bioressource Technol. 1998;63:205-211.
6. Garg VK, Kaushik P. Vermistabilization of textile mill sludge spiked with poultry droppings by an epigeic earthworm *Eisenia foetida*. Biores. Technol. 2005;96:1063–1071.
7. Alves WL, Passoni AA. Compost and vermicompost of urban solid waste in Licania tomentosa (Benth) seedlings production to arborization. Pesqui. Agropecu. Bras. 1997;32(10):1053–1058.
8. Logsdon G. World wide progress in vermicomposting. Biocycle. 1994;35:63–65.
9. Singh A, Sharma S. Composting of a crop residue through treatment with microorganisms and subsequent vermicomposting. Biores. Technol. 2002; 85:107–111.
10. Arancon NQ, Edwards CA, Atiyeh RM, Metzger JD. Effects of vermicomposts produced from food waste on greenhouse peppers. Bioresource Technology. 2004; 93:139-144.
11. Atiyeh RM, Arancon N, Edwards CA, Metzger JD. Influence of earthworm-processed pig manure on the growth and yield of greenhouse tomatoes. Bioresource Technology. 2000;75:175-180.

12. Arancon NQ, Edwards CA, Bierman P. Influences of vermicomposts on field strawberries: Part 2. Effects on soil microbiological and chemical properties. Bioresource Technology. 2005;97:831-840.
13. Jarecki MK, Chong C, Voroney RP. Evaluation of compost leachates for plant growth in hydroponic culture. Journal of Plant Nutrition. 2005;28:651–667.
14. Tejada M, Gomez I, Hernandez T, Garcia C. Utilization of vermicomposts in soil restoration: Effects of soil biological properties. Soil Science Society of America Journal. 2010;74(2):525-532.
15. Ismail SA. Vermicology: The biology of earthworms. Orient Longman Ltd, Chennai, India. 1997;92.
16. Gutierrez-Miceli FA, Oliva-Llaven MA, Nazar PM, Ruiz-Sesma B, Alvarez-Solís JD, Dendooven L. Optimization of vermicompost and worm-bed leachate for the organic cultivation of radish. Journal of Plant Nutrition. 2011;34(11):1642-1653.
17. Pant AP, Radovich TJK, Hue NV, Talcott ST, Krenek KA. Vermicompost extracts influence growth, mineral nutrients, phytonutrients and antioxidant activity in pak choi (*Brassica rapa cv.* Bonsai, Chinensis group) grown under vermicompost and chemical fertilizer. Journal of Science of Food and Agriculture. 2009;89(14):2383-2392.
18. Merrill R, McKeon J. Organic teas from compost and manures. Organic farming research foundation project report. Cabrillo community college, Aptos, CA; 1998.
19. Bio-Cycle Staff. Compost teas in agriculture. BioCycle, JG Press, Inc., Emmaus, PA. 1996;65.
20. Brinton WF. The control of plant pathogenic fungi by use of compost teas. Tiodynamics. 1995;12-15.
21. Weltzein HC. Biocontrol of foliar fungal diseases with compost extracts. In: Microbial Ecology of Leaves. J. H. Andrews and S. S. Hirano (eds) Springer Verlag, New York, N.Y. 1991;430-450.
22. El-Shinawy MZ, Abd-Elmoniem EM, Abou-Hadid AF. The use organic manure for lettuce plants grown under NFT condition. Acta. Hort. 1999;491:315-318.
23. Wilson GCS. The physico- chemical and physical properties of horticultural substrate. Acta Hort. 1983;150:19-32.
24. Raul IC. Measuring physical properties. Rutgers Cooperative Extension. New Jersey Agriculture Experiment Station. New Jersey University; 1996.
25. Inbar Y, Chen Y, Hadar Y. Journal of Environmental Quality. 1993;22:875-863.
26. Allen SE. Chemical Analysis of Ecological Materials. Black-Well, Oxford. 1974;565.
27. Chapman HD, Pratt PF. Methods of analysis for soil, plant, and water. University of California, Division of Agric Sci; 1961.
28. Food and Agriculture Organization (FAO). Micron Nutrients and the Nutrient Status of Soils. FAO Soils Bulletin. 1980;48:242-250.
29. Watanabe FS, Olsen SR. Test of an ascorbic acid method for determining phosphorus in water and Na HCO_3 extracts from soil. Soil Sci. Soc. Amer. Proc. 1965; 29:677-678.
30. Hand P, Hayes WA, Frankland JC, Satchell JE. The vermicomposting of cow slurry. Pedobiologia. 1988;31:199–209.
31. Domínguez J, Aira M, Gomez-Brandon M. Vermicomposting: earthworms enhance the work of microbes. In: Insam H, Franke-Whittle I (eds) Microbes at work. Springer, Berlin. 2010;93-114.
32. Chan PLS, Griffiths DA. The vermicomposting of pre-treated pig manure. Biological Wastes. 1988;24:57-69.
33. Edwards CA, Burrows I. The potential of earthworm compost as plant growth media. In: Edwards, C.A. and E. Neuhauser (eds), Earthworms in Waste and Environmental Management. SPB Academic press, The Hague, The Netherlands. 1988;21-32.
34. Anjana SU, Muhammad I, Abrol YP. Are nitrate concentrations in leafy vegetables within safe limits? Current Science. 2007;92(3).
35. EFSA (European Food Safety Authority), 2010. Statement on possible public health risks for infants and young children from the presence of nitrates in leafy vegetables. The EFSA Journal. 2010; 8(12):1935.
36. Tognetti C, Laos F, Mazzarino MJ, Hernandez MT. Composting vs. vermicomposting: A comparison of end product quality. Compost Science & Utilization. 2005;13:6-13.
37. Lazcano C, Arnold J, Tato A, Zaller JG, Domíngues J. Compost and vermicompost as nursery pot components: effects on tomato plant growth and morphology. Spanish Journal of Agricultural Research, 2009;7(4):994-951.

38. Tringovska I, Dintcheva T. Vermicompost as substrate amendment for Tomato transplant production. Sustainable Agriculture Research. 2012;1(2):115-122.
39. Orozco SH, Cegarra J, Trujillo LM, Roig A. Vermicompostsing of coffee pulp using the earthworm *Eisenia fetida*: Effects on C and N contents and the availability of nutrients. Biology and Fertility of Soils. 1996;22:162-166.

5

Seedling Emergence of *Blighia sapida* (K. D. Konig) Seeds Subjected to Different Lengths of Storage and Pretreatments

Olufunke O. Olayode[1*] and Theodora N. Osuji[1]

[1]Department of Forestry, Wildlife and Fisheries Management, Ekiti State University, Ado-Ekiti, Nigeria.

Authors' contributions

This work was carried out in collaboration between the both authors. Author OOO designed the study and supervised author TNO in carrying out the experiment. Author TNO wrote the first draft while author OOO restructured the manuscript and edited it appropriately. Both authors managed the literature searches and approved the final manuscript.

Editor(s):
(1) Manuel Esteban Lucas-Borja, Castilla La Mancha University, School of Advanced Agricultural Engineering, Department of Agroforestry Technology and Science and Genetics, Spain.
Reviewers:
(1) Edward Missanjo, Malawi College of Forestry and Wildlife, Dedza, Malawi.
(2) Rosilda Mara Mussury Franco Silva, Universidade Federal da Grande Dourados, Brazil.

ABSTRACT

Aims: The study investigated the influence of different lengths of storage at room temperature and pretreatments on seedling emergence of *Blighia sapida* seeds.
Study Design: Randomized complete block design with four replicates was used.
Place and Duration of Study: The experiment was conducted at the nursery of Faculty of Agricultural Sciences, Ekiti State University, Ado-Ekiti, Nigeria for about three months.
Methods: Seeds of *Blighia sapida* under five lengths of storage at room temperature being Fresh, 7 days, 14 days, 21 days and 28 days were sown in germination boxes filled with topsoil after subjection to four pretreatments thus: Control, Hot water, Soaking in water at room temperature and Scarification. Cumulative germination count was done daily till no further germination was observed for about a week.
Results: Seeds sown Fresh and pretreated by Soaking in water at room temperature had the highest germination percentage of 96.7% followed by those under Control with germination

**Corresponding author: Email: funkefaboy@yahoo.com*

percentage of 76.7%. Seeds stored for 7 days had germination percentage of 33.3% under Control and Scarification whereas the ones subjected to Soaking in water at room temperature produced 40% germination percentage. Hot water pretreatment gave 0% germination under the different lengths of storage. In seeds stored for 14 days, 3.3%, 6.7% and 13.3% germination percentages were observed for Control, Soaking in water at room temperature and Scarification respectively. However, seeds stored for 21 days and 28 days gave 0% germination for the different pretreatments.

Conclusion: Fresh seeds of *B. sapida* are the most suitable for sowing whereas they can also be stored at room temperature for 7days to obtain appreciable viable seeds beyond which viability cannot be guaranteed even with the use of different pretreatments.

Keywords: Seed storage; room temperature; germination; recalcitrant.

1. INTRODUCTION

Seed storage is an essentiality in supplying good quality seeds for a planting programme whenever needed [1,2], whereas planting programme is rather inevitable considering many factors. Such factors include high rate of deforestation in developing countries, indiscriminate logging activities and climate change. Seed is the common propagation material for most tropical tree species [3] because propagation from seed is considered inexpensive and usually effective, and is therefore a viable method for *ex-situ* conservation of plant genetic resources [4]. This may be one of the reasons why propagation of tropical tree species still relies heavily on availability of seeds. Furthermore, the proper handling of seeds through the processes of ripening or maturation, collecting, processing, storage, dormancy, stratification and germination is required for success in producing a new seedling [5]. It has also been discovered that seed maturation does not correspond many times with seedling production thereby making seed storage a very important part of seed handling.

The period of storage is often limited by the technical and physiological storage potentials, that is, the length of time seeds of a particular species will survive under the available storage conditions. Therefore, in order to maintain viability over a prolonged period, it is important that the optimal storage environment for the species is met, as much as possible. However, even under the best storage conditions, some species will only be viable for a very short time nevertheless deterioration in such species may be delayed by adopting the best possible storage conditions, but long term storage is not possible for those species. Even where fruiting is regular and abundant every year, it is rather cheaper to collect surplus seeds to cover for several years rather than collecting seeds every year [6]. The importance of seed storage has been recognized ever since human beings began to domesticate plants. The length of successful seed storage depends on both the objectives and the species concerned, but the overall objective is to help maintain viable seeds from one growing season to the next; that is for months or even years depending on when the seeds will be needed.

Many tropical fruit tree species exhibit uncertain storage behavior. Seeds have traditionally been grouped into two according to their physiological storage potentials viz. recalcitrant and orthodox. Even with these two broad groups, variations exist among each group. Likewise, viable period for seeds of some species can be prolonged by certain storage media. [7] said other storage media such as cold storage rooms and cryopreservation could prolong viability of seeds but some of these are not readily available or are limited by one factor or another in a developing country like Nigeria. Also appropriate seed pretreatments are required for nursery establishment and production of maximum number of quality seedlings with minimum cost, time and labour [8].

Blighia sapida (K.D. Konig) has been classified as a neglected and underutilized species which signify that its potentials both as an income generating species as well as its contribution to human well being have been grossly underexploited. This neglect places it in danger of continued genetic erosion and disappearance. *Blighia sapida* commonly called ackee apple is a tropical fruit belonging to the Sapindaceae family. It is a woody perennial multipurpose fruit tree species native to the Guinean forests of West Africa but has traversed the Atlantic Ocean making the Caribbean its home. *Blighia sapida* occurs naturally from Senegal to Cameroon and Equatorial Guinea, and possibly also in Gabon. It is commonly planted in its natural area of distribution, as fruit tree and ornamental shade tree. It has been introduced in many other tropical countries and in some subtropical regions such as Florida (United States) and is widely cultivated as fruit and ornamental tree in India and tropical America. It had already been introduced in tropical America by the end of the 18th century, and has since become locally naturalized [9]. Although so far overlooked by researchers in Nigeria, the tree is highly valued by farmers and is an important component of traditional agroforestry systems in some other countries. Furthermore, the seed storage behavior of *B. sapida* is said to be uncertain [10] whereas the benefits derivable from *B. sapida* are quite enormous. It is said to be a nice ornamental tree, especially when decorated with the brightly coloured fruits. It is also considered useful for planting to improve soil fertility and reduce erosion through its large rooting system [11]. Boiled arils are an ingredient of a popular traditional dish in Jamaica, together with salt fish. In West Africa, arils are sometimes eaten raw, fried or roasted while canned ackee is exported primarily to the United kingdom and Canada [12]. Considering the diverse benefits derivable from *B. sapida*, this study on its length of storage at room temperature and pretreatments to aid its seed germination becomes important. The use of storage at room temperature is to give the knowledge of when *B. sapida* seed will still retain its viability without access to any sophisticated means of storage. Also, the pretreatments adopted in this study are those that are socially acceptable and extra precaution may not be needed to apply them. Thus, this study aimed at evaluating the viability of *B. sapida* seeds subjected to different lengths of storage at room temperature and assessment of its early growth.

2. MATERIALS AND METHODS

2.1 Fruit Collection, Depulping and Seed Weight Determination

Mature fruits of *Blighia sapida* were collected from Oye-Ekiti, Ekiti State, Nigeria (latitude 7°8'N and longitude 5°3'E). These were carefully depulped using hand to avoid seed damage. Six hundred (600) seeds of *Blighia sapida* were then selected based on their good morphological characteristics. These were divided into five parts with each part containing 120 seeds. The five parts stand for the different lengths of storage for *Blighia sapida* seeds being Fresh, 7 days storage, 14 days storage, 21 days storage and 28 days storage represented as LS1, LS2, LS3, LS4, and LS5 respectively. Storage was done at room temperature by spreading the seeds on trays in the laboratory. Forty seeds were randomly selected under each length of storage and weighed on sensitive weighing balance in four replicates.

2.2 Effect of Pretreatments on the Germination of *Blighia sapida* Seeds

The pretreatments used for aiding germination in *B. sapida* seeds were Control, Hot water, Water at room temperature and Scarification represented as PT1, PT2, PT3 and PT4 respectively. In Control, no form of pretreatment was applied while for hot water, water was allowed to boil before soaking the seeds in it and the water was allowed to cool with the seeds for 24 hours. In the case of water at room temperature, the seeds were soaked in water for 24 hours before sowing. Scarification was done using sand paper to inflict scar on the seeds thereby forcing the seed coat open to encourage imbibition of water and gas.

Seeds under each length of storage were further sub-divided into four parts with each part subjected to the different pretreatments and each had thirty seeds which were sown in germination trays previously filled with topsoil in three replicates. Watering was done daily in the morning using finely-meshed watering can. Germination count was done daily until no further germination was observed for about a week. Germination was taken to have occurred when the plumule emerged on the soil surface.

2.3 Statistical Analysis

The data collected were subjected to two-way Analysis of Variance (ANOVA) and the various means were separated using Duncan's Multiple Range Test (DMRT).

3. RESULTS

3.1 Seed Weight

Mean seed weight values showed significance among the different lengths of storage of *Blighia sapida* when subjected to Analysis of Variance (ANOVA) at $\alpha = 0.05$. However, when Duncan's Multiple Range Test (DMRT) was used to separate the means, it revealed that LS1 was significantly different ($p<0.05$) from LS2, LS3, LS4 and LS5. Contrariwise, LS2 and LS3 were not significantly different ($p>0.05$) from each other but different from those of LS4 and LS5.

Furthermore, LS4 and LS5 did not differ from each other at p>0.05 (Table 1).

Table 1. Differences in mean seed weight of *B. sapida* from the different lengths of storage

Length of storage (days)	Seed weight (g)
0	36.40[a]
7	27.50[b]
14	26.03[b]
21	22.73[c]
28	22.47[c]

Note: Means with the same letter are not significantly different

3.2 Germination of Ackee Apple Seeds

3.2.1 Germination of seeds sown fresh using different pretreatments

Germination began on the 11th Day after Sowing (DAS) for seeds subjected to PT3 and was completed on the 19th DAS whereas it began on the 12th DAS for seeds subjected to PT1 and PT4, and was completed on the 20th DAS. Germination percentages of 76.7%, 0%, 96.7% and 73.3% were recorded for PT1, PT2, PT3 and PT4 respectively (Table 3). Fig. 1 shows the cumulative number of seedlings for *B. sapida* seeds sown fresh.

3.2.2 Germination of seeds sown after 7 days of storage subjected to different pretreatments

Germination started on the 12th DAS for seeds subjected to PT3 and was completed on the 20th DAS whereas at the 13th DAS, germination began for seeds subjected to PT1 and PT4 and was brought to an end on the 20th and 19th DAS respectively. Percentage germination of 33.3%, 0%, 40.0% and 33.3% were recorded for PT1, PT2, PT3 and PT4 respectively (Table 3). Fig. 2 reveals the cumulative number of seedlings for *B. sapida* seeds sown after 7 days storage.

3.2.3 Germination of seeds sown after 14 days storage under different pretreatments

Germination began on the 13th DAS for seeds under PT4 and was completed on the 18th DAS. Whereas germination which began on the 14th DAS for seeds under PT1 and PT3 came to an end on the 14th and 17th DAS respectively. Germination percentages of 3.3%, 0%, 6.7% and 13.3% were recorded for PT1, PT2, PT3 and PT4 respectively (Table 2). Fig. 3 indicates the cumulative number of seedlings for *B. sapida* seeds sown after 14 days storage.

3.2.4 Germination of seeds sown after 21 days and 28 days storage subjected to different pretreatments

None of the seeds germinated all through the period of the experiment even under the different pretreatments. Therefore, germination percentage was 0% for the two lengths of storage under the various pretreatments (Table 2).

3.2.5 Analysis of Variance (ANOVA) results for germination rate of *Blighia sapida* across lengths of storage and pretreatments

ANOVA result showed no significance at $\alpha = 0.05$ for lengths of storage, pretreatments and; interaction between storage and pretreatments at 11 DAS. Although at 12 DAS, ANOVA result indicated significance for lengths of storage, whereas, it revealed no significance for pretreatments and; interaction between storage and pretreatments. Furthermore, at 13 DAS, 14 DAS up to 21 DAS there was significant difference for lengths of storage, pretreatments and; interaction between storage and pretreatments.

Table 2. Mean germination percentage of *Blighia sapida* seeds subjected to different lengths of storage and pretreatments

Lengths of storage	Germination percentage (%)				Germination begins (DAS)				Germination ends (DAS)			
	PT1	PT2	PT3	PT4	PT1	PT2	PT3	PT4	PT1	PT2	PT3	PT4
LS1	76.7	0	96.7	73.3	12	0	11	12	19	0	19	20
LS2	33.3	0	40.0	33.3	13	0	12	13	20	0	20	19
LS3	3.3	0	6.7	13.3	14	0	14	13	14	0	17	18
LS4	0	0	0	0	0	0	0	0	0	0	0	0
LS5	0	0	0	0	0	0	0	0	0	0	0	0

**DAS means days after sowing*

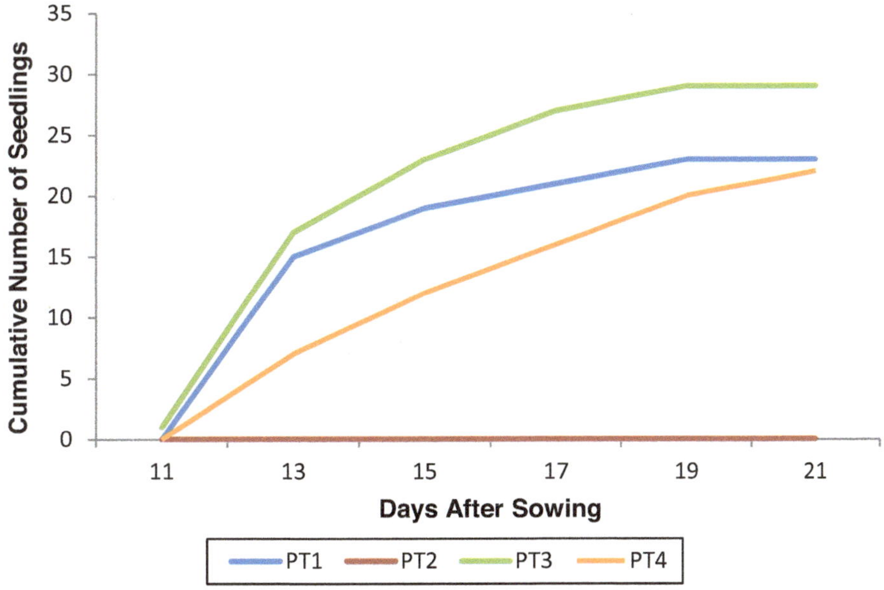

Fig. 1. Cumulative germination of *Blighia sapida* seeds sown Fresh under different pretreatments

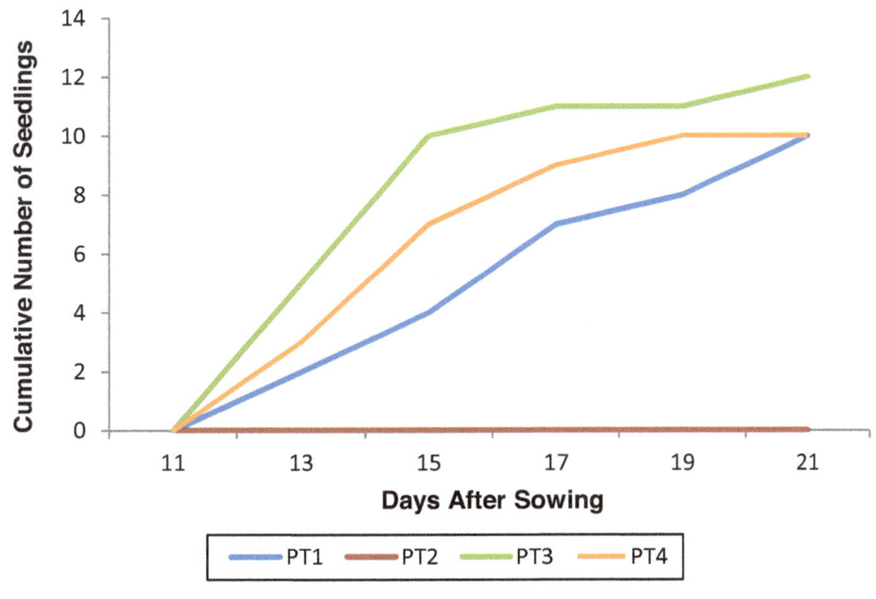

Fig. 2. Cumulative germination of *Blighia sapida* seeds sown after 7 days storage under different pretreatments

When Duncan's Multiple Range Test (DMRT) was used to separate the mean germination for the different lengths of storage, no significance ($p>0.05$) was shown at the 11th DAS for Fresh (LS1), 7 days of storage (LS2) and 14 days of storage (LS3). However at the 12th DAS, LS2 and LS3 were different ($p<0.05$) from LS1 but they were not different ($p>0.05$) from each other. Furthermore, DMRT showed significant difference ($p<0.05$) among LS1, LS2 and LS3 from 13th DAS till the experiment was concluded on the 21st DAS (Table 3).

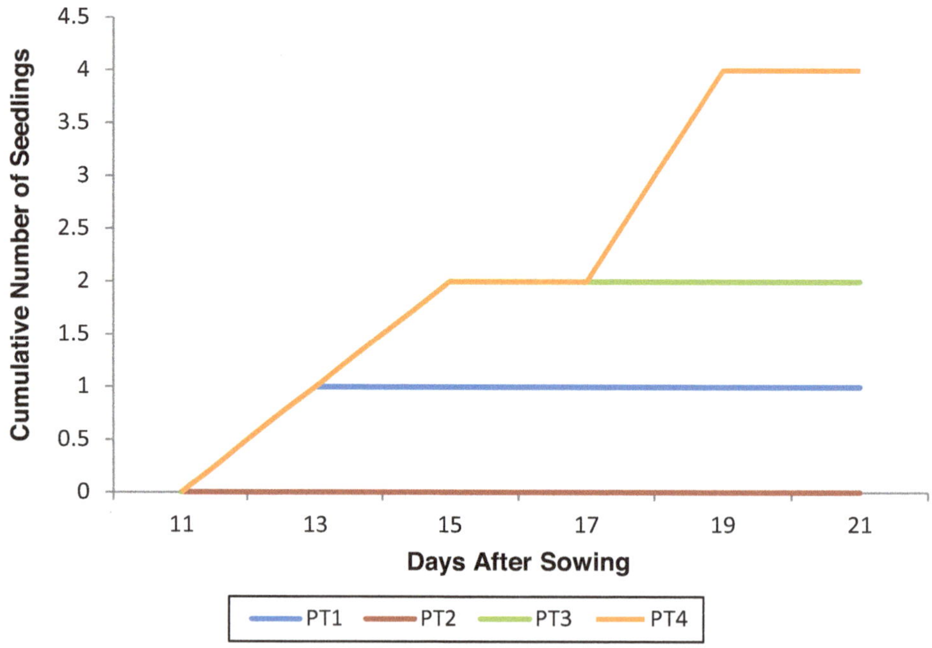

Fig. 3. Cumulative germination of *Blighia sapida* seeds sown after 14 days storage under different pretreatments

Table 3. Mean germination rate of *Blighia sapida* seeds across the different lengths of storage

DAS	Lengths of storage		
	LS1	**LS2**	**LS3**
11	0.08^a	0.00^a	0.00^a
12	1.67^a	0.17^b	0.00^b
13	3.42^a	0.83^b	0.08^c
14	4.00^a	1.25^b	0.25^c
15	4.50^a	1.75^b	0.33^c
16	4.83^a	2.00^b	0.33^c
17	5.33^a	2.25^b	0.42^c
18	5.67^a	2.33^b	0.80^c
19	6.00^a	2.42^b	0.58^c
20	6.17^a	2.67^b	0.58^c
21	6.17^a	2.67^b	0.58^c

Note: Means with the same letter across rows are not significantly different
**DAS means days after sowing*

3.2.6 Mean germination rate of *Blighia sapida* seeds across the various pretreatments

The mean values for PT1, PT2, PT3 and PT4 were not significantly different ($p>0.05$) from one another when they were subjected to DMRT at 11 DAS. However, at 12 DAS, the mean value for PT3 differed significantly ($p<0.05$) from PT2 whereas, it did not differ ($p>0.05$) from PT1 and PT4, likewise PT2 did not differ ($p>0.05$) from PT1 and PT4. Also, at the 13th DAS, 15th DAS and 16th DAS, DMRT showed that PT1 was significantly different ($p<0.05$) from PT3 and PT4, all of which differed from PT2. Furthermore, at the 14th DAS there was no significant difference ($p>0.05$) among the pretreatments while at the 20th and 21st DAS, only the mean value of PT2 was significantly different from those of other pretreatments (Table 4).

Table 4. Mean germination rate of *Blighia sapida* seeds across the various pretreatments

DAS	Pretreatments			
	PT1	PT2	PT3	PT4
11	0.00[a]	0.00[a]	0.11[a]	0.00[a]
12	0.67[ab]	0.00[b]	1.00[a]	0.78[ab]
13	1.89[ab]	0.00[c]	2.44[a]	1.44[b]
14	2.44[b]	0.00[d]	3.33[a]	1.56[c]
15	2.67[ab]	0.00[c]	3.78[a]	2.33[b]
16	3.11[ab]	0.00[c]	3.89[a]	2.56[b]
17	3.22[b]	0.00[c]	4.44[a]	3.00[b]
18	3.44[b]	0.00[c]	4.56[a]	3.44[b]
19	3.56[a]	0.00[b]	4.67[a]	3.78[a]
20	3.78[a]	0.00[b]	4.78[a]	4.00[a]
21	3.78[a]	0.00[b]	4.78[a]	4.00[a]

Note: Means with the same letter across rows are not significantly different
**DAS means days after sowing*

4. DISCUSSION

There was reduction in seed weight as lengths of storage increased which indicated reduction in moisture content of the seeds. Also, it was realized that Fresh seeds which had the highest weight were the first to germinate in comparison with other lengths of storage. This result agrees with that of [13] who observed that the progeny of *Dialium guineense* Willd with the highest seed weight were the first to germinate.

Considering the different lengths of storage, seeds sown fresh had the highest germination rate above the other lengths of storage while onset of germination increased with increasing length of storage. The result of this study agrees with that of [7] who reported prompt germination in seeds of *Dacryodes edulis* G. Don sown fresh in comparison with seeds stored for different lengths of time. Similarly, seeds sown fresh had the highest germination percentage while this reduced with increasing lengths of storage till no germination was observed in seeds stored for longer duration. This result still agrees with that of [7] who observed that *D. edulis* seeds sown fresh produced the highest germination percentage while this reduced till no germination was observed with increasing lengths of storage at room temperature. [14] observed that seeds are affected by desiccation and are considered short-lived. However, not all tree seeds are affected by desiccation but only those that fall under the category of being recalcitrant. It is generally known that tropical recalcitrant seeds cannot be dried without causing damage to the endospermic unit of the seed. Thus, germination of seeds in this group dried prior to sowing should not be expected [2]. Likewise, [15] reported that fresh seeds with a certain level of moisture content (15-90%) normally germinate with a germination percentage of 80-90% but that even a small decrease in moisture content will lead to a significant decrease in seed germination. They further observed fresh seeds as having the highest germination percentage than those stored for different lengths of time.

Furthermore, under the various pretreatments, the highest germination percentage was observed for seeds soaked in water at room temperature in both seeds sown Fresh and the ones stored for 7 days prior to sowing followed by Control. Although under 14 days length of storage, Scarification produced the highest germination percentage (13.3%) that was not appreciable followed by water at room temperature (6.7%). However, this result is not in agreement with the work of [16] who reported that soaking Prosopis seed in water at room temperature is generally ineffective in breaking dormancy. In the case of *B. sapida* used in this study, soaking in water at room temperature for 24 hours produced the highest germination percentage. It can then be inferred that seeds of different tree species differ in their response to various pretreatments. Furthermore, hot water gave 0% germination percentage across the assessment period and this could be as a result of the fact that the embryo of *B. sapida* seeds cannot withstand hot water pretreatment. Whereas, in the case of Prosopis seed [16] reported that hot water pretreatment is suitable in breaking its seed dormancy. Moreover, the various pretreatments used in this study produced very low germination percentage in seeds of *B. sapida* stored for 14 days while they did not aid germination at all in seeds stored for

21 days and 28 days. The loss of moisture earlier observed in *B. sapida* seeds as lengths of storage increased must have negatively affected their viability. This result disagrees with that of [17] who observed that viability of *Mimosa foliolosa* Benth seeds can be maintained for at least three years under artificial storage conditions. Their study suggested alleviation of seed dormancy with storage time which they attributed to probable seed coat softening under artificial condition.

5. CONCLUSION

The difference observed in germination percentages of *B. sapida* from the five different lengths of storage used in this experiment showed that Fresh seeds of *B. sapida* are the most suitable for sowing whereas they can also be stored at room temperature for 7 days to obtain appreciable viable seeds beyond which viability cannot be guaranteed even with the use of different pretreatments. Also, soaking seeds of *B. sapida* in water at room temperature is the best pretreatment to aid its germination. It is therefore recommended that seeds of *Blighia sapida* be sown as soon as collection is made but not stored beyond 7 days. Where this does not correspond with planting time, they can be raised in the nursery first before eventual planting out. Moreover, considering the fact that *B. sapida* seeds are desiccation sensitive, it can be concluded that they are recalcitrant.

COMPETING INTERESTS

Authors have declared that no competing interests exist.

REFERENCES

1. Siddique AB, Wright D. Effects of different seed drying methods on moisture percentage and seed quality (viability and vigour) of Pea seeds (*Pisum sativum* L.) Pakistan Journal of Agronomy. 2003;2(4): 201-208.
2. Onyekwelu JC, Fayose OJ. Effect of storage methods on the germination and proximate composition of *Treculia africana* seeds. Conference on International Agricultural Research for Develompent. University of Kassel-Witzenhausen and University of Göttingen, October 9-11; 2007.
3. Bowes BG. A Colour Atlas of Plant Propagation and Conservation. Newyork Botanical Garden Press, the University of Wisconsin – Madison; 1999.
4. Abirami K, Rema J, Matthew PA, Srinivoisan V, Hamza S. Effect of different propagation media on seed germination, seedling growth and vigour of nutmeg (*Myristica fragrans* Houtt). Journal of Medicinal Plants Research. 2010;4(19): 2054-2058.
5. SP629 growing trees from seed. The University of Tennessee Agricultural Extension Service. Accessed 25 May 2014.
   Available:http://trace.tennessee.edu/utk_agexfores/90
6. Schmidt L. Guide to handling of tropical and subtropical forest seed. Danida Forest Seed Centre. 2000;511. ISBN: 87-982428-6-5.
7. Olayode OO, Kolawole FS. Seed viability of *Dacryodes edulis* (G. Don) H.J. Lam subjected to different lengths of storage. Journal of Agriculture and Sustainable Research. 2013;13(2):16-20.
8. Hossain MA, Arefin MK, Khan BM, Rahman MA. Effects of seed treatments on germination and seedling growth attributes of Horitaki (*Terminalia chebula* Retz) in the nursery. Research Journal of Agricultural and Biological Sciences. 2005;1(2): 135-141.
9. PROTAbase (Plant Resources of Tropical Africa). Accessed 2 May 2014.
   Available:http://database.prota.org/PROTAhtml/Blighia%20sapida_En.htm
10. Available:www.kew.org/science-conservation/millenium-seed-bank/seed-research/blighia-sapida
    Accessed 2 September 2015.
11. Burkill HM. The useful plants of West Tropical Africa. 2nd Edition. Families S–Z, Addenda. Royal Botanic Gardens, Kew, Richmond, United Kingdom. 2000;5.
12. Centers for Disease Control (CDC). Toxic hypoglycemic syndrome: Jamaica, 1989-1991. MMWR Morb Mortal Wkly Rep. 1992;41:53-5.
13. Olayode OO, Gbadamosi AE. Seed sources and pre-treatments effects on the emergence of velvet tamarind (*Dialium guineense* Willd) seedlings. Journal of Sustainable Forestry. 2009;28:895-903.
14. Maurya RP, Lewis DM, Chandler JStA. Studies on the propagation of Jamaican Ackee (*Blighia sapida* L.) by air-layering. HortScience. 2013;48(10):1298-1300.

15. Pangou SV, De Zoysa N, Maury G, Masanga A, Pangou-Yoka CM. Seed storage behaviour of some indigenous valued tropical tree species from Congo (Africa). International Research Journal of Agricultural Science and Soil Science. 2011;1(8):307-313.
16. Pasiecznik NM, Harris PJC, Tavares JdeP, Cassama M. Pretreatment of prosopis seeds to break dormancy. International Tree Crops Journal. 1998;9(3):187-193.
17. Silveira FAO, Negreiros D, Ranieri BD, Silva CA, Araújo LM, Fernandes, GW. Effect of seed storage on germination, seedling growth and survival of *Mimosa foliolosa* (Fabaceae): Implications for seed banks and restoration ecology. Tropical Ecology. 2014;55(3):385-392.

6

Soil Acidity Management by Farmers in the Kenya Highlands

E. M. Muindi[1*], E. Semu[2], J. P. Mrema[2], P. W. Mtakwa[2] and C. K. Gachene[3]

[1]Department of Crop Science, Pwani University, Kenya.
[2]Department of Soil Science, Sokoine University of Agriculture, Tanzania.
[3]Department of Land Resource Management, University of Nairobi, Kenya.

Authors' contributions

This work was carried out in collaboration between all authors. Author EMM designed the study, wrote the protocol, involved in data collection, data analysis, interpretation and writing the first draft of the manuscript. Authors ES, JPM, PWM and CKG guided on study design, data collection, evaluation and interpretations. All authors read and approved the manuscript.

Editor(s):
(1) Mirza Hasanuzzaman, Department of Agronomy, Faculty of Agriculture, Sher-e-Bangla Agricultural University, Bangladesh.
Reviewers:
(1) Ade Onanuga, Dalhousie University, Halifax, Canada.
(2) Zakka Israila Yashim, Ahmadu Bello University Zaria, Nigeria.
(3) María Daniela Chavez, National Institute for Agricultural Technology, Argentina.

ABSTRACT

Declining soil fertility attributed to soil acidity is a major soil productivity problem in sub-Saharan Africa. A study was carried out in nine counties across the Kenya highlands, namely Meru, Embu, Kerugoya, Nyeri, Kiambu, Kinangop, Siaya, Busia and Eldoret, where the problems associated with soil acidity are prominent. The study aimed at assessing farmers' awareness of soil acidity, and establishment of common acidity management practices following administration of structured questionnaires. From the information gathered through personal interviews via questionnaires, <37% of the farmers were attached to a farmers training group in all study sites; among them, <4% were aware of soil acidity problems and <8% had carried out chemical analysis of their soils. The farmers who had applied lime at least once on their farms were <3% in all sites. Most farmers (>80%) used both inorganic fertilizers and manure on their farms, with the majority using DAP, CAN and farmyard manure. On cultural soil fertility management, choice of subsequent crop was

Corresponding author: Email: e.muindi@pu.ac.ke

dictated by sustainability rather than cropping system like rotation. There was a significant ($P<0.05$) negative relationship between livestock keeping and soil fertility management, with <30% of the farmers returning crop residues back to the farm. Most of them fed crop residues to their livestock. Only 8% of the farmers incorporated crop residues into the soil. There was a significant ($P \leq 0.05$) positive correlation between education level and inorganic fertilizer use in crop production. Farmer's age and maize yields correlated negatively with each other. Additionally, farmers' training programmes and frequencies positively influenced choice of inorganic fertilizers and levels of application. Training is therefore one of the most significant issues affecting soil fertility management in the Kenya highlands. To further enhance the understanding of soil acidity and fertility management in Kenya highlands, farmers training should be prioritized.

Keywords: Inorganic fertilizers; soil fertility; soil fertility management; soil acidity; Kenya highlands; farmers training.

1. INTRODUCTION

Soil acidity is a major problem for agricultural productivity worldwide [1,2]. Acid soils account for about 4 billion ha of the total world land area [3]. This is 30% of the total world land area and 58% of land suitable for agriculture, inhabited by 73% of the world's population. Most acid soils are found in South and North America, Asia and Africa, due to extensive weathering and leaching. The acid soils occupy 29% of the total land area in the sub-Saharan Africa (SSA) zone and 13% of Kenyan land area [4,5]. Soil acidity is associated with hydrogen (H), aluminium (Al), iron (Fe) and manganese (Mn) toxicities to plant roots and corresponding deficiencies of plant available P, molybdenum (Mo), calcium (Ca), magnesium (Mg) and potassium (K) [6,7], which negatively affect soil fertility and productivity.

The agricultural sector is the backbone of the Kenyan economy [8]. It contributes 25% of the total GDP and employs about 75% of the national labour force [9]. Although Kenya's food policy emphasizes self-sufficiency, food insecurity is a big problem [10]. More than 10 million Kenyans (nearly one third of the population) are chronically food insecure [8,9] due to declining food production and non-uniform distribution of agricultural products. Crop production in Kenya has been on the decline over the years, leading to importation of food [9]. The low food production has been attributed to declining soil fertility, poor crop management practices, poor post harvest handling procedures [11] and uneven and inadequate natural precipitation. Plant nutrient depletion from soils is greatest when no or only minimal quantities of nutrient inputs are added to the soils under crop production [12] to replenish those removed by plants or crop harvests. Studies carried out in Kenya indicate that the amount of chemical inorganic fertilizers used for maize production by most farmers is below the recommended rates [13,14]. This is attributed to the lack of knowledge on the effects of inorganic fertilizers on crop yields, lack of capital due to the low wealth status of the farmers or lack of the appropriate agricultural inputs.

Maize (*Zea mays* L.) is the staple food crop for the majority of Kenyans [15]. However, maize production has been declining over the years [16]. Both commercial and small-scale farmers grow the crop but in most areas the production is constrained by low soil nitrogen and phosphorus availability and high soil acidity [17-21] and inadequate natural precipitation in some areas. Phosphorus availability in acid soils is greatly influenced by aluminium ion concentration in the soil. Unfortunately, most improved maize germplasms and landraces used by Kenyan farmers are sensitive to the high (>20%) Al saturation and high exchangeable Al (>2 cmol kg^{-1}) commonly encountered in many acid soils of the Kenya highlands [22,23]. Use of calcitic and dolomitic liming materials which ameliorate Al toxicity, raising soil pH and promoting increased base saturation and soil extractable P [2,24-26] is, therefore, important in management of P in these soils.

Soil acidity and fertility can be managed through the use of tolerant plant germplasms; improved agronomic, cultural and biological practices and use of inorganic fertilizers. Some of these management practices are known to farmers, who undertake them routinely within their farming protocols. However, some of the relevant management practices are not undertaken due to various reasons including lack of credit to purchase the required inputs, knowledge on the importance of lime, improved crop varieties tolerant to soil acidity constraints, and inadequate amount of organic materials [25]. Several studies have been undertaken in Kenya to assess the best management practices for soil acidity amelioration. Research findings by [27-30] indicated that integrated plant nutrient management (IPNM) by combining inorganic and organic fertilizers improves maize yields. Additionally, [23,31,32] reported some success in development of P - use efficient maize germplasms which are tolerant to Al toxicity in acid soils.

The many studies undertaken on soil acidity and fertility status in Kenya have concentrated on nutrient management issues [24,30,33]. Limited research has focused on farmers' knowledge, attitudes and current adoption levels of the management recommendations. Farmers' knowledge of the cause of the problem and potential solutions is often the first step towards identifying and designing appropriate strategies for any successful management practice. In addition, their perception of how well a technology performs is understandably one of the key factors influencing their decision to adopt new technologies [34,35]. It is also generally recognized that the technology adoption process may be hindered by failure to incorporate views and perceptions of intended users during the design or development process. For the effective adoption of soil fertility management, soil fertility gaps identification and addressing of farmers' views and perceptions is paramount. The information generated would be useful in guiding future soil acidity management programme designs for the Kenya highlands. Therefore, the objective of the research was to evaluate farmer's awareness of soil acidity and assess the current soil fertility management practices in the Kenya highlands.

2. METHODOLOGY

2.1 Location of the Study Sites

Sites were selected to represent the major agro-ecological areas with acid soils in the Kenyan highlands where the staple food crop, maize, is grown east and west of the Rift Valley [5,36]. The study sites had been under continuous crop cultivation for years with minimal attention to the state of the art maize production practices. Descriptions of the selected sites are shown in Table 1. From the information in Table 1, Meru, Embu, Kerugoya, Nyeri, Kiambu and Kinangop represented the highlands east of the Great Rift Valley while Siaya, Busia and Eldoret represented the highlands west of the Great Rift Valley.

2.2 Administration of the Questionnaire

A systematic sampling procedure [37] was used to select farmers who acted as responders in the nine study sites. The farmers were selected after a distance of 500 metres apart, resulting in a sample size of 20 farmers per site and 180 for the whole study (9 counties). Data were obtained from the sampled farmers through interviews and structured questionnaires. The Farmers' survey was undertaken in October/December, 2013, using a single - visit survey approach [38]. Data collected included information on farmers' socio - economic status, soil acidity and fertility management practices.

2.3 Data Analysis

The social data was coded and analyzed by using the SPSS software, version 17 [39].

Table 1. Description of the study areas

County	Site	GPS of sites		Elevation (m)	Soil type
Meru	Kaguru	00 05 S	037 39 E	1460	Humic Nitisols
Embu	Kavutiri	00 25 S	037 30 E	1700	Ando- humic Nitisols
Kerugoya	Inoi	01 04 S	036 47 E	1539	Ando- humic Nitisols
Nyeri	Chehe	00 25 S	037 10 E	1920	Nitisols
Kiambu	Githunguri	01 03 S	036 45 E	1720	Nitisols
Kinangop	Magumu	00 46 S	036 35 E	2691	Eutric Nitisols
Eldoret	Turbo	00 38 N	035 10 E	1835	Chromic Acrisols
Siaya	Sega	00 13 N	034 14 E	1228	Orthic Acrisols
Busia	Bumala	00 16 N	034 12 E	1248	Orphic Ferralsols

Soil type information sourced from FURP, 1988

Descriptive analysis was applied whereby frequencies of scores were computed. Dependency tests were also conducted to find out if there were any relationships between the various variables addressed by the questionnaires.

3. RESULTS

3.1 General Information about Respondents Gender and Education Levels

More than 50% of the respondents were men except for Kiambu and Kinangop counties where female respondents were more than their male counterparts (Table 2). The education levels of the respondents varied greatly from site to site. More than 50% of the respondents in Meru, Embu, and iaya had only attained primary school education, except in Kerugoya and Nyeri where most of the respondents had tertiary education.

3.2 Membership to Farmers Training Groups

Most of the farmers in some of the research areas did not belong to any particular agricultural training group (Fig. 1), except for Kerugoya, Nyeri and Eldoret where 80%, 80% and 40%, respectively, of the farmers were members. On average, less than 37% of the interviewed farmers were members of a training group. It was found out that most farmers were not interested in farmers' training groups because of unavailability of regular training programs.

3.3 Seasonal Choice of Type of Crop to Grow

The criteria governing the choice of crops to be grown per season varied among sites and individuals (Fig. 2). The majority of farmers in Meru (82%), Embu (80%) and Eldoret (85%) chose crops depending on level of sustainability - whether the crops would be able to meet their food needs or not, while majority of farmers in Kerugoya (60%), Nyeri (60%), Kinangop (60%) and Siaya (50%) chose crops depending on their level of profitability.

3.4 Soil Acidity Perceptions or Awareness among Farmers

Most farmers were not aware of soil acidity problems (Fig. 3). Kiambu, Kinangop and Busia respondents were totally unaware of soil acidity problems while less than 15% were aware in the other areas. Only 20% of the farmers in Meru and Embu had carried out physical and chemical analyses for their soils at least once while the rest of the respondents had never analysed the soils of their farms. On average, < 4% of all the interviewed farmers were aware of soil acidity and less than 8% had carried out nutrient analysis on their soils.

3.5 Soil Acidity Management through Liming

Farmers reported minimal use of lime in soils (Fig. 4). Only 10% of the farmers from Embu and Siaya and 5% of farmers in Nyeri and Eldoret had used lime.

3.6 Inorganic Fertilizer and Manure Use

Use of inorganic fertilizers and manures varied from site to site (Fig. 5). More than 75% of the respondents used both inorganic fertilizers and manures on their farms, except in Kiambu and Eldoret counties where only 60% of respondents applied both.

Table 2. Respondents gender and education levels

County	Gender (% of respondents)		Education level (% of respondents)		
	Female	Male	Primary	Secondary	Tertiary
Meru	33.3	66.7	60	25	15
Embu	33.3	66.7	60	25	15
Kerugoya	33.3	66.7	10.5	36.8	52.6
Nyeri	33.3	66.7	10.5	36.8	52.6
Kiambu	55	45	45	25	20
Kinangop	55	45	45	25	20
Eldoret	30	70	21.1	42.1	26.3
Siaya	45	55	52.6	15.8	15.8
Busia	40	60	26.3	36.8	36.8

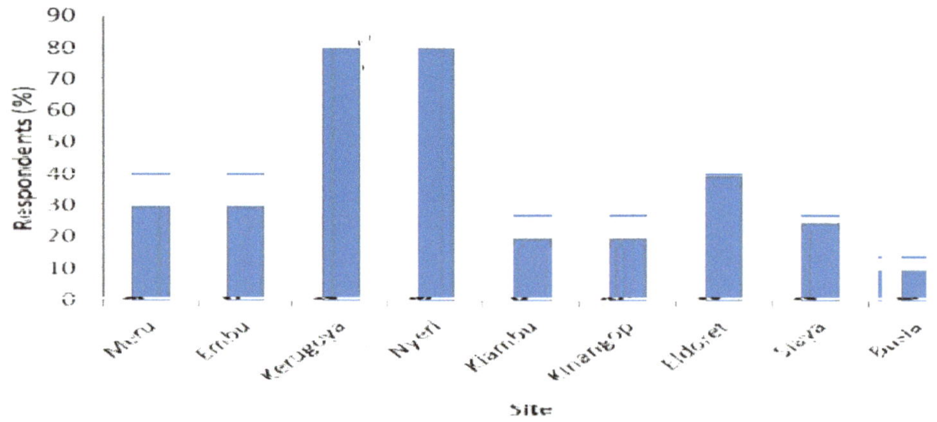

Fig. 1. Percentage of farmers in each site who are members of some training groups

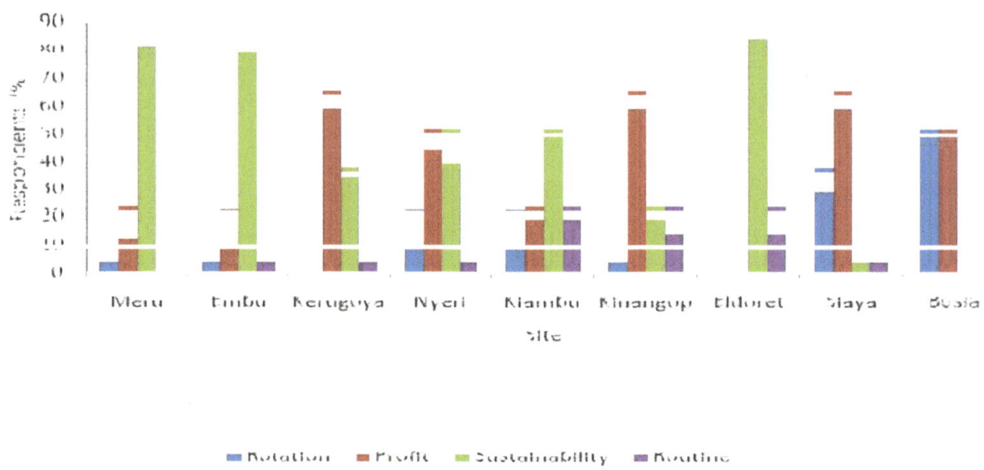

Fig. 2. Criterion applied by farmers when choosing the type of crop to plant per season in the nine sites

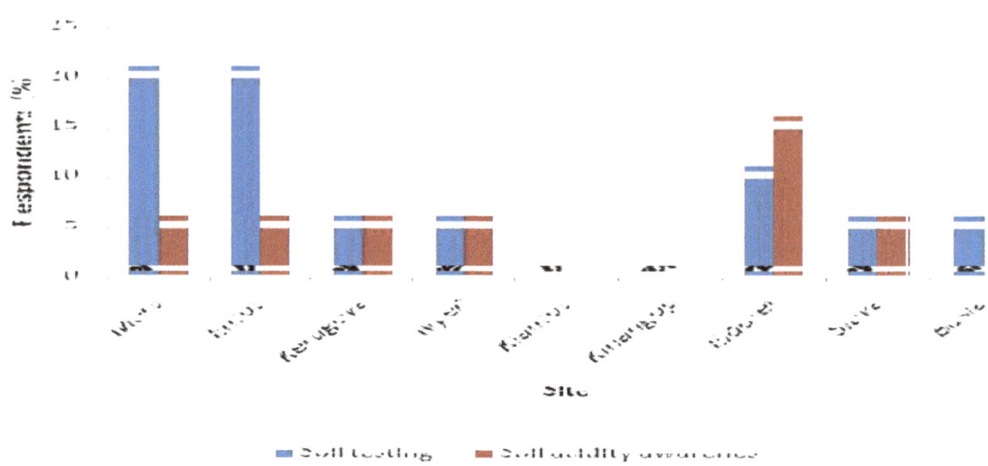

Fig. 3. Percentage of farmers having soil acidity knowledge and have tested their soils in the nine sites

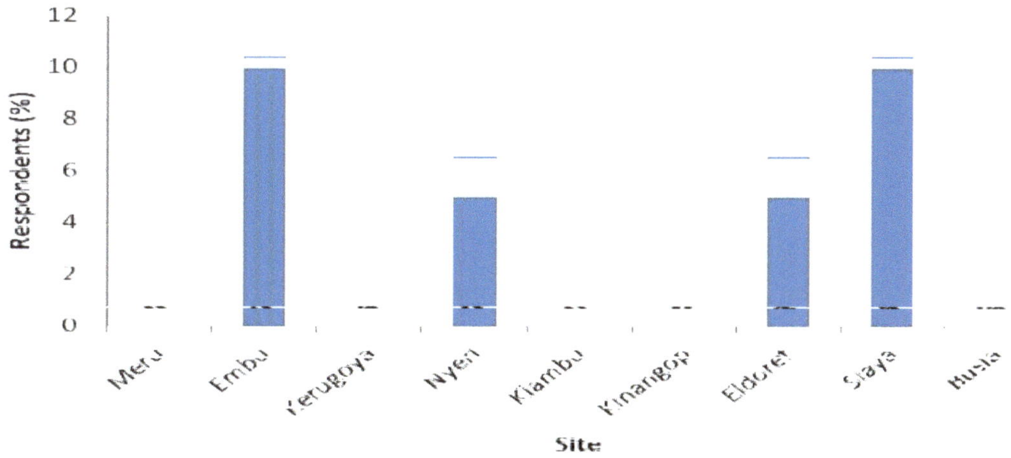

Fig. 4. Percentages of farmers applying lime in their farms at the nine sites

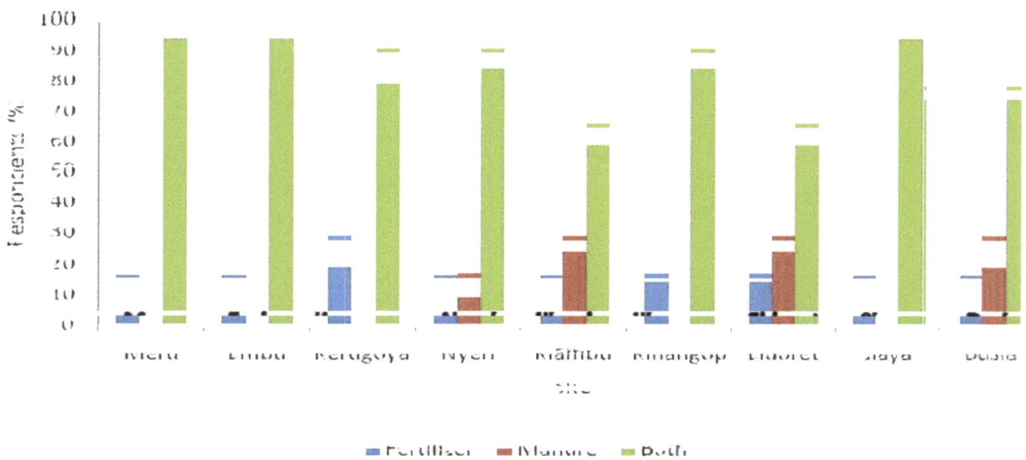

Fig. 5. Percentages of farmers applying manures or inorganic fertilizers in their farms at the nine sites

3.6.1 Manure type and sources

Farmers used only farmyard or compost manure (Fig. 6). More than 75% of the manures were from own sourcing, with only few (<1%) farmers sourcing the manures from relatives and friends and others purchasing. The proportion of farmers using farmyard manure was >70%, with <10% using compost manure.

3.6.2 Choice of inorganic and organic fertilizers in the Kenya highlands

Farmers' choices of types of inorganic fertilizers and organic fertilizers (manures) was dictated by availability, advice from agricultural extension staff, what they are used to, and advice from other farmers (Table 3). About of 50% and 74% chose inorganic fertilizer, and manure, respectively, depending on local availability, while 30% and 21% of the farmers chose inorganic fertilizer and manure type depending on what they were used to. Most farmers used inorganic fertilizers and manures mainly for soil fertility enhancement, hence expecting crop yield increases. More than 50% of the farmers had benefited from inorganic fertilizer and manure use, with <50% of the farmers indicating that sometimes the costs of production were higher than the income. The inorganic fertilizers commonly used by farmers were diammonium phosphate (DAP) [$(NH_4)_2$ HPO_4], calcium ammonium nitrate (CAN) [$Ca\ (NO_3)_2.NH_4.NO_3$], and triple super phosphate (TSP) [$Ca\ (H_2PO_4)_2$] while the common manure types were farmyard manure (FYM) and compost.

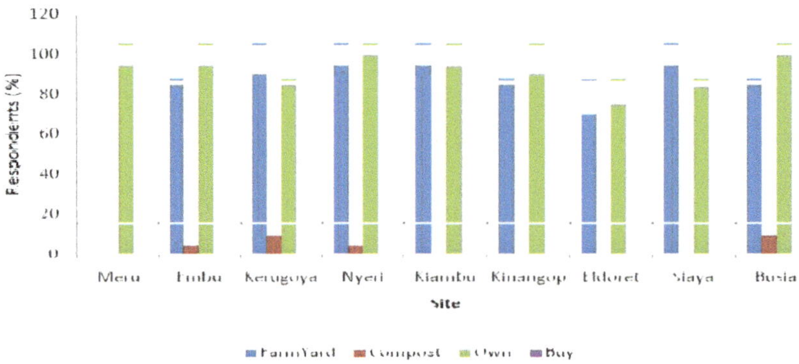

Fig. 6. Sources and types of manures applied by farmers at the nine sites

Table 3. Factors determining inorganic fertilizer and manure choice in the Kenya highlands

Question	Reasons	% respondents	
		Inorganic fertilizer	Manure
Types choice	Availability	50	74
	Advice by extension staff	18	5
	Used to	30	21
	Others	2	0
Why apply	Fertility enhancement	52	5
	Increase yield	45	90
	Others	3	5
Availability of benefit	Yes	60	50
	No	40	40
Acquisition of profit	Yes	28	20
	No	20	27
	Sometimes	32	23
	Don't know	20	30

3.7 Crop Residues Utilization by the Farmers

Methods of crop residue disposal varied greatly (Fig. 7) whereby 70% of the farmers interviewed claimed that they fed all the crop residues to their livestock, 9.4% utilized it for compost manure preparation, 0.5% burned the residues while 2.2% sold it to other progressive farmers. Only 8.3% of the farmers left residues on the farm for soil fertility improvement. For example, more than 85% of the farmers in Meru, Kerugoya and Kiambu fed the crop residues to their livestock while 5% of the farmers in Eldoret disposed of their crop residues by burning. Busia had the highest number of farmers (35%) who left crop residues on the farm to be incorporated into the soil.

3.8 Relationship between Farmers' Practices and Crop Yields

A significant negative relationship was observed between farmers' age and inorganic fertilizer use levels (Table 4). A positive relationship was observed between farmer education level, average maize yield and inorganic fertilizer use levels. Additionally, inorganic fertilizer use was positively correlated with the frequency by which farmers participated in training programmes.

4. DISCUSSION

From the farmers' responses to the questionnaire, most farmers complained of lack of structures for establishment of regular training platform as the reason they did not belong to any farmers' training group. According to [40], training of farmers is mostly through agricultural officials such as extension or field officers. The knowledge acquired by farmers enables them to learn, internalize, evaluate, try and adopt new technologies [41]. However, [40] indicated that the extension officers must be knowledgeable and credible enough to win farmers' trust. This is in agreement with farmers' responses because they indicated that extension officers were knowledgeable but the problem was that they were too few to effectively meet farmers' demand for training and follow up.

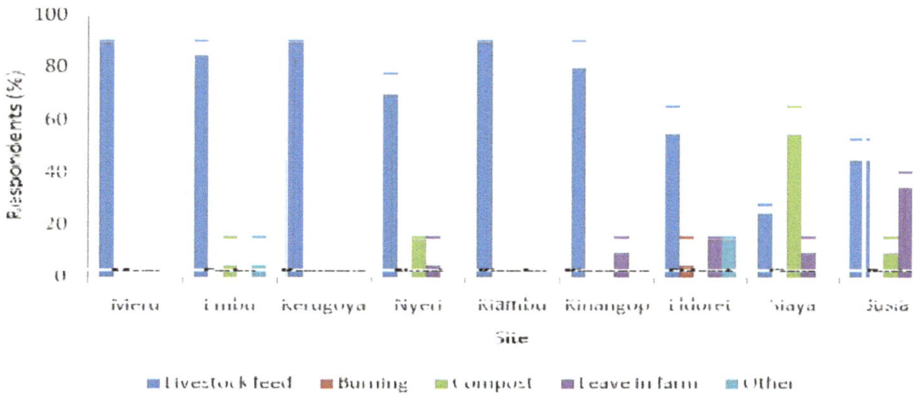

Fig. 7. Disposal of crop residues by farmers from the nine sites

Table 4. Correlations between farmers soil management practices and crop yield

Pair	R	P-value
Farmers age- Inorganic fertilizer use	-0.30**	0.002
Acreage- Average maize yield	0.29**	0.006
Education- Average maize yield	0.24*	0.025
Education- Inorganic fertilizer use	0.23*	0.024
Farmers training- Inorganic fertilizer use	0.40*	0.016
Maize yield- Inorganic fertilizer use	0.31**	0.005

**Significant at 0.05 level; **Significant at 0.01 level*

The farmers (<10%) who had used lime at least one time complained that they were not aware of soil acidity problems, did not vividly know their importance and also that lime was not readily available in the local agricultural input shops. These findings are similar to those from a review conducted by [25] who indicated that very few farmers in Kenya were aware of the soil acidity problem. They further stated that most farmers did not use lime either due to lack of knowledge about its importance or lack of liming materials in the market, or because the liming activity was labour intensive and expensive for smallholder farmers.

The preference of DAP as opposed to other nitrogen sources, despite its acidifying effect on the acid soils, could be attributed to the cost of the N sources available in the market. Studies carried out by [42] to evaluate the cheapest source of N in the Kenyan market for farmers' use indicated that mono-ammonium phosphate (MAP) and DAP were cheaper N sources compared to NPK inorganic fertilizers like 17:17:17, 20:20:0 or 23:23:0. The farmers claimed that they preferred FYM over other organic sources of fertilizer because it is readily available locally, is easy to handle and is cheaper than other manures. Similar findings reported by [30,43] indicated that most smallholder farmers in Western Kenya preferred FYM as compared to other organic sources of plant nutrients whose costs could not be offset by the yields obtained.

Education level influences farmers' access to information as well as their ability to understand technical aspects of innovations which largely affects production decisions [44]. The positive correlation between educational level, maize yields and inorganic fertilizer use is in agreement with the findings of [45], which showed that educated farmers adopted recommended inorganic fertilizer rates with ease, leading to improved crop yields and low levels of environmental pollution. It was also established by [46] that education influenced the rate of adoption of developed technologies or innovations, hence positively influencing the resulting output.

The significant positive correlation between inorganic fertilizer use and participation in farmer trainings could be attributed to access of information and knowledge on handling of inorganic fertilizers. Baah et al. [47] observed that farmers who were members of farmers' associations had better access to inorganic fertilizer information that contributed to the adoption of appropriate inorganic fertilizer use practices. Similarly, [48] attributed non-adoption of some agricultural technologies by farmers to inadequate information and sketchy knowledge, lack of awareness of the technologies and lack of follow-up by extension staffs, which are adequately catered for in training forums.

The significant (P=0.001) negative correlation (-0.30) between farmers' age and inorganic fertilizer use might be attributed to fear of the unknown. The results of this study were in agreement with [49], who observed that farmers who were above 39 years were most likely to have lower adoption rates of new technologies, because older people fear the risk of unexpected events whilst young farmers tend to be more flexible in their decisions to adopt new ideas and technologies more rapidly. Similarly, [48,50] reported that adoption of improved technologies was associated with age, education, farm size, income, exposure, scientific orientation, knowledge level and training received.

5. CONCLUSIONS

The results have revealed that farmers' in the study sites have limited knowledge on soil acidity problems, and management strategies. Additionally, farmers' access to information on new technologies is constrained by failure to participate in farmers' trainings. Strategies should, therefore, be established which would encourage farmers' participation in agricultural training groups to improve their agricultural knowledge base. The approach to be used could be the participatory rural appraisal (PRA), participatory learning and action research (PLAR) or participatory agro-ecosystem management (PAM). This will enhance levels of understanding and adoption of improved soil acidity and fertility management practices, hence improve crop yields and production across the Kenya highlands.

ACKNOWLEDGEMENTS

The authors are grateful to the Alliance for a Green Revolution Africa (AGRA) for funding the study. We also acknowledge with gratitude, the agricultural extension officers, Ministry of Agriculture, Livestock and Fisheries, Kenya for technical advice, guidance and assistance in the whole study process.

COMPETING INTERESTS

Authors have declared that no competing interests exist.

REFERENCES

1. Van Straaten P. Agrogeology: the use of rocks for crops. Enviroquest Limited, Cambridge. 2007;440.
2. Brady CN, Weil RR. *The* nature and properties of soils. 14th Ed; Pearson Prentice Hall, New Jersey. 2008;975.
3. Von Uexkull HR, Mutert E. Global extent, development and economic impact of acid soils. Pl & So. 1995;171:1-15.
4. Eswaran H, Reich P, Beigroth F. Global distribution of soils with acidity. In soil plant interactions at Low pH. (Edited by Moniz, A.C.) Brazilian Soil Science Society, Vicosa. 1997;159-164.
5. Kanyanjua SM, Ireri L, Wambua S, Nandwa SM. Acid soils in Kenya: Constraints and remedial options. KARI Technical Note. 2002;11:24.
6. Giller KE, Wilson KJ. Nitrogen fixation in Tropical cropping systems. CAB International, Wallingford. 1991;313.
7. Jorge RA, Arrunda P. Aluminium induced organic acids exudation by roots of aluminium tolerant maize. Phytochemistry. 1997;45(4):675-681.
8. Government of Kenya. National Food and Nutrition Security Policy. Nairobi: Government Printer, Nairobi, Kenya. 2011;82.
9. MAFAP. Review of food and agricultural policies in Kenya. MAFAP Country Report series, FAO, Rome, Italy. 2013;196.
10. Ombaka DM. Of Kenya's eaters and eatists: Hunger as a development and social justice challenge. J Soc Welf Hum Righ. 2014;2(1):107-129.
11. Nyoro J, Ayieko M, Muyanga M. The compatibility of trade policy with domestic policy interventions affecting the grains sector in Kenya. In: Proceedings of FAO workshop trade and policy for food products conducive to development. 1-2 March, Rome, Italy. 2007;434.
12. Smaling EMA, Nandwa SM, Janssen BH. Soil fertility is at stake. In Replenishing Soil fertility in Africa. American Society of Agronomy and Soil Science Society of America, Madison. 1997;47-62.
13. Ruigu GM, Schulter M. Fertilizer and seeds requirements in Kenya from 1981/83 to

1990/91. World Bank Working Paper. World Bank, Washington D.C. 1990;305.
14. Mugunieri GL, Nyangito HO, Mose LO. Agronomic and socioeconomic factors determining maize yield response to fertilizers in Western Kenya. In: Proceedings of the African Crop Science Conference. 13 - 17 January, Pretoria, South Africa. 1997;1465-1471.
15. Wesonga CJ, Sigunga DO, Musandu AO. Phosphorus requirements by maize varieties in different soil types of Western Kenya. Afric Cr Sc J. 2008;16(2):161-173.
16. Sigunga DO, Wandahwa PJ. Land and soil resources and their management for sustainable agricultural production in Kenya: Current position and future challenges. Egert J Sci Techn. 2011;11: 66-86.
17. Sanchez PA. Soil fertility and hunger in Africa. Sc. 2002;295:2019-2020.
18. Muhammad L, Underwood E. The maize agricultural context in Kenya. In: Risk assessment of genetically modified organisms: A case study of Bt maize Kenya. (Edited by Andow DA, Hilbeck A.) CABI Publishing, Wallingford. 2004;21-56.
19. Kiiya WW, Mwonga SM, Obura RK, Ngugi JG. Effect of incorporation of legumes on selected soil chemical properties and weed growth in a potato cropping system at Timboroa, Kenya. Afric J Agric R. 2010; 5(17):2392-2398.
20. Obura PA, Darrell A, Schulze G, John A, Okalebo R, Caleb B, Othieno O, Cliff B, Johnston T. Characterization of selected Kenyan acid soils. In: Proceedings of the 19th World Congress of Soil Science, Soil Solutions for a Changing World. 1 - 6 August, Brisbane, Australia. Published on DVD. 2010;9-12.
21. Kisinyo PO. Constraints of soil acidity and nutrient depletion on maize (*Zea mays* L.) production in Kenya. Ph.D thesis, Moi University, Kenya. 2011;1-79.
22. Schulze D, Santana D. A report on laboratory analysis of soils from acid soils-maize growing areas of Central and Western Kenya, McKnight Foundation, USA, Project. First Year Progress Report. (EMBRAPA, Purdue and Cornell Universities (USA) and Moi University (Kenya). 2003;107.
23. Ligeyo DO. Genetic analysis of maize (*Zea Mays* L.) tolerance to aluminium toxicity and low phosphorus stress and development of synthetics for use in acid soils of Western Kenya. Ph.D. Thesis. Moi University, Eldoret, Kenya. 2007;1-68.
24. Kisinyo PO, Othieno CO, Gudu SO, Okalebo JR, Opala PA, Maghanga JK, Ngetich WK, Agalo JJ, Opile RW, Kisinyo JA, Ogola BO. Phosphorus sorption and lime requirements of maize growing acid soils of Kenya. Sus Agr Res. 2013;2(2): 116-123.
25. Kisinyo PO, Opala PA, Gudu SO, Othieno CO, Okalebo JR, Palapala V, Otinga AN. Recent advances towards understanding and managing Kenyan acid soils for improved crop production. Afr J Agr Res. 2014a;9(31):2397-2408.
26. Kisinyo PO, Othieno CO, Gudu SO, Okalebo JR, Opala, PA, Ng'etich WK, Nyambati RO, Ouma EO, Agalo JJ, Kebeney SJ, Too EJ, Kisinyo JA, Opile WR. Immediate and residual effects of lime and phosphorus fertilizer inorganic fertilizers on soil acidity and maize production in Western Kenya. Exp Agr. 2014b;50(1):128-143.
27. Smaling EMA, Nandwa SM, Prestele H, Roetter R, Muchena FN. Yield response of maize to fertilizers and manure under different agro-ecological conditions in Kenya. Agric Eco Env.1992;41:241-252.
28. Palm CA, Myers RJK, Nandwa SM. Combined use of organic and inorganic sources for soil fertility maintenance and replenishment. In Replenishing Soil Fertility in Africa. American Society of Agronomy and Soil Science Society of America, Madison. 1997;193-218.
29. Jama B, Swinkels RA, Buresh RJ. Agronomic and economic evaluation of organic and inorganic sources of phosphorus in Western Kenya. Agron J. 1997;89:597-604.
30. Opala PA, Okalebo JR, Othieno CO. Comparison of effects of phosphorus sources on soil acidity, available phosphorus and maize yields at two sites in Western Kenya. Arch Agron S Sci. 2013; 59(3):327-339.
31. Matonyei KT. Acid soil tolerance studies on selected maize breeding lines from Kenya. Master of Philosophy thesis, presented to Moi University, Kenya. 2010;1-80.
32. Ouma E, Ligeyo D, Matonyei T, Agalo J, Were B, Too E, Onkware, Gudu S, Kisinyo P, Nyangweso P. Enhancing maize grain yield in acid soils of Western Kenya using aluminium tolerant germplasm. J Agric Sc Techn. 2013;3:33-46.

33. Opala PA. Comparative effects of lime and organic materials on selected soil chemical properties and nutrient uptake by maize in acid soils. Arch Appl Sc Res. 2011;3(1): 96-107.
34. Ajayi OC, Akinnifesi FK, Sileshi G, Chakeredza S. Adoption of renewable soil fertility replenishment technologies in the southern African region: Lessons learnt and the way forward. Nat Res Forum. 2007;31(4):306-317.
35. Kumbhar MI, Makhijani HB, Panhwar KN, Mughal S, Abbasi NA. Study of extension teaching methods adopted through crop maximization project: A case study of Sindh province. J Bas Appl Sc. 2015;11: 300-303.
36. Sombroek WG, Braun HM, van de Pouw. Exploratory soil map and agro-climatic zone map of Kenya. Scale 1:1000, 000. Exploratory Soil Survey Report No. E1. Kenya Soil Survey. 1982;60.
37. Banning R, Camstra A, Knottnerus P. Sampling theory. Sampling design and estimation methods. Statistics Netherlands. The Hague. 2012;87.
38. Thompson SK. Theory of sample surveys. New York: John Wiley and Sons. 1997; 301.
39. SPSS. SPSS Statists for Windows, Version 17.0. Chicago: SPSS Inc. technology. Am J Agric Econ. 2008;66: 312-320.
40. Feder G, Slade R. The acquisition of information and the adoption of new technology. Amer J Agric Econ. 2004;66: 312-320.
41. Caswell MK, Fuglie C, Ingram SJ, Kascak C. Adoption of agriculture production practices: Lessons learned from the US. US Department of Agriculture, Agriculture Economic Report No.792. Washington DC. 2001;12.
42. Kanyanjua SM, Ayaga GO. A guide to choice of mineral inorganic inorganic fertilizers in Kenya. KARI Technical Note. 2006;17:1-17.
43. Opala PA, Jama BJ, Othieno CO, Okalebo JR. Effects of phosphate fertilizer application methods and nitrogen on maize yields in Western Kenya. An agronomic and economic evaluation. Exp Agr. 2007; 43:477-487.
44. Rahman S. Environmental impact of modern agricultural technology diffusion in Bangladesh: An analysis of farmers' perceptions and their determinants. J Env Mn. 2003;68:183-191.
45. Zhou Y, Yang H, Mosler H, Abbaspour KC. Factors affecting farmers decisions on fertilizer use: A case study for the Chaobai watershed in Northern China. The J Sus Dev.2010;4(1):80-102.
46. Olagunju FI, Salimonu KK. Effect of adoption pattern of fertilizer technology on small scale farmers productivity in Boluwaduro Local Government. W Rur Obs. 2010;2(3):23-33.
47. Baah F, Anchirinah V, Amon-Armah F. Soil fertility management practices off cocoa farmers in the Eastern Region of Ghana. Agric & Biol J N Amer. 2011;2(1):173-181.
48. Ajayi MT, Oloruntoba A. Assessment of factors affecting farmers' adoption and utilization of major agricultural technologies developed by International Institute of Tropical Agriculture (IITA). J Agri F& Soc Sc. 2007;5(1):20-28.
49. Teklewold H, Dadi L, Yami A, Dana N.Determinants of adoption of poultry technology: A double-hurdle approach. Liv Res Rur Dev. 2006;18(40):1-3.
50. Zanu HK, Antwiwaa A, Agyemang CT. Factors influencing technology adoption among pig farmers in Ashanti region of Ghana. J Agric Techn. 2012;8(1):81-92.

7

Comparison of Leaf Traits and Branching Patterns between *Acacia tortilis* (Forsk.) Hayne *subsp. raddiana* (Savi) Brenan, *Balanites aegyptiaca* (L.) Del. and *Ziziphus mauritiana* Lam. Seedlings Originated from the Sahel

Fidèle Tonalta Ngaryo[1], Ampa-Kande Badiatte[2], Venceslas Goudiaby[3,4]* and Léonard-Élie Akpo[2]

[1]Department of Biology, University Adam Barka of Abéché, Faculty of Sciences and Technology, B.P 1173, Abéché, Tchad.
[2]Department of Crops Sciences, University Cheikh Anta Diop, Faculty of Sciences and Technology, B.P 5005, Dakar, Sénégal.
[3]Département des Sciences Biologiques, Université du Québec à Montréal, Case Postale 8888, succursale Centre-ville, Montréal (Québec) H3C 3P8, Canada.
[4]Wageningen University, Forest Ecology and Forest Management Group, P.O. Box 47 6700AA Wageningen, The Netherlands.

Authors' contributions

This work was carried out in collaboration between all authors. Author LÉA designed the study, and wrote the protocol. Authors FTN and AKB conducted the measurements. Author FTN wrote the first draft of the manuscript. Authors VG and FTN managed the literature searches. Author VG performed the statistical analyses and discussed the conclusion. All authors read and approved the final manuscript.

Editor(s):
(1) Monica Rosa Loizzo, Department of Pharmacy, Health Sciences and Nutrition of University of Calabria, Italy.
Reviewers:
(1) Gabriel Muturi, Kenya Forestry Research Institute, Kenya.
(2) Bahaa El Din Bastawy Mekki, National Research Centre, Egypt.

Corresponding author: Email: venceslas.goudiaby@uqat.ca

ABSTRACT

Aims: The aim of this study was to compare the leaf traits and branching patterns between *Acacia tortilis*, *Balanites aegyptiaca* and *Ziziphus mauritiana* seedlings, three species occurring in the same range in the Sahel, a semi-arid tropical area of Africa.

Study Design: Seeds used in this experiment were collected from a semi-arid area (16°20'N, 15°25'W), from species growing in their natural range in the Sahel. Seeds were after transferred at the experimental site where they were germinated and grown in a common garden experiment, on an area of 1300 m².

Place and Duration of Study: The experimental site was located at around 400 km south of the seed origin (17°33'N, 14°55'W). The experiments were carried out between October 2002 and October 2004.

Methodology: The seedlings were watered three days a week from the beginning of the experiment until July 2004. Thereafter, seedlings were assessed at 11 months, 16 months old, and at 24 months after the seed germination date. In the assessment, we measured the length (L_{GU}) and the number (N_{GU}) of growth units, the number of nodes (N_{node}), and leaves (N_{leaf}), and the single (A_{leaf}) and total ($A_{foliage}$) leaf areas.

Results: Following the water stress, N_{leaf} and A_{leaf} only decreased in *A. tortilis*, to tightly control the transpiration. A_{leaf} increased in *B. aegyptiaca* and *Z. mauritiana*, due to persistent leaves flushed before the water stress. $A_{foliage}$, N_{GU} and N_{node} were generally consistent, while L_{GU} decreased in *B. aegyptiaca*. The defoliation in dry season despite the watering suggested an endogen control, allowing the species to escape harsh conditions in their natural range.

Conclusion: The study do not support the hypothesis according to which species naturally coexisting are likely to display similar trend in leaf traits and branching patterns in response to drought. The study has been limited to three weeks of water stress and deserves to be extending to provide more insight into traits pattern for a longer period of water stress.

Keywords: Drought; growth units; leaf area; phenology; tropical area.

1. INTRODUCTION

The Sahel is one of the driest savanna ecozones with high inter-annual rainfall variability, ranging from 200 to 400 mm, high potential evaporation of about 2000 mm year^{-1} and a deep groundwater of more than 30 m [1]. The rainy season spans from July to September, and the dry season from October to June. The mean annual maximum temperature is between 35-37°C with a peak up to 48°C, and the minimum between 13-15°C [2]. Rainfall is one of the main limiting factors, explaining the highly variable recruitment potential generally occurring in semi-arid savannas [3]. Leaf traits are associated with resistance to drought [4] and can be used to assess species vulnerability to water shortage.

Acacia tortilis (Forsk.) Hayne, *Balanites aegyptiaca* (L.) Del., and *Ziziphus mauritiana* Lam. are multipurpose tree species naturally occurring in the Sahel, with leaf traits adapted to frequent water stress. They are used by the local population as fodder for livestock, as food and herbal medicine. The genus *Acacia* is particularly known to produce commercial gum, generating substantial income for local population [5]. In search for the sustainability for these species in the context of advance aridification due to climate change, there is a growing interest to better know the functional and morphological traits allowing these species to cope with drought regularly occurring in these areas. Their phenology is closely linked with climate including drought and can be used to provide insight into species response to drought.

Several studies on mature Sahelian native species [6-8] focused on qualitative phenology based on visual assessment of leafing patterns in relation with climate. Hence, *A. tortilis* is a semi-evergreen [7,9], whereas *B. aegyptiaca* (L.) Del. is an evergreen species [9,10]. *Z. mauritiana* Lam. is sometimes considered as semi-evergreen [8,9], or deciduous [11]. Leaves generally flush at the end of the dry season, when the water availability at the soil surface is completely depleted for *A. tortilis* [7,9], *B. aegyptiaca* [8] and *Z. mauritiana* [8,9]. While *A. tortilis* [7] and *Z. mauritiana* [8,12] have their maximum leafing in the earlier rainy season and lasts until the middle of the dry season, *B. aegyptiaca* remains green all year round [8].

Water stress reduces the duration of full canopy stage for *A. tortilis* [7], while we did not find such evidence in *B. aegyptiaca* and *Z. mauritiana*. These patterns were described in mature trees and may not necessarily be translated into seedlings because the direction of the responses may be changing according to the species ontogeny. Water stress significantly reduces leaf area in *A. tortilis* seedlings [13], leaf area and leaf weight in *B. aegyptiaca* [14], and leaf area in *Ziziphus* sp. [15], suggesting that stress conditioning enhances adaptive traits for drought tolerance [14]. However, there are still uncertainties about the comparative traits patterns in response to drought among *Acacia tortilis*, *B. aegyptiaca* and *Z. mauritiana*. Species naturally coexisting in the same area are believed to display similar strategies to cope with water stress, which allow them to adapt to similar habitats and form the niche differentiation [16]. In this study we aimed to know the extent to which *Acacia tortilis*, *B. aegyptiaca* and *Z. mauritiana* differ in response to same stress.

2. MATERIALS AND METHODS

2.1 Study Site, Seeds Collection and Climate

2.1.1 Site description

A. tortilis, *B. aegyptiaca* and *Z. mauritiana* naturally occur in the Sahel ecozone of Senegal. Their seeds were collected near Souilène village (16°20'N, 15°25'W) located 20 km south of Dagana where there is a weather station that is used as reference for the zone (Fig. 1). The dry season spans from October to June, and the rainy season from July to September. Annual rainfall averages 280 mm. The dry season can be subdivided into a hot dry season in May with average maximum and minimum temperatures of 41 and 22.8°C respectively, and a cool dry season in January with respective maximum and minimum temperatures of 31 and 14.8°C (annual average = 28.8°C). The soils are sandy with a deep water table [7].

2.1.2 Seeds collection

The collected seeds were germinated and their seedlings transferred in the experimental site located in the Institut de Recherche pour le Développement (IRD) of Bel-Air in Dakar 17°33'N, 14°55'W, at about 400 km south of the seed origin (Fig. 1).

Soils of the experimental site are tropical ferruginous, non-washed [17], slightly basic (pH = 8.35) resting on sandy sedimentary substratum, and contain 0.92% of organic matter, 0.53% of organic carbon and 0.46% of nitrogen [18]. The cold Canary current and the maritime trade winds moderate the climate [1]. Total annual rainfall is 376.3 mm, lowest temperatures are noted in January-February (20.7°C), and highest in September-October (27.7°C).

2.1.3 Climate

In 2002, 2003, and 2004, monthly mean maximum temperatures were around 28°C and usually occurred in rainy season while monthly mean minimum temperatures (21°C) were noted in the dry season around January and February (Fig. 3). Also, monthly mean vapor pressure displayed a minimum in the dry season (18×10^{-1}hPa) and a peak in the rainy season (30×10^{-1}hPa). In 2002, total annual rainfall was 151 mm but the seedlings were transplanted in the late rainy season, exactly the first week of October when was recorded only 4 mm of rain before the end of the rainy season. In 2003 the rainy season spanned from June to October with total amount of rainfall of 385 mm. In 2004, 234 mm of rainfall were recorded between June and October when dropped a last significant rain of 4 mm. The seedlings consecutively remained almost 3 weeks without additional water.

2.2 Experimental Design and Measurements

In order to obtain a uniform and high germination rate, pretreatments were performed on seeds before sowing for germination. The seeds of *A. tortilis* were pre-treated with sulphuric acid for 60 minutes and soaking in cold water for 10 to 15 minutes. Seeds of *B. aegyptiaca* and *Z. mauritiana* seeds were pre-treated by removing the impermeable seed coats and soaking in cold water for 24 hours. In August 2002, the pretreated seeds were germinated in plastic bags filled with sand and soil mixture, at a rate of one seedling per plastic bag in a nursery located in the Centre National de Recherches Forestières (CNRF / ISRA, Dakar). The germination was followed by a nursery acclimation phase (September 2002, end of rainy season), and two months from the germination date the seedlings were transplanted in the experimental site of IRD.

The experiment was carried out in an open area of 1300 m² (65 m x 20 m). To account for mortality in the course of the experiment, we planted more seedlings than we really need for the experiment (Fig. 2). All seedlings were planted at a spacing of 2 m by 1.3 m to reduce competition during the experiment [19,20]. In total, 96 seedlings of *A. tortilis*, 72 seedlings of *B. aegyptiaca*, and 84 seedlings of *Z. mauritiana* were planted in the experimental site (Fig. 2).

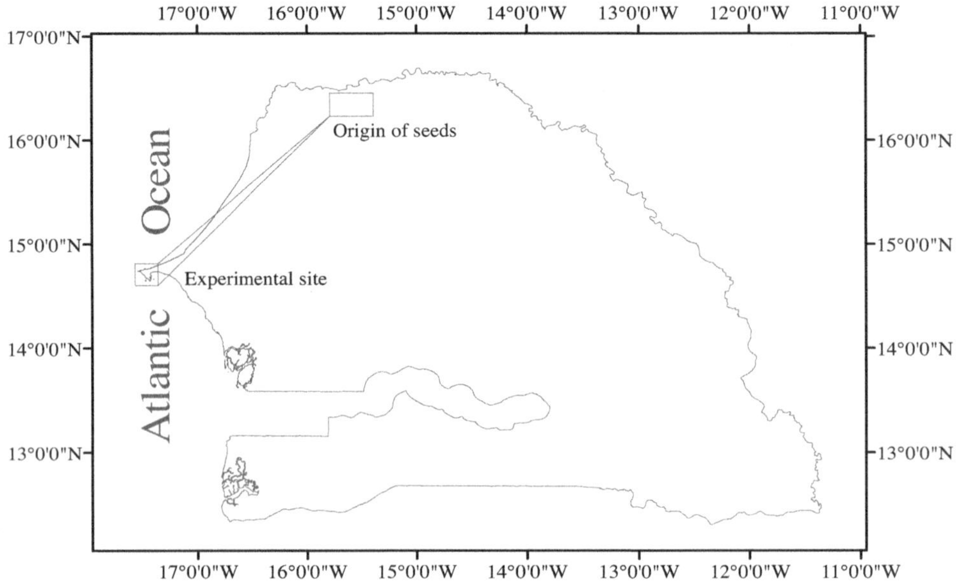

Fig. 1. Origin of the seeds of *A. tortilis, B. aegyptiaca* and *Z. mauritiana* from their natural stands, and location of experimental site approximately (Dakar, Senegal), 400 km south of the seeds origin

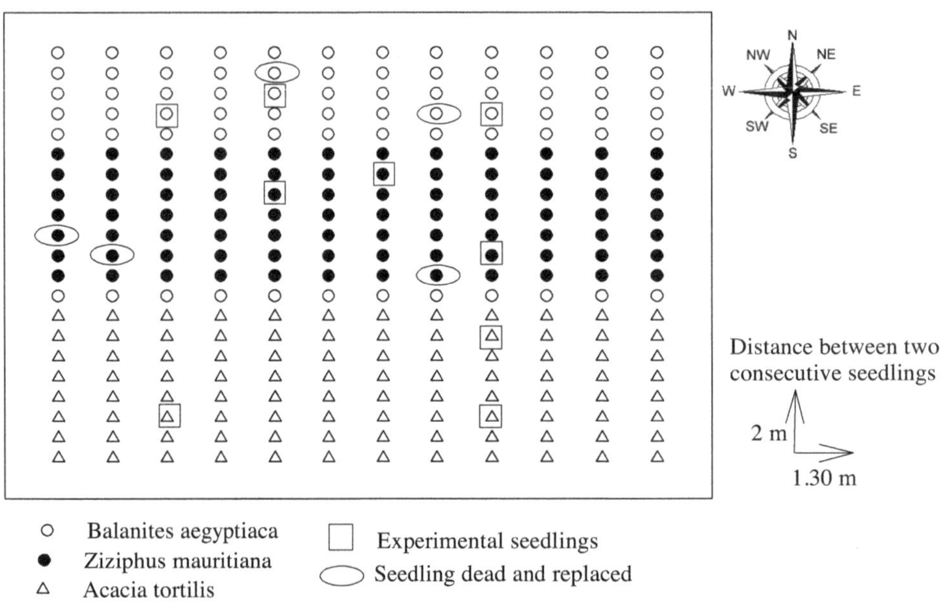

Fig. 2. Experimental design in an open area located in the Institut de Recherche pour le Développement (IRD) of Bel-Air, Dakar 17°33'N, 14°5 5'W. Two month old seedlings of *A. tortilis, B. aegyptiaca* and *Z. mauritiana* were directly transplanted in the soil, after germination and acclimation in a nursery

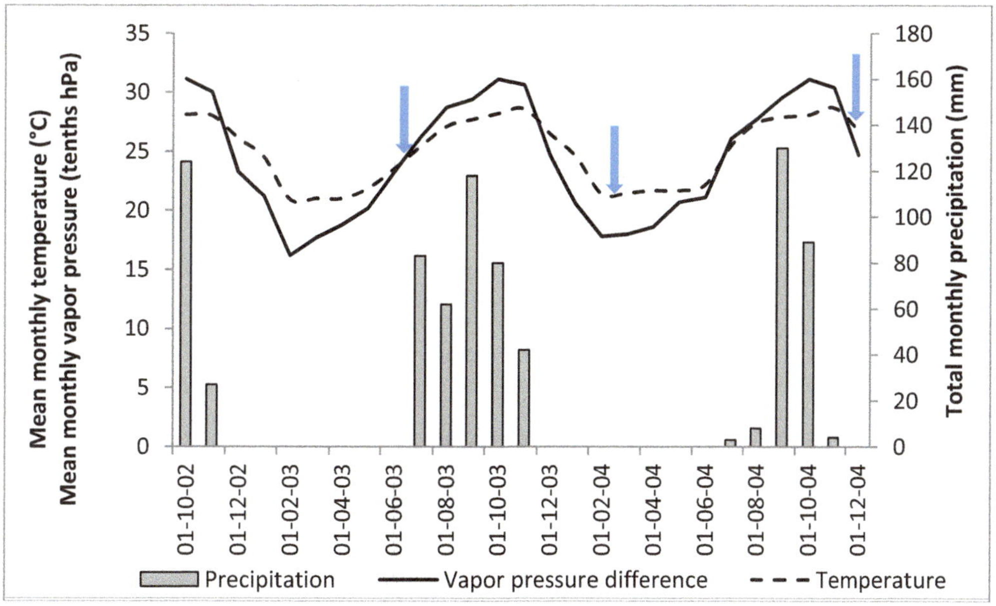

Fig. 3. Total monthly rainfall, average monthly temperature and vapor pressure of the experimental site (17°33'N, 14°55'W). The reference weather station is located in Dakar-Yoff, at 12 km far as a crow flies from the experimental site. Data came from the monthly climatic data for the World 2002–04 (WMO-NOAA, 2002–04). The arrows indicate the measurement dates

In order to build models to estimate leaf areas, leafy twigs were destructively sampled randomly in three individuals different from the measured seedlings. The sampled twigs were stored in small refrigerated containers until they were brought to the laboratory where the projected leaf areas were measured using a leaf area meter (Model AM1, Delta-T Devices, Cambridge, UK). Linear regression models were consecutively performed then used to estimate the areas of the leaves followed in the non-destructive measurements.

On 3 month old seedlings, non-destructive measurements were carried out from October 2002 to October 2004, on 3 seedlings per species selected randomly. In tree architecture model, when an episode of continuous twig elongation is followed by an interruption of growth, it produces a morphological marker of growth interruption. A Growth Unit (GU) is the section of twig delimited by two consecutive morphological markers of growth interruption [21]. A growth unit is constituted of a succession of nodes and internodes and is defined by its order in the branching ramification pattern. Hence, along a given leafy axis, we defined a GU of order 1 as the one closer to the main stem. It bears a GU of order 2, followed by a GU of order 3 and so one, until the tip of the leafy axis. In the followed seedlings, all the different orders of growth units (GU) were numbered and marked from the top of the branches and progress downwards until the main stem. The seedlings were watered three days a week from the beginning of the experiment until July 2004 (beginning of rainy season) when the watering was withheld, making the rains the only source of water provision, until the end of the experiment at 26 months from the germination date. The seedlings were measured at three different dates. The first measurement was carried out in 11 month old seedlings (M_{11}), the second measurement in 16 month old (M_{18}), and the third one in 26 month old seedlings (M_{26}) (Fig 3).

In every measurement date, we counted the number of nodes and leaves of the GU. Since *B. aegyptiaca* has simple leaves and *Z. mauritiana* is made up only of two leaflets, the leaf length and the maximum width perpendicularly to the central rib were measured. *A. tortilis* has compound leaves so that number of leaflets was also counted. Single leaf area was estimated using the regression models obtained by the destructive sampling, to derive an average value for the GU. Total number of leaves and leaf area of a given seedling were obtained by adding the values of all the GU of the seedling. Total length of GU was obtained by summing up all the GU of the seedling.

2.3 Statistical Analyses

A two-way analysis of variance was performed using the GLM procedure of the SAS software package version 9.3, according to the model:

$$Y_{ijk} = \beta_0 + \beta_1 S_i + \beta_2 GU_{ij} + \beta_3 M_{ijk} + \beta_4 SM_{ijk} + \varepsilon_{ijk}$$

Where Y is the dependent variable and represents the total number of leaves, average single leaf area, total leaf area, total number of GU, total GU length, and total number of nodes per seedling. The subscripts i, j, and k represent the species, GU rank, and month respectively. The fixed effects of the model are represented by S, GU, and M respectively for the species, GU, and month. β_0 is the intercept, β_1 to β_4 the estimates for the fixed effects, and ε_{ijk} is the error term ($\varepsilon_{ijk} \sim N(0;\sigma^2)$). Interaction between S, M and GU was not considered due to low replicate of GU ending with insufficient degree of freedom. An effect was considered significant at P<0.05 based on a type III F-test on the fixed effects. When a fixed effect was significant, paired-comparisons between species, and month were performed with a t-test (P<0.05) using the LSMEANS statement and PDIFF option of SAS software.

3. RESULTS

3.1 Leaf Area Estimation Models

The linear regression models predict well the single leaf area for *A. tortilis*, *B. aegyptiaca* and *Z. mauritiana* (Table 1).

3.2 Total Number of Leaves per Seedling

The species (F = 30.32, P<0.001) and the month (F = 5.06, P = 0.014) had significant effects on the total number of leaves (N_{leaf}) while the interaction species × month was not significant (Table 2).

For *A. tortilis*, N_{leaf} were 96, 66 and 63 in M_{11}, M_{18} and M_{26}, respectively (Fig. 4). Significantly lower N_{leaf} were noted in M_{18} (P = 0.05) and M_{26} (P = 0.03) relatively to M_{11}, while no difference was noted between M_{18} and M_{26}. For *B. aegyptiaca* and *Z. mauritiana* N_{leaf} hardly varied for the duration of the experiment. *A. tortilis* had all the time significantly higher N_{leaf} than *B. aegyptiaca* (P<0.001) and *Z. mauritiana* (P<0.001). The water stress reduced N_{leaf} by 5% in *A. tortilis* and 83% in *B. aegyptiaca*, whereas an increase of 62% was noted in *Z. mauritiana*.

3.3 Average Single Leaf Area

The species, the month and the interaction between species and month had significant effects on the average single leaf area (A_{leaf}) as shown in Table 2. *A. tortilis* displayed a quite constant A_{leaf} of about 0.31 cm^2 in average over the three measurement dates (Fig. 4). For *B. aegyptiaca*, A_{leaf} were 0.56, 1.38 and 1.72 for M_{11_wet}, M_{18_wet} and M_{26}, respectively. Relatively to M_{11_wet}, significant increase of A_{leaf} was noted in M_{26}, (P = 0.008), but no difference was observed between M_{18_wet} and M_{26}. For *Z. mauritiana*, A_{leaf} were 0.30, 0.33 and 2.29 for M_{11}, M_{18} and M_{26}, respectively. In M_2, A_{leaf} significant increased relatively to both M_{11} (P<0.001) and M_{18} (P<0.001). The three species showed similar A_{leaf} in M_{11}. In M_{18}, only *A. tortilis* displayed significantly lower A_{leaf} than *B. aegyptiaca* (P = 0.031).

3.4 Total Leaf Area of Seedlings

Only the month had a significant effect (Table 2) on the total seedling leaf area ($A_{foliage}$). The average $A_{foliage}$ were respectively 26.92 and 28.24 cm^2 in *A. tortilis* and *B. aegyptiaca*, with no significant fluctuations over the three measurement dates (Fig. 4). For *Z. mauritiana*, $A_{foliage}$ were 9.05, 2.21 and 74.57, in M_{11}, M_{18} and M_{26}, respectively. A significant increase was noted in M_{26} relatively to M_{11} (P = 0.013). Overall, $A_{foliage}$ did not significantly vary between species.

3.5 Total Number of Growth Units per Seedling

Growth units differed significantly among species (Table 2). Total number of growth units (N_{GU}) slightly decreased in *A. tortilis* and increased in *B. aegyptiaca*, and *Z. mauritiana*, but the effects of water stress were not significant. N_{GU} had average values of 10 in *A. tortilis*, 21 in *B. aegyptiaca*, and 14 in *Z. mauritiana*. The three species showed no significant difference in M_{11}. Following the water stress, *A. tortilis* had significantly lower N_{GU} than *B. aegyptiaca* (P = 0.005) and *Z. mauritiana* (P = 0.037), while all other differences between species were insignificant.

3.6 Total Length of Growth Unit per Seedling

The species, GU, and month had significant effects (Table 2) on the total length of GU (L_{GU}). L_{GU} was in average 21.67 cm in *A. tortilis* with no significant variations (Fig. 5). For *B. aegyptiaca*, L_{GU} were 25, 79, and 47 in M_{11}, M_{18} and M_{26}, respectively. L_{GU} significantly increased from M_{11} to M_{18} ($P<0.001$), and decreased from M_{18} to M_{26} ($P = 0.024$). For *Z. mauritiana*, L_{GU} were 9, 54, and 26 for M_{11}, M_{18} and M_{26}, respectively. A significant increase was noted between M_{11}, and M_{18} ($P = 0.005$). In M_{11}, L_{GU} did not vary between the three species. However, *A. tortilis* had significantly shorter GU than *B. aegyptiaca* in both M_{18} ($P = 0.004$) and M_{26} ($P = 0.04$).

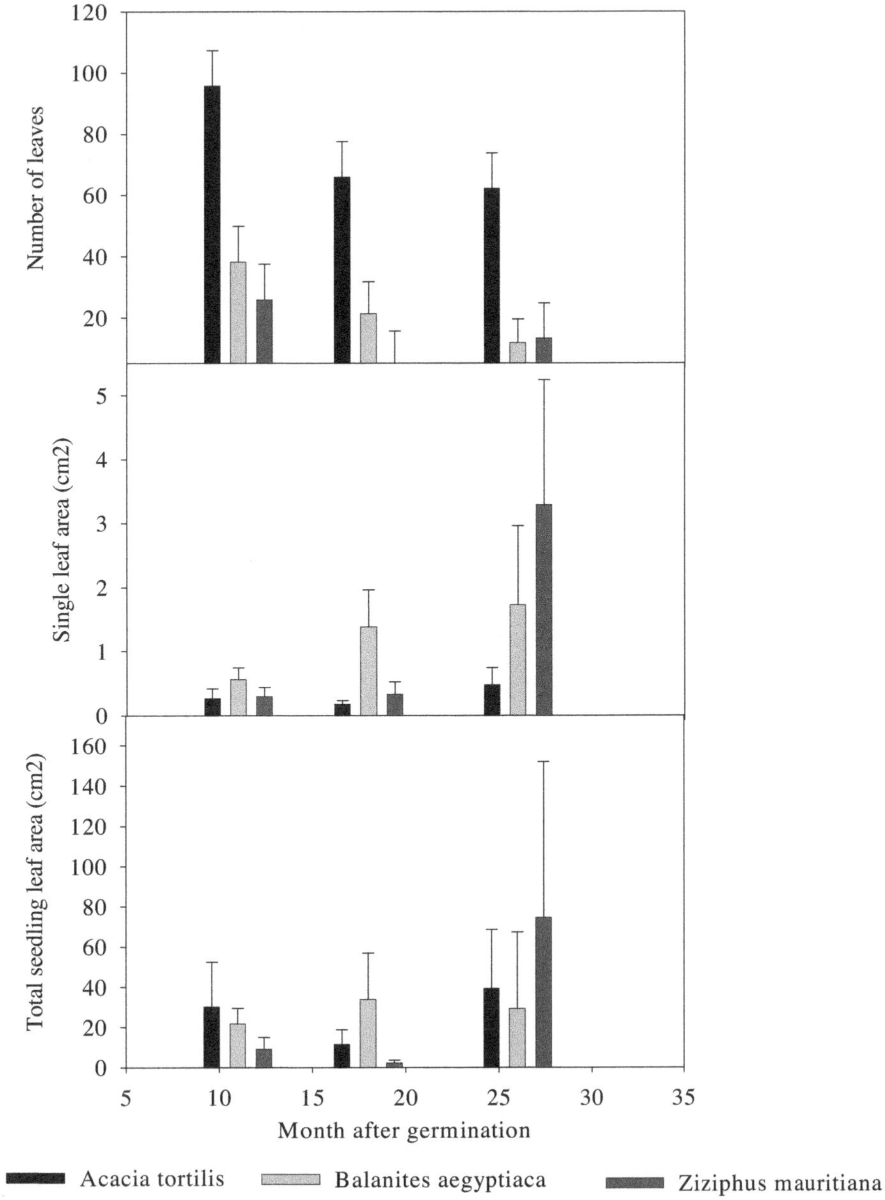

Fig. 4. Number of leaves, single leaf area, and total leaf area followed on seedlings from their germination to 26 month old in *A. tortilis, Z. mauritiana*, and *B. aegyptiaca* in a semi-controlled experiment. Seedlings were watered from the start of the experiment until month 23; when the watering was withheld at the beginning of the rainy season until month 26. Error bars represent standard errors

Table 1. Linear model of assessment of the single leaf area (A_{leaf}, cm^2) of 24 month old seedlings of *A. tortilis*, *B. aegyptiaca* and *Z. mauritiana*

Species	Prediction model of A_{leaf}	N	R^2	P
A. tortilis	$0.164 \times (L_{leafstalk}) + 0.164 \times (N_{leaflet}) - 0.523$	228	0.61	P<0.001
B. aegyptiaca	$1.040 \times (L_{leaf}) + 1.999 \times (W_{leaf}) - 4.639$	109	0.84	P<0.001
Z. mauritiana	$1.493 \times (L_{leaf}) + 3.437 \times (W_{leaf}) - 5.759$	130	0.96	P<0.001

Note: N = total number of leaves, $L_{leafstalk}$ = length of the leafstalk (cm), $N_{leaflet}$ = number of leaflets, L_{leaf} = leaf length (cm) and W_{leaf} = leaf width (cm), the R^2 were 0.61, 0.84, and 0.96 in A. tortilis, B. aegyptiaca and Z. mauritiana, respectively. These high R^2 associated with significant p-values indicated a good prediction capability of the models

Table 2. Two-way ANOVA of the effects of Species, GU rank, month and the interactions Species × Month on the leaf and growth unit traits in a semi-controlled experiment on seedlings of *A. tortilis*, *B. aegyptiaca* and *Z. mauritiana*

Source	df	Anova SS	Mean square	F value	P>F
Number of leaves per seeding					
Species	2	25610.40	12805.20	30.32	< 0.001
GU rank	6	2770.94	461.82	1.09	0.392
Month	2	4174.05	2087.02	4.94	0.015
Species × Month	4	668.57	167.14	0.40	0.810
Average single leaf area					
Species	2	9.10	4.55	6.38	0.006
GU rank	6	4.11	0.68	0.96	0.471
Month	2	18.05	9.02	12.65	< 0.001
Species × Month	4	11.69	2.92	4.10	0.011
Total leaf area per seedling					
Species	2	215.40	107.70	0.09	0.911
GU rank	6	3050.17	508.36	0.44	0.848
Month	2	8655.11	4327.55	3.72	0.038
Species × Month	4	6663.05	1665.76	1.43	0.252
Total number of growth units per seedling					
Species	2	772.76	386.38	3.87	0.0313
GU rank	6	8737.37	1456.23	14.58	< 0.001
Month	2	35.55	17.77	0.18	0.838
Species × Month	4	675.04	168.76	1.69	0.177
Total length of growth units per seedling					
Species	2	6968.01	3484.01	6.53	0.004
GU rank	6	51363.02	8560.50	16.05	< 0.001
Month	2	12193.38	6096.69	11.43	< 0.001
Species × Month	4	1176.29	294.07	0.55	0.700
Total number of nodes per seedling					
Species	2	2058.50	1029.25	12.03	<0.001
GU rank	6	14840.63	2473.44	28.92	< 0.001
Month	2	3300.21	1650.11	19.29	< 0.001
Species × Month	4	194.56	48.64	0.57	0.688

Note: GU = Growth unit

3.7 Total Number of Nodes per Seedling

Significant effects of species, GU and month were noted in the number of nodes (N_{node}). For *A. tortilis*, N_{node} were 25, 43 and 44 in M_{11}, M_{18} and M_{26}, respectively (Fig. 5), with a significant positive shift in M_{18} (P = 0.019) and M_{26} (P = 0.01) compared to M_{11}. *B. aegyptiaca* had 6, 36 and 36 nodes in M_{11}, M_{18} and M_{26}, with significant increase in M_{18} (P<0.001) and M_{26} (P<0.001) relatively to M_{11}. For *Z. mauritiana*, N_{node} were 7, 25, and 26 in M_{11}, M_{18} and M_{26}, respectively. However, relatively to M_{11}, N_{node} significantly increased in M_{18} (P<0.008) and M_{26}

(P<0.006). *A. tortilis* had significantly lower N_{node} than *B. aegyptiaca* (P = 0.05) in M18. Following the water stress, *A. tortilis* showed significantly lower N_{node} compared to *B. aegyptiaca* (P = 0.005) and *Z. mauritiana* (P = 0.037).

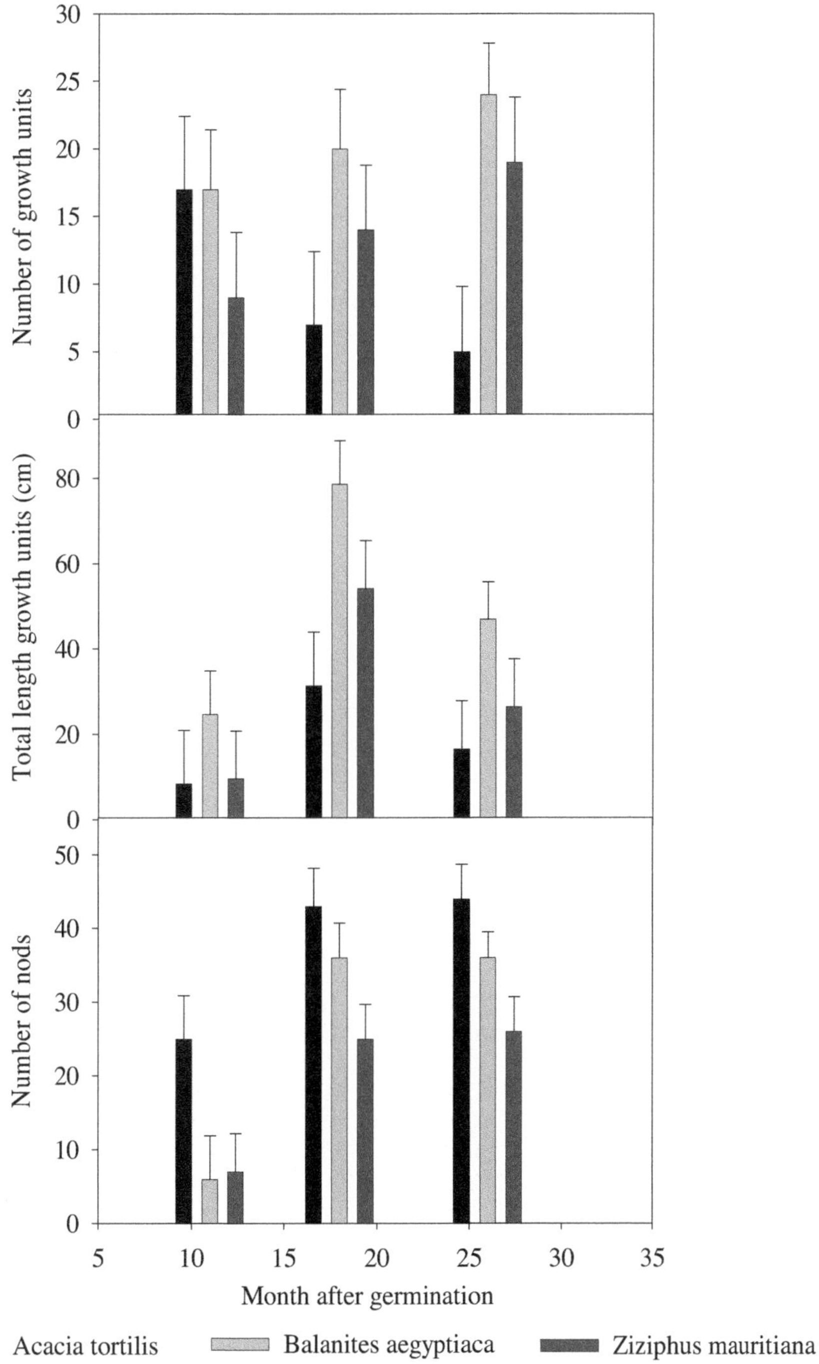

Fig. 5. Number of growth units, their total length, and the number of nodes followed on seedlings from their germination to 26 month old in *A. tortilis, Z. mauritiana*, and *B. aegyptiaca* in a semi-controlled experiment. Seedlings were watered from the start of the experiment until month 23; when the watering was withheld before the rainy season until month 26. Error bars represent standard errors

4. DISCUSSION

This study aimed to compare leaf traits and branching patterns between *A. tortilis*, *B. aegyptiaca* and *Z. mauritiana* seedlings, three species naturally occurring in the same area. Two main information can be drawn from this study. The first one pertains to the patterns displayed within one species, and the second one lays on the difference between species in their response to drought.

4.1 Variation within Species

Despite the favorable soil water status due to the watering, all the species shed their leaves, even slightly, in the dry season, but not translating into total leaf area reduction. In their natural range, a systematic defoliation also occurs in hot dry season for *A. tortilis* [7], and *Z. mauritiana* [9]. *B. aegyptiaca* also shed its leaves in hot dry season but defoliation overlaps with the next leaf flush [9]. In addition, a systematic defoliation of *A. tortilis*, grown in the Negev where it was introduced, occurs when soil water status is locally favorable, corresponding to the hot dry season of their origin in Africa [22]. The defoliation at least for *A. tortilis* may therefore be related to an endogen control, and the same mechanism may have played for *B. aegyptiaca* and *Z. mauritiana*, as well. In their natural range where drought is real at this time of the year, this could be seen as an escape strategy because maintaining lower foliage in hot dry season is conducive to some extent to tight regulation of leaf transpiration. Hence, soil water availability is seldom, if ever, to be the only variable explaining leaf shading in dry season.

B. aegyptiaca and *Z. mauritiana* increased their growth unit lengths in the middle of the dry season in response to favorable soil water availability while *A. tortilis* showed a stronger seasonality of its GU elongation. This also indicates that *B. aegyptiaca* and *Z. mauritiana* elongate their twigs in dry season to be able to respond to soil water experimentally increased in this period. Following the water stress, the GU length only decreased for *B. aegyptiaca* suggesting the active phase of GU elongation in dry season and the plastic behavior to water stress of this species. Based on GU patterns, *Z. mauritiana* displayed higher tolerance to stress as shown by its consistent GU elongation despite water stress. The greater number of GU and nodes in *A. tortilis* consecutively to water stress resulted in reduced internode distances, ending up with a greater number of nodes per unit of leafy axis. The study illustrates the statement according to which, measurement of shoot or twig is an effective way to assess fine scale plant response to environmental stimuli [23].

4.2 Variation between Species

The water stress was followed by a marked decrease in single leaf area for *A. tortilis*, but an increase for *B. aegyptiaca* and *Z. mauritiana*, however not resulting in significant variation in total leaf area. However, the slight defoliation generally observed, was enough to maintain a consistent total leaf area despite the significant change in single leaf area in the three species. *A. tortilis* had generally more leaves than *B. aegyptiaca* and *Z. mauritiana* all the time, even before the stress was applied. Earlier studies have reported leaf area decrease due to water stress in *A. tortilis* [13], *B. aegyptiaca* [24] and *Ziziphus* [15]. Water stress has been shown to reduce leaf area through reduced leaf initiation, leaf size and leaf production rate [13]. We also observed in *A. tortilis* a reduced leaf production rate, even if it was not significant. The advantage of having higher number of leaves associated with lower size single leaf areas is to allow more plasticity in leaf surface area regulation. Small leaves ensure for convective heat loss that improves cooling in hot, dry conditions, and a tight control of transpiration [25].

Unexpectedly, *Z. mauritiana* had greater single leaf areas following the water stress. *B. aegyptiaca* also positively reacted to water stress, but in comparison with first measurement date and not the second. The greater single leaf area can be explained by the fact that leaves earlier flushed when the trees were not stressed still persisted when the water stress occurred. To be able to increase their single leaf area while maintaining a consistent total leaf area following the water stress, *B. aegyptiaca* and *Z. mauritiana*, have to shed proportionally their leaves to balance their single leaf area increase. After the rainy season of 2004 with 234 mm of rainfall, the seedlings remained 3 weeks without additional water in soil. *B. aegyptiaca* and *Z. mauritiana* may have kept their leaves flushed during this previous rainy season or sooner while the seedlings were still watered. Three weeks of water stress seemed to be too short for *B. aegyptiaca* and *Z. mauritiana* to shed their leaves, showing that they can withstand water stress, longer than *A. tortilis*.

4.3 What can we Learn so Far?

A. tortilis, *B. aegyptiaca* and *Z. mauritiana* naturally occur in the same range characterized by harsh climate conditions with frequent drought episodes. They have sclerophillous leaves and can withstand severe water stress. Two clear patterns emerge from this study. *A. tortilis* bears more leaves which are also smaller than those of their counterparts, as a strategy to cope with soil water depletion. *B. aegyptiaca* and *Z. mauritiana* withstand water stress by keeping larger single leaf area, because they might have kept their greater size leaves flushed whether during the previous rainy season or earlier when the seedlings were still watered. In addition, *B. aegyptiaca* decreased its GU length following the water stress. Overall, larger plasticity in response to drought was observed in *A. tortilis* compared to *B. aegyptiaca* and *Z. mauritiana*. In addition, within a given species, the same seedlings were followed during the course of the experiment, at different ages. However it is unlikely that the age of the seedlings influenced the leaf size, since individual leaf area was shown to have a modest increase with plant size [26]. In a typical common garden experiment, many provenances are grown in a common area to study the phenotypic variability. In our experiment, the seeds which have been germinated were rather collected in a restricted area where the species naturally coexist. The differences in leaf traits and branching patterns may therefore not be attributed to a genetic variability leading to a phenotypic divergence within the species.

5. CONCLUSION

In the Sahel, a semi-arid tropical area of Africa where we collected the geminated seeds used in this experiment, soil water availability is the main environmental driving factor. *A. tortilis*, *B. aegyptiaca* and *Z. mauritiana* naturally coexist in these areas and the understanding of how they cope with water shortage is essential to implement guidelines for a more efficient assisted recruitment. This study provides to these purposes important information on leaf traits and branching patterns related to soil water availability. In contrast with what we hypothesized, species naturally coexisting do not necessarily displayed similar trend in leaf traits and branching patterns in response to drought. Although straightforward, the study was limited to three weeks of water stress. Perspectives for future studies suggest extending the drought period.

ACKNOWLEDGEMENTS

Project administration and funding were provided by Université Cheikh Anta Diop, Dakar, Senegal. The research was conducted in the Laboratory of plant ecology of the Institut de Recherche pour le Development (IRD).

COMPETING INTERESTS

Authors have declared that no competing interests exist.

REFERENCES

1. Le Houérou HN. Definition, geographical limits, contacts with other ecoclimatic zones. In: The grazing land ecosystems of the African Sahel, Springer Berlin Heidelberg; 1989.
2. Lawesson JE. Sahelian woody vegetation in Senegal. Vegetatio. 1990;86:161-174.
3. Couteron P, Kokou K. Woody vegetation spatial patterns in a semi-arid savanna of Burkina Faso, West Africa. Plant Ecol. 1997;132(2):211-227.
4. Deines JM, Hellmann JJ, Curran TJ. Traits associated with drought survival in three Australian tropical rainforest seedlings. Aust. J. Bot. 2011;59(7):621-629.
5. Ngaryo FT, Goudiaby VA, Dagbenonbakin GD, Agbangba EC, Diatta S, Akpo LE. Modeling the production of Arabic gum (*Acacia senegal* (L.) Willd) based on the proper tapping characteristics in the semi arid Sahel of Chad. International Journal of Science and Advanced Technology. 2011; 1(8):24-30.
6. Grouzis M, Sicot M. A method for the phenological study of browse populations in the Sahel: The influence of some ecological factors. In: Browse in Africa, the current state of knowledge, Le Houérou éd. ILCA, Addis Abeba; 1980.
7. Do FC, Goudiaby VA, Gimenez O, Diagne AL. Diouf M, Rocheteau R, Akpo LE. Environmental influence on canopy phenology in the dry tropics. Forest Ecol. Manag. 2005;215:319-328.
8. De Bie S, Ketner P, Paasse M, Geerling C. Woody plant phenology in the West Africa savanna. J. Biogeogr. 1998;25(5):883-900.
9. Seghieri J, Do CF, Devineau JL, Fournier A. Phenology of woody species along the climatic gradient in West Tropical Africa. In: Phenology and Climate Change, Dr. Xiaoyang Zhang (Ed.), In Tech; 2012.

10. Akpo LE. Phenological interactions between tree and understory herbaceous vegetation of a sahelian semi-arid savanna. Plant Ecol. 1997;131(2):241-248.
11. Mishra RK, Upadhyay VP, Bal S, Mohapatra PK, Mohanty RC. Phenology of species of moist deciduous forest sites of Similipal biosphere reserve. Lyonia. 2006; 11(1):5-17.
12. Grice C. Ecology in the management of Indian Jujube (*Ziziphus mauritiana*). Weed Science. 1998;46(4):467-474.
13. Otieno DO, Schmidt MWT, Adiku S, Tenhunen J. Physiological and morphological responses to water stress in two *Acacia* species from contrasting habitats. Tree Physiol. 2005;25(3): 361-371.
14. Elfeel AA, Warrag El, Musnad HA. Response of *Balanites aegyptiaca* (L.) Del. seedlings from varied geographical source to imposed drought stress. Discov. Innov. 2007;184:319-325.
15. Arndt SK, Clifford SC, Wanek W, Jones HG, Popp M. Physiological and morphological adaptations of the fruit tree *Ziziphus rotundifolia* in response to progressive drought stress. Tree Physiol. 2001;21(11):705-715.
16. Xu F, Guo W, Xu W, Wei Y, Wang R. Leaf morphology correlates with water and light availability: What consequences for simple and compound leaves? Prog. Nat. Sci. 2009;19(12):1789-1798.
17. Diagne Y. Boisements et paysages dans la ville de Dakar. Mémoire de maîtrise, UCAD, FLSH, Dakar; 1988. French.
18. Ngaryo FT, Akpo LE, Goudiaby, VA, Badiatte AK, Logbo J, Diouf M. Estimation de la surface foliaire de jeunes plants de trois essences sahéliennes: *Acacia tortilis*, *Balanites aegyptiaca* et *Ziziphus mauritiana* en conditions semi naturelles (Dakar, Sénégal). Annales de l'Université de N'Djaména. Série C N°2:27-41. French; 2007.
19. Goudiaby V, Brais S, Grenier Y, Berninger F. Thinning effects on jack pine and black spruce photosynthesis in Eastern boreal forests of Canada. Silva Fennica. 2011; 45(4):595-609.
20. Goudiaby V, Brais S, Berninger F, Schneider R. Vertical patterns in specific volume increment along stems of dominant jack pine (*Pinus banksiana*) and black spruce (*Picea mariana*) after thinning. Can. J. For. Res. 2012;42(4):733-748.
21. Barthélémy D, Caraglio Y. Plant architecture: A dynamic, multilevel and comprehensive approach to plant form, structure and ontogeny. Ann. Bot. 2007; 99:375-407.
22. Halevy G, Orshan G. Ecological studies on *Acacia* species in the Negev and Sinai. II. Phenology of *Acacia raddiana*, *A. tortilis* and *A. gerrardii negevensis*. Israël J. Bot. 1973;22:120-138.
23. Naydenov K, Tremblay F, Bergeron Y, Goudiaby V. Effect of active charcoal on the primary growth of jack pine seedlings, Influence du charbon actif sur la croissance primaire des plantules de pin gris. Can. J. For. Res. 2006;36(3):761-767.
24. Elfeel A, Warrag E, Musnad H. Response of *Balanites aegyptiaca* (L.) Del. seedlings from varied geographical source to imposed drought stress. Discov. Innov. 2007;184:319-325.
25. Yates MJ, Anthony Verboom G, Rebelo AG, Cramer MD. Ecophysiological significance of leaf size variation in *Proteaceae* from the Cape Floristic Region. Funct. Ecol. 2010;24(3):485-492.
26. Price CA, Wright IJ, Ackerly DD, Niinemets U, Reich PB, Veneklaas EJ. Are leaf functional traits 'invariant' with plant size and what is 'invariance' anyway? Funct. Ecol. 2014;1-14.

8

Allelopathic Effect of Three Wild Species on Seed Germination and Seedling Growth of *Vicia faba*, *Hordeum vulgare* and *Triticum aestivum*

Azza Ibrahim Ebid[1*]

[1]*Department of Biology, Faculty of Science, Princess Nora bint Abdulrahman University, Saudi Arabia.*

Author's contribution

The sole author designed, analyzed and interpreted and prepared the manuscript.

Editor(s):
(1) Daniele De Wrachien, Department of Agricultural and Environmental Sciences of the State University of Milan, Italy.
Reviewers:
(1) Rezzoug Waffa, Ibn Khaldoun University, Algeria.
(2) M. Angeles Calvo Torras, Universidad Autónoma de Barcelona, Spain.
(3) Abdel-Galil Mohamed Abdel-Galil, Field Crops Research Institute, Egypt.

ABSTRACT

Many wild and cultivated grasses, exhibit allelopathy to exclude the associated species by reducing their regeneration. This process involving secondary metabolites produced by plant influence the growth and development of agricultural and biological system. This research was to determine the allelopathic potential of aqueous extracts of different wild species (*Calligonum comosum* L, Her, *Rhanterium epapposum* Oliv and *Rhazya stricta* Decne) on seed germination and seedling growth of three crops (*Vicia faba*, *Hordeum vulgare* and *Triticum aestivum*). Effect of water extracts from dry aboveground plant biomass in concentration of 10% (100 g/L) was examined under laboratory conditions in petri dishes. Germination indicators (Germination Percentage (GP) and germination start) were evaluated. The results showed that all the extracts significantly decreased germination percentage, plumule and radicle length of seedlings. However, leaf extracts of *R. stricta Decne* had greater inhibitory potential and reduced seedling growth. The significant allelopathic effect remained up to 10 days. The extract had strong inhibitory effect to root elongation of seedling of plants. Differences in sensitivity between plant cultivars were documented. Based on the study

**Corresponding author: Email: azzaebid2002@yahoo.com*

results, stem, leaves and flowers residues of (*C. comosum* L, Her, *R. epapposum* and *R. stricta* Decne) showed a negative allelopathic effects on plant growth and should be eliminated from the field.

Keywords: Allopathic; wild plants; Vicia faba; Hordeum vulgare; Triticum aestivum.

1. INTRODUCTION

Allelopathy is the interactions among plants, bacteria, fungi and algae with the living organisms in a certain environment. Interaction that are mediated by production of secondary metabolites release by plants or microorganisms into the environment, are stimulatory or inhibitory and produce biological changes [1,2]. Allelochemicals usually are called secondary plant product or waste products of the main metabolic pathways in plants. These chemicals are the major factor in regulating the structure of plant communities [3,4]. Allelochemicals interference must be species-specific to explain why nonindigenous species dominate an invaded community while they normally do not reach high dominance in their native community. Most research on allelopathy has focused on the effect of allelochemicals as weed control [5-9]. Identification of allelopathic potential of wild plants on growth and yield of crops leads to better understanding of their negative impact on plants which will help to control invasive plants. However, [10] reported that the difficult to find noticeable effect in the established communities because of possible adaptation of co-existing species to allelochimicals released by competitors. *Calligonum comosum* L, Her, *Rhanterium epapposum* Oliv and *Rhazya stricta* Decne are a wild plants, which naturally grows on the plains and hilly areas of Saudi Arabia and neighboring countries.

Calligonum comosum L, Her, locally known as (Alarta) belongs to the botanical family Polygonaceae (Fig. 1). It is a plant of tropical and subtropical regions and is wide spread in Saudi Arabia and Arabian Peninsula. Several studies have confirmed the anti-bacterial activity and benefits of *C1. comosum* leaves in the treatment of certain skin diseases [11,12,13].

Rhanterium epapposum Oliv, locally known as (Alarfaj) belongs to the botanical family Asteraceae (Fig. 2). It is a C_3 desert shrub that can form monotonous stands covering vast areas of north-eastern Arabia [14]. It grows as a perennial woody shrub approximately 80 cm high with many stems branching, small and narrow leaves and straw- yellow flowers. It is grow mainly in deep sandy soil [15].

Fig. 1. Alarta (*Calligonum comosum L,* Her)

Fig. 2. Alarfaj (*Rhanterium epapposum* Oliv)

Rhazya stricta Decne, locally known as (Alharmal) belongs to the botanical family Apocynaceae (Fig. 3). It is a desert evergreen perennial sub-shrub. It is poisonous, thus not browsed by livestock. It is a valuable medicinal plant as it contains several chemical compounds and has ethno-medicinal uses world-wide. Despite its medicinal importance, *R. stricta* is a troublesome plant as it invades deteriorated rangelands. *R. stricta* species have harmful effects on crops including reduced seed germination [16,17,18].

Fig. 3. Alharmal (*Rhazya stricta* Decne)

The aim of this study was to determine the possible allelopathic effects of commonly distributed wild species in Saudi Arabia (*C. comosum* L, Her, *R. epapposum* Oliv and *R. stricta* Decne) on germination and seedling growth of *Vicia faba*, *Hordeum vulgare* and *Triticum aestivum*.

2. MATERIALS AND METHODS

2.1 Collection of Plant Material and Preparation of Extracts

The profile of three wild plants used in this study is shown in (Table 1). Fresh plant material used in this study were: *C. comosum* L, Her, *R. epapposum* Olive and *R. stricta* Decne collected from the north-east of Riyadh city, in "Dahna" area in Saudi Arabia. The fresh leaves and roots were cut into small pieces, washed in distilled water and air-dried in 75% shade at room temperature. After drying, the plant materials were ground to powder and extracted with hot water, 10 g of the powdered sample was suspended in 100 ml of distilled hot water. The samples were then shaken and allowed to settle at room temperature for 24 hrs in hot water. The solutions were filtered thought a double layers of muslin cloth followed by a Whatman No. 1 filter paper, the pH values were adjusted to 7 with 1.0 M HCl, these were kept in refrigerator at 4 C° until further use [19]. The osmotic concentrations of bioassay solutions were less than 0.1 Mpa and hence not considered a factor affecting germination [20]. Seeds of *Vicia faba*, *Hordeum vulgare* and *Triticum aestivum* were collected from the different localities of study area, sterilized by 0.3% sodium hypochlorite, rinsed by distilled water and shade dried again on the filter paper in the laboratory at room temperature for 7 days according to [21].

2.2 Preliminary Phytochemical Screening

The plant materials were screened for the presence of different classes of secondary metabolites including alkaloids, flavonoids, phenols, saponins and tannins using previously described methods [22].

2.3 Germination and Growth Bioassays

Two layers of Whatman No. 1 filter paper were placed in 90 mm diameter glass petri dishes. In each petri 20 seeds were placed and 10 mm of each plant extract added. A check treatment was assigned with distilled water and let at room temperature. Starting from the first day after experiment set on, germinated seeds were counted and removed daily. A seed with 0.5 cm of radical was considered germinated. The plume and radicle lengths of seedlings were measured on 10 day after treatment (DAT).

Experiment designed was Randomized Complete Block design with four replicates and experiment repeated twice and the percentage of germination was calculated.

2.4 Germination and Growth Indicators

Percentage of Germination (PG) and germination start compounds (GS) were calculated according to the following extract formulas:

1) Percentage of germination = 100(n/N).
2) Germination start = The long time between seed.
3) Where N is the number of seeds germinated on day i and T is the number of days from sowing.

Table 1. Used part of the selected plants

Botanical name	Family	Local name	Part used	Extract dry weight (g ml^{-1})
Calligonum comosum L'Her	Polygonaceae	Alarta	Stem	10
Rhanterium epapposum Oliv	Asteraceae	Alarfaj	Flower	10
Rhazya stricta Decne	Apocynaceae	Alharmal	Leaves	10

3. RESULTS AND DISCUSSION

3.1 Phytochemical Analysis

Freshly prepared extracts were subjected to preliminary phytochemical screening for various constituents. The results (Table 2) revealed the presence of phenols, polyphenols and alkaloids in *C. comosum*, *R. epapposum* and *R. stricta* Decne, Tannis and saponins were also detected in the extracts except in the extract of R. *epapposum*. The chemical compounds such as Tannins, Trepenoids, Phenylpropans, Acetogenins, Terpenoids, Steroids, Alkaloids, may play an important role in plant-plant interferences [1, 23, 24].

3.2 Effect of Aqueous Extracts of Wild Plants on Germination of *Vicia faba, Hordeum vulgare and Triticum aestivum* Growth

The allelopathic effect of aqueous extracts of *C. comosum*, *R. epapposum* and *R. stricta* Decne on the germination percentage of *Vicia faba, Hordeum vulgare and Triticum aestivum* after 5 days after treatments (DAT) are shown in Table 3. After 5 days of treatment, significant germination reduction was observed with all extracts ($P \leq 0.05$). The *R. stricta* Decne extracts exhibited higher negative effective on seed germination of 22% and 30% for seed of *Vicia faba and Triticum aestivum*, except *Hordeum vulgare,* there was complete failure of seed germination of *R. stricta* Decne aqueous extract. All aqueous extracts of wild plants significantly reduced seed germination compared with distill water control treatments (Table 3).

The inhibition of germination was found strong in extract of *R. stricta* Decne, was the most potent inhibitor of seed germination; in agreements with [18] found that, allelopathic effects of *R. stricta* Decne aqueous leachate from leaves, stems and follicles reduced the germination of *A. leucoclada*, *L. scindicus* and *S. villosa*.

The probable reason for *R. stricta* Decne potent allelopathic activity may be due to the presence of many bioactive secondary metabolites in this plant, particularly phenolics and alkaloids (Table 2). According to [25,17] the extract of *R. stricta* Decne plant contain alkaloids, that may be cause significant reductions in the germination of the seed of many plants [16]. In addition, [26] demonstrated that leaf aqueous extracts of *R. stricta* Decne plant exhibited significant inhibitory effects on germination of lettuce seeds.

Moreover, allolopathic potential could be attributed to the bioactive compounds such as phenolic and flavonoid contents [27,28,29]. From preliminary screening it was found that leaf extract of *R. stricta* Decne had the strongest allelopathic effect on seed germination compared to the stem extract of *C. comosum* and flower extract of *R. epapposum*. The results are agreed with [30] they reported that the most inhibitory effect of allelopathic plants produced by leaf extract. Moreover, [31] found that high degree of inhibition occurred with leaves extracts at the highest concentrations in all tested crop plants. In addition, [32] also reported that the inhibitory allelopathic potential of leaf extract was more powerful than of other vegetative parts. Furthermore, negative impact of allelopathic compounds from the aqueous extract of *C. comosum* stem was predictable, reduction of the low germination percentage and the decreased of Plumule and radical length detected when compared with the control (Tables 3, 4 and 5). [33] reported that the genus *Calligonum* revealed the presence of new lipoxygenase flavonoids, alkaloids, proteins, tanins, steroids, phenols and terpenoids [34]. Moreover, [35], reported that the *Calligonum polygonoides* extract at concentrations (75 g/L) inhibited seed germination of *Echinochloa crus-galli* by about 60.42%, while it inhibited the root and shoot growth by about 99.12% and 67.10%, respectively.

In this study, the indirect association between lower seed germination and allelopathic inhibition may have resulted to inhibition of water uptake

Table 2. Phytochemical composition of selected plant species

Treatment	Phenols	Polyphenols	Tannins	Saponins	Alkaloids	Flavonoids
Calligonum comosum L, Her	+	+	+	+	+	+
Rhanterium epapposum Oliv	+	+	-	-	+	+
Rhazya stricta Decne	+	+	+	+	+	+

(+): Present; (-): Absent

Table 3. Effect of different plant extracts on germination percentage after 5 days

Treatment	*Vicia faba*	*Hordeum vulgare*	*Triticum aestivum*
*Calligonum comosum L'*Her	47±0.35	60±0.40	70±0.60
Rhanterium epapposum Oliv	40±0.65	50±0.25	60±0.59
Rhazya stricta Decne	22±0.58	NG	30±0.33
Control	90±0.45	95±0.32	85±0.51
$LSD_{0.05}$	3.3	4.1	3.7

**: mean value ± standard error. Different superscript letters indicate values significantly lower than the respective control (P ≤ 0.05). NG= No Germination*

Table 4. Effect of different plant extracts on the plumule length (cm) after 10 days

Treatment	*Vicia faba*	*Hordeum vulgare*	*Triticum aestivum*
*Calligonum comosum L'*Her	0.36±0.02	0.08±0.40	1.68±0.03
Rhanterium epapposum Oliv	3.55±0.53	0.51±0.06	2.03±0.08
Rhazya stricta Decne	0.24±0.49	NG	0.33±0.63
Control	5.24±0.40	8.03±0.14	5.25±0.06
$LSD_{0.05}$	0.3	0.1	0.2

**: mean value ± standard error. different superscript letters indicate values significantly lower than the respective control (P ≤ 0.05). NG= No Germination*

Table 5. Effect of different plant extracts on the radical length (cm) after 10 days

Treatment	*Vicia faba*	*Hordeum vulgare*	*Triticum aestivum*
Calligonum comosum L, Her	2.93±0.12	3.14±0.17	2.56±0.85
Rhanterium epapposum Oliv	1.45±0.32	1.53±0.00	1.25±0.02
Rhazya stricta Decne	0.14±0.19	NG	0.06±0.25
Control	4.54±0.53	10.66±0.03	5.76±0.44
$LSD_{0.05}$	0.2	0.1	0.2

**: mean value ± standard error. Different superscript letters indicate values significantly lower than the respective control (P ≤ 0.05). NG= No Germination*

and enzyme activity. Although enzyme activity was not analyzed in this study, this observation is in accordance to the study of [36] reported that, increasing the concentration of aqueous leaf extracts significantly inhibited the water uptake by germination of *Rhanterium epapposum* seeds.

Moreover, [1] reported that many enzymatic functions important to plants are inhibited by the presence of allelochemicals, such as protease which plays an important role in the hydrolysis of proteins during germination. Usually, the allelochemicals inhibit seed germination by blocking hydrolysis of nutrients reserve and cell division and cause significant reductions in the growth of plumule and radical of various crops [37,38].

3.3 Effect of Aqueous Extracts of Wild Plants on Length of Plumule and Radicle of *Vicia faba, Hordeum vulgare and Triticum aestivum* Growth

The allelopathic effect of wild plants extracts on plumule growth of *Vicia faba, Hordeum vulgare* and *Triticum aestivum* are shown in (Table 4). The leaves extracts of *R. stricta* Decne were the most effective extract as it inhibit plumule length of seedling was about 0.24 and 0.33 cm for plumule of *Vicia faba and Triticum aestivum*, and the *Hordeum vulgare,* there was failure of seed germination of *R. stricta* Decne aqueous extract. Generally, all extracts of the tested wild plants significantly reduced the plumule length at depended manner. The allelopathic activities of water extracts from tested wild plant on radical growth of *Vicia faba, Hordeum vulgare* and *Triticum aestivum* seedling are shown in (Table 5). The extracts of all wild plants showed inhibition of radical length. The extract of *R. stricta* Decne was the most effective as the radicle length was 0.14 and 0.06 cm for radicale of *Vicia faba and Triticum aestivum*, and the *Hordeum vulgare,* there was failure of seed germination of *R. stricta* Decne aqueous extract. Plumule and radical length was significantly reduced by plants extracts when compared with control treatments (Tables 4 and 5). The radicle length was relatively more sensitive to autotoxic allelochemicals from extracts than plumule length. These results are in agreement with [36] reported that water extracts of allelopathic plants were more pronounced effects on radicle growth than Plumule growth. Because it is likely that roots are the first to absorb the allele-chemicals or autotoxic-compounds from the environment.

4. CONCLUSION

The results concluded that the wild plants plays an important role in the formation of its natural habitats as it contains the allele-chemical compounds that enable the plant to compete with other species. Moreover, it may change communities when recycled as a green manure in the soil for increasing organic matter in agro-ecosystems, where it inhibits crop growth and production. Moreover, the present results indicated that higher concentrations of extracts of tested species could be used for weed control. Since the experiment was conducted only in Petri dishes, further studies in pots or under field conditions are necessary to fully determine plants allelopathic potential effects. Also needed to determine the influence of seasonal and cultivar variations, and to identify the active compounds of autotoxicity and allelopathy effects.

COMPETING INTERESTS

Author has declared that no competing interests exist.

REFERENCES

1. Rice EL. Manipulated ecosystems: Roles of allelopathy in Agriculture, In: allelopathy, (Rice EL Ed.) 2nd Ed, Academic Press, Orlando. 1984;8-73.
2. Heidarzade A, Pirdashti H, Esmaeili M. Quantification of allelopathic substances and inhibitory potential in root exudates of rice (*Oryza sativa*) varieties on Barnyard grass (*Echinochloa crus-galli* L.). Plant Omics Journal. 2010;3:204-209.
3. Aliotta G, Cafiero G, Otero AM. Weed germination, seedling growth and their lesson from allelopathy in agricultural. In: Reigosa MJ, Redrol N, Gonzales L (eds.) Allelopathy: A physiological process with ecological implications. Dordrecht, The Netherlands: Springer Publisher. 2006; 285-299.
4. Narwal SS. Allelopathy in crop production. Scientific Publishers, Jodhpur. 1994;288.
5. Mattner SW. The impact of pathogens on plant interference and allelopathy In: Inderjit, Mukerji KG (eds.). Allelochemicals: Biological control of plant pathogens and diseases. Springer, Dordrecht, The Netherlands; 2006.
6. Smith AE, Martin LD. Allelopthatic characteristics of 3 cool-season grass species in the forage ecosystem. Agronomy Journal. 1994;86:243-246.
7. Soyler D, Canihos E, Temel N, Hajyzadeh M. Determination of chemical fungicide against soil borne fungal diseases of capers (*Capparis ovata* Desf. var. herbacea) during early stages. Pakistan Journal of Agriculture Science. 2012;49: 345-348.
8. Newman EI, Rovira AD. Allelopathy among some British grassland species. Journal of Ecology. 1975;63:727-737.
9. Tajuddin Z, Shaukat SS, Siddiqui IA. Allelopathic potential of Solanum forskalii Dunal: A tropical weed. Pakistan Journal of Biological Science. 2002;5:866-868.
10. Harper JL. The population biology of plants. Academic Press, London, UK; 1977.
11. Riadh H, Imen F, Abdelmajid Z, Sinda F. Detection and extraction of *Calligonum comosum*, a medicinal plant from arid region of Tunisia. African Journal of Traditional Complementary and Alternative Medicine. 2011;8 (3):322-327.
12. Liu XM, Zakaria MNM, Islam MW, Radhakrishnan R, Ismail A, Chen HB, Chan K, Al-Attas A. Anti-inflammatory and anti-ulcer activity of *Calligonum comosum* in rats. Journal Fitoterapia. 2001;72:487-491.
13. Ebid AI. Anti-bacterial activity of folk medicinal plant extracts of Saudi Arabia on isolated bacteria. Journal of Applied Life Sciences International. 2015;3:49-54.
14. Brown G. Vegetation ecology and biodiversity of degraded desert areas in noth-eastern Arabia-Habilitation thesis. Rostock University. 2001;90-92.
15. Halwagy R, Moustafa AF, Kamel SM. On the ecology of the desert vegetation in Kuwait. J. Arid Environment. 1982;5:95-107.
16. Assaeed AM, Al-Doss AA. Allelopathic effects of *Rhazya stricta* on seed germination of some range plant species. Annals Agric. Sci., Ain Shams Univ. 1997;142:159-167.
17. Gilani SA, Kikuchi A, Shinwari Z, Khattak Z, Watanable K. Phytochemical, pharmacological and ethnobotanical

studies of *Rhazya stricta* Decne. Phytother Res. 2007a;21:301-307.
18. Al-Otaibi MO, Assaeed AM, Azza E. Allelopathic effect of aqueous extract of *Rhazya stricta* Decne on seed germination of some range plant species. Journal of Saudi Society for Agricultural. 2014; 13(2a):278- 295.
19. Rice EL. Allelopathic effect of *Andropogon virginicus* and its persistence in old field. American Journal Botany. 1972;59:752-755.
20. Hegazy AK. Allelopathic effect of *Glossonema edule* in Qatar. Allelopathy Journal. 1997;4:133-138.
21. Uremis I, Arslan M, Uludag A. Allelopathic effect of some *Brassica* species on germination and growth of cut leaf ground cherry (*Physalis angulate* L.). Journal of Biological Science. 2005;5(5):661-665.
22. Harborne JB. Phytochemical methods. London. Chapman and hall, ltd. 1973;49-188.
23. Johnston A. Some factors effecting germination, emergence and early growth of three range grasses. Canadian Journal Plant Sceience. 1961; 41: 59-70.
24. Atiqur Rahman, Mossa JS, Al-Said MS, Al-Yahya MA. Medicinal plant diversity in the flora of Saudi Arabia 1: A report on seven plant families. Fitoterapia. 2004;75:149-161.
25. Gilani SA, Shinwari ZK, Watanabe KN. Monograph on *Rhazya stricta*. Plant Genetic Resources Monograph Series, 1. Mimatsu Corp., Tokyo, Japan. 2007b;81.
26. Hani MA, Elgamal H, Kamel S, Karl P, Karlheinz S. Triterpenoid saponins from *Zygophyllum* spcies. Phytochemistry. 1995;40:1233-1236.
27. Hassanean HA, El-Hamouly MMA, El-Moghazy SA, Bishay DW. 14-Decarboxy quinovic and quinovic acid glycosides from *Zygophyllum album*. Phytochemistry. 1993;33:667-670.
28. Debouba M, Balti R, Hwiwi S, Zouari S. Antioxidant capacity and total phenol richness of *Cistanche violacea* hosting *Zygophyllum album*. International Journal of Phytomedicine. 2012;4:399-340.
29. Turk MA, Tawaha AM. Inhibitory effects of aqueous extracts of black mustard on germination and growth of lentil. Pakistan Journal of Agronomy. 2002;1(1):28-30.
30. Al-Zahrani HS, Al-Robai SA. Allelopathic effect of *Calotropis procera* leaves extract on seed germination of some plants. JKAU: Sci. 2007;19:115-126.
31. Tefera T. Allelopathic effects of *Parthenium hysterophorus* extracts on seed germination and seedling growth of *Eragrostis tef* (Zucc.) Trotter. Journal of Agronomy and Crop Science. 2002;188(5): 306-310.
32. Yawer MA, Ahmed E, Malik A, Ashraf M, Rasool MA, Alza W. New lipoxygenase in habiting constituents from *Calligonum polygonoides*. Chemical Biodivers. 2007;4(7):1578-1585.
33. Rasmussen JA, Rice EL. Allelopathic effects of *Sporobolus pyramidatus* on vegetation patterning. Am. Midl. Nat. 1971;86:309-325.
34. Samejo MQ, Memon S, Bhanger MI, Khan KM. Preliminary phytochemicals screening of *Calligonum polygonoides* Linn. Journal of Pharmacy Research. 2011;4(12):4402.
35. El-Amier YA, Abdullah TJ. Allelopathic effect of four wild species on germination and seedling growth of *Echinocloa crus-galli* (L.) P. Beauv. International Journal of advanced research. 2014;2:287-294.
36. Chung IM, Miller DA. Natural herbicide potential of alfalfa residue on selected weed species. Agronomy Journal. 1995; 87:920-925.
37. Irshad A, Cheema ZA. Influence of some plant water extract on the germination and seedling growth of Barnyard Grass (*E. crus-galli* (L) Beauv). Pakistan Journal of Science Ind. Res. 2004;43(3):222-226.
38. Ogbe FMO, Gill LS, Iserhien EOO. Effects of aqueous extracts of *C. odorata* L. on radical and plumule growth and seedling height of Maize, *Z. mays* L. Compositae Newsletters. 1994;25:31-38.

9

Effect of Carbendazim Exposure and Vitamin E Supplementation in African Giant Rats

A. O. Omonona[1*] and T. A. Jarikre[2]

[1]Department of Wildlife and Ecotourism Management, Faculty of Agriculture and Forestry, University of Ibadan, Nigeria.
[2]Department of Veterinary Pathology, Faculty of Veterinary Medicine, University of Ibadan, Nigeria.

Authors' contributions

This work was carried out in collaboration between both authors. Author AOO designed the study, wrote the protocol and wrote the first draft of the manuscript. Author TAJ managed the literature searches, analyses of the study performed the structural equation modelling and discuss the conclusion. Both authors read and approved the final manuscript.

Editor(s):
(1) Anonymous.
Reviewers:
(1) Myrene Roselyn D'souza, Mount Carmel College, Bangalore, India.
(2) Anonymous, Pakistan.

ABSTRACT

Aim: Carbendazim is a broad spectrum benzimidazole and metabolite of benomyl used to control farm pests. Unregulated use of pesticides inadvertently affects the ecosystem and loss of biodiversity with precipitous decline of some wildlife species. We investigated the toxicity of carbendazim in African giant rat AGR on exposure and the ameliorative effect of tocopherol (vitamin E).

Methodology: The AGR were randomly divided into four experimental groups of 4 animals each. Group A was exposed to carbendazim only; B- carbendazim + vitamin; C- vitamin only and D- blank (control). Assessment was done clinically, microhaematocrit (erythrogram) and hemocytometric (Leucogram) methods. Cholinesterase (AchE) and markers of oxidative stress were quantified, and tissue changes examined microscopically.

Results: There was decrease in body weight of the AGRs, abortion after 23 days of exposure to carbendazim and significant changes in the erythrogram ($p<0.05$). Malondialdehyde MDA increased significantly ($p<0.05$) in group A. Vitamin E supplementation reduced MDA level

**Corresponding author: Email: ao.omonona@gmail.com*

significantly ($p<0.05$). Also, remarkable decrease in acetylcholinesterases in the pesticide intoxicated rats ($p<0.05$). The liver showed diffuse hepatocellular degeneration in carbendazim exposed group but which was comparatively reduced in the Vitamin E "supplementated" AGR.
Conclusion: Strict measures are needed to monitor intensive agricultural practices in the use of chemicals for pest control.

Keywords: African giant rat; antioxidant; carbendazim; pesticides; toxicity.

1. INTRODUCTION

In many parts of the world dramatic changes in farming practices have occurred. Not only have farms grown larger but there is also an increase in specialization in agriculture with the efficient use of costly farm machine, fertilizers, and irrigation and soil improvement techniques [1]. The resultant system of intensive agriculture has less use for crop rotation which is known to be a useful way of controlling pest, in view of this, there is an increase in the use of pesticides thereby increasing the release of chemical toxicant into the environment which could cause devastating health effects such as endocrine disruption, carcinogenicity, mutagenic effect and also cause ecosystem degradation by contamination along food chain and eventually loss of biodiversity [2,3].

Wild animals have been used as a resource throughout the course of human existence [4]. Wildlife resources play conspicuous role in the rural as well as national economics of tropical Africa [5]. The populations of numerous wildlife species have undergone a precipitous decline in the past century, resulting in significant reduction for many and extinction or near - extinction of others [6]. Many of these problems have been attributed directly to habitat loss and over exploitation resulting majorly from increasing human activities.

African giant rat AGR belongs to the family Nesomyidea, they are also referred to as pouched rat because of their large cheek pouches [7]. They are one of the wild rodent of the genus Cricetomys growing up to 0.9 meters including their tail which makes up half of their body length [8]. They are found inhabiting different places in Africa in the savanna around the edges of forest and in mountain up to 3500 m above sea level. The AGR is a solitary animal, with nocturnal habits and prefers to live in burrows. It feeds on large varieties of plant and animal foods, such as tubers, grains and nuts [9]. The continued globalization, human population growth and associated landscape changes due to habitat loss and other human induced landscape changes further enhances the interface between wildlife and human thereby facilitating additional disease emergence [10]. There are many sources of environmental toxicity, organic or inorganic like the use of pesticides which include insecticides, fungicides, rodenticides and herbicides [11]. Benzimidazole fungicides are broad spectrum fungicides, used for pre and post harvest treatment and for control of a wide range of pathogens [12].They readily penetrate plants through the roots and leaves and can directly enter natural water by drainage. Most of these compounds persist in the environment after application some even remain for years. Carbendazim is a widely used broad spectrum benzimidazole fungicides and a metabolite of benomyl. It is used to control plant disease in cereals and fruit [13]. Studies have shown that high doses of carbendazim cause infertility and destroy the testicles of laboratory animals [14]. It is of major concern due to its suspected hormone disrupting effect and action by inhibiting the development of spindle formation at mitosis [12].

The process of oxidation damages cell membranes and other structures including cellular protein, lipids and DNA. However, reactive oxygen species formation can increase in environmental stress [15,16]. Oxidation can be accelerated by stress, sunlight, pollution and other factors [17]. Anti-oxidants inhibit the oxidation of other molecules or neutralizes the effect of free radicals. Tocopherol, a class of organic chemical compounds is fat soluble anti-oxidant and also noted to have many other functions in the body [18,19]. Five tocopherols are identified by their prefixes as alpha, beta, gamma, delta and epsilon. Alpha tocopherol is an important lipid soluble anti oxidant in the glutathione peroxide pathway and it protects cell membrane from oxidation by reacting with radicals produced in the lipid peroxidation chain reaction.

In assessing the general health condition of animals, biomedical changes, histological and hematological changes can be used to determine the health condition of a particular organism [20,21]. In this study, the effect of carbendazim exposure and the ameliorative effect of tocopherol (vitamin E) were assessed on African giant rat.

2. MATERIALS AND METHODS

2.1 Study Animals

The African giant rats (*Cricetomys gambianus*) were purchased from a local market in Ibadan, Oyo state, Nigeria. A total of sixteen animals were purchased. The animals were housed in the Experimental animal house of the Faculty of Veterinary medicine, University of Ibadan and stabilized for 2 weeks. During stabilization, vitalite was added in their water for the first three days in order to reduce the effect of stress, after which they were dewormed using albendazole (shanuzole®). The animals were fed with compounded concentrate (Crude protein 14%, Fat 7%, Crude fibre 10%, Calcium 1%, Available phosphorus 0.35%, Metabolisable energy 2550 kcal/kg) and clean water (20 cl) was supplied ad-libitum. The experiment protocol complied with the ethical guidelines of the University of Ibadan.

2.2 Source of Pesticide

The pesticide used was forcelet, a preventive, and synthetic agricultural fungicide which contains 50% carbendazim, belonging to the toxicity class IV.

2.3 Antioxidant (vitamin E)

Vitamin E of the brand name Eviol, di- alpha-tocopheryl acetate manufactured in Greece by GAP SA was used. Each tablet contains 100 mg of tocopherol. Dosage used was 25 mg/kg of the animal for 14 days (pre-treatment).

After stabilization, the animals were weighed and were then divided into four experimental groups (A to D) comprising four animals each. Two groups were pretreated with vitamin. Thus, group A was given carbendazim only; B- carbendazim + vitamin; C- vitamin only and D- blank (control).

2.4 Sampling

After four weeks of exposure all the rats were weighed using sensitive weighing scale (AMW-PN-201C) and blood samples were collected for analysis accordingly.

The rats were sedated with 0.2 ml Ketamine Chloride administered intra-peritoneally. 2 ml of blood was collected from peri-orbital sinus into lithium heparinised bottles for hematology and blank bottles for serum needed for antioxidant assay. This was aliquot into Eppendorf tubes using Pasteur pipette and stored at -20°C until analyzed). Afterwards, six rats were necropsied (2 from each group) for gross and microscopic tissue examination.

2.5 Hematologic, Cholinesterase, Oxidative assays

The Packed cell volume PCV, Hemoglobin concentration, Red cell count was estimated by the microhaematocrit method, Leucocyte and Platelet counts were determined using the haemocytometric method as described by Schalm [22].

Cholinesterase (AchE) was quantified as Butyryl Cholinesterase activity using an *In vitro* diagnostic kit (Fortress Diagnostics, BXC0801), following manual procedure. In brief, 1000 µl of the working reagent and 20 µl of the sample were pipetted into the test tube, mixed and initial absorbance read after 90 seconds incubation at the assay temperature. Similarly, 20 µl of the sample was incubated with 1000 µl of buffer and 200 µl of substrate before reading absorbance activity. Wavelength was set at 405 nm, 37°C and 1cm cuvette light path.

$$\text{Activity} = \Delta\text{Abs/min} \times 55000 \text{ [serum start]}$$
$$\text{Activity} = \Delta\text{Abs/min} \times 65800 \text{ [substrate start]}$$

The determination of Malondialdehyde (MDA) was carried out using the thiobarbituric acid (TBA) test as described by Vidyasagar et al. (2004). The concentration of MDA was read from standard calibration curve in nanomoles of MDA per ml of serum.

$$\text{Concentration of MDA} = \text{absorbance at 540 nm}/1.56\text{x}10\ 5\ m^{-1}cm^{-1}$$

Superoxide Dismutase (SOD) activity was estimated using the method of Misra & Fridovich [23]. In brief, 1 ml of serum was diluted in 9 ml of distilled water. Aliquot of 0.2 ml was added to 2.5 ml of 0.05 M phosphate buffer at pH 10.2 to equilibrate in the spectrophotometer. The reaction was started by the addition of freshly prepared 0.3 ml of adrenaline. Reference cuvette contained 2.5 ml of phosphate buffer, 0.3 ml of substrate (adrenaline) and 0.2 ml of water. The increase in absorbance was monitored every 30 second for 150 seconds and expressed in Unit/mg protein.

%Inhibition=

increase in absorbance for substrate X 100/increase in absorbance of blank

Catalase (CAT) activity was estimated using method of Sinha [24]. Briefly, dichromate was reduced to chromic acetate when heated in the presence of hydrogen peroxide (H_2O_2). Chromic acetate produced was measured colorimetrically at 570 nm. Result was expressed in µmol/mg protein.

2.6 Pathology

Samples from the liver, kidney and brain were processed routinely for histopathology. Slides were evaluated using the Olympus light microscope (CX21) attached to a digital computerized camera (AmScope, MU900).

2.7 Statistical Analysis

The data were presented in Mean±SEM, subjected to one-way ANOVA, and followed by Bonferoni test for multiple comparisons using SPSS 16. P values <0.05 were considered significant.

3. RESULTS

There were no behavioral changes observed in the animals, but there was a decrease in body weight of the animals in the group A and B. There were no changes in the external morphological features of the animals. During the period of exposure, there was delivery of a malformed feotus by one of the animals in group A after 23 days of exposure to carbendazim.

The hematological parameters of the animals after four weeks of exposures shows the values of the packed cell volume, hemoglobin count, red blood cell count, white blood cell count, the platelet count, the absolute values of the lymphocytes, the neutrophils, eosinophils and the mean cell volume (MCV) and the mean cell hemoglobin concentration (MCHC) among the groups (Table 1).

There was significant differences in the packed cell volume, the hemoglobin concentration and the red blood cell counts ($p<0.05$).

The increases in Malondialdehyde (MDA) was significant ($p<0.05$) in the pesticide intoxicated rats compared to control. Vitamin E supplementation reduced MDA level significantly ($p<0.05$).

There was a sharp remarkable decrease in acetylcholinesterase levels in the pesticide intoxicated rats ($p<0.05$). Vitamin E supplementation normalized the AchE levels comparable to that in control (Table 2).

Pesticide intoxication reduced superoxide dismutase (SOD) activity in the exposed rats ($p<0.05$). Supplementation with Vitamin E later increased SOD activity in the rats ($p<0.05$). Carbendazim also reduced catalase activity.

Grossly, the vital organs appeared normal in the pesticide exposed and control groups except moderate pulmonary congestion. Histopathologically, there was severe diffuse hepatocellular swelling in carbendazim exposed group (Plate A & B). The severity of hepatocellular injury was reduced in the rats with vitamin E. Pulmonary congestion in rats exposed to carbendazim (Plate C) and normal lungs in control (Plate D). There was also neuronal necrosis and demyelination in neuropil of rats exposed to carbendazim (Plate E).

4. DISCUSSION

This study ascertained the fact that carbendazim exposure has effect on the values of hematological parameters of African giant rat. Carbendazim has been observed by various researchers to have a wide range of effect on variety of species ranging from invertebrates like annelids, mollusks, crusteceans, nematodes and vertebrates. Changes in the body weight agrees with the observation [14,25] that exposure of carbendazim to Wister rat for 30 days led to a decrease in body weight of the animals.

Carbendazim caused foetal abortion, this was also observed by Culik et al. [26], who reported carbendazim induced reproductive disorder in adult male, female and the foetus, his result showed that the foetus from pregnant rat exposed to carbendazim showed a lot of abnormalities from skeletal defect, visceral and soft skin and tissue malformation to dead foetus.

Table 1. The hematological parameters of AGR exposed and non-exposed to CBZ and Vit E

Group	PCV	HB	RBC	WBC	PLT	LYM	NEUT	MON	EOS	MCV	MCHC
A	34.0±1.0*	11.4±0.2*	5.02±0.32	6.05±0.20	1.03±0.03	4.37±0.25	1.47±0.47	0.12±0.1	0.1±0.0	50.2±20.0	33.5±0.4
B	29.75±4.33*	9.78±1.49*	4.74±0.76*	5.19±0.36	0.89±0.01	3.64±0.28	1.32±0.16	0.8±0.0	0.14±0.0	63.1±2.2	32.8±0.5
C	45.75±1.59	15.45±0.63	7.82±0.32	5.15±0.52	0.88±0.12	3.52±0.47	1.44±0.23	0.11±0.0	0.08±0.0	58.6±0.9	33.8±0.1
D	44.0±0.58*	14.73±0.29*	7.50±0.32*	6.18±1.61	2.64±1.38	4.03±0.98	1.96±0.64	0.11±0.0	0.08±0.1	58.7±0.5	33.5±0.3

**Values with superscript are significant at 0.05; Packed Cell Volume- PCV (%), Haemoglobin Concentration-Hb (g/dl), Red Blood Cell-RBC (*10^3/µL), White Blood Cell-WBC (*10^3/µL), Platelet Count-Plate (*10^5/ µL), Lymphocytes-Lympho, Neutrophils-Neutro, Monocytes-Mono, Eosinophils-Esino, Mean Cell Volume-MCV (fl), Mean Cell Haemoglobin Concentration-MCHC (pg)*

Table 2. The biochemical parameters of AGR exposed and non-exposed to CBZ and Vit E

Group	MDA	CAT	S0D	CHOLIN
A	24.7±0.19*	1.90±1.29*	2.60±0.6*	4.01±0.92*
B	19.95±15.46	2.16±0.84	2.68±0.11	6.10±2.0
C	11.96±20.02*	3.45±1.00	2.95±0.34	9.08±1.09
D	3.96±1.43*	3.36±0.25*	2.77±0.33*	9.71±6.46*

*AchE (CHOLIN)- Acetylcholinesteerase; MDA- Malonyl aldehyde; SOD- Superoxide dismutase; CAT- Catalase.. Values with different superscripts in row are statistically different and * are significant at 0.05*

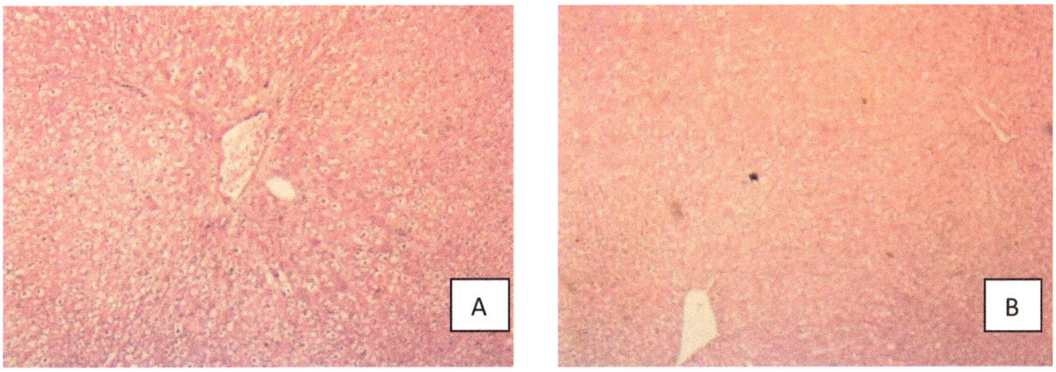

Plate A & B. Photomicrograph of the liver- A. Hepatocellular degeneration in carbendazim exposed rats. B. Normal hepatocytes in control. HE x100

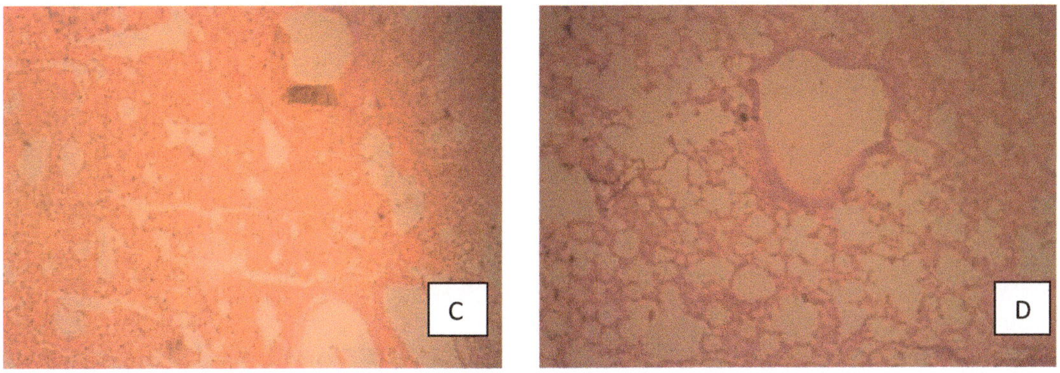

Plate C & D. Photomicrograph of the lungs- C. Pulmonary congestion in rats exposed to carbendazim. D. normal alveoli in control HE x100

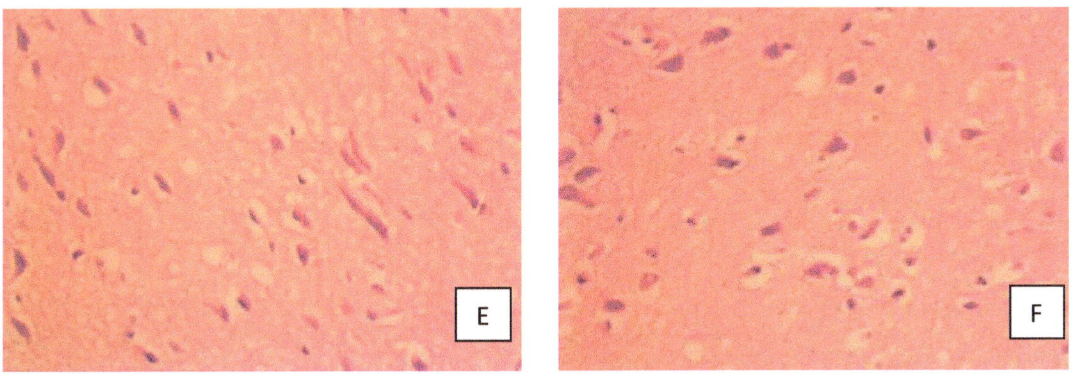

Plate E & F. Photomicrograph of the brain- A. neuronal necrosis and demyelination in rat exposed to carbendazim. B. Normal neurons in control. HE x100

This study has ascertained the fact that carbendazim has been showed to have effect on the values of hematological parameter of Africa giant rat, specifically in having reductive effect in the packed cell volume, the hemoglobin concentration and red blood cell count. Although studies have shown that it reduces the white blood cell count, lymphocytes count and the neutrophils count but from this study the white blood cell count and the lymphocytes count did not seem to be affected by carbendazim exposure as the mean values obtained from the white blood cell count and the lymphocytes count for all the groups agrees with the values given by

Oyewale et al. [27]. Oxidative damaging effect of carbendazim on the reproductive ability of the affected animal has been confirmed by this work as was observed in the aborted foetus in one of the animals exposed to carbendazim, the effect of carbendazim on the platelet count can't be really ascertained as no values of the standard platelet count has been recorded for Africa giant rat. Carbendazim has been shown to have a reduction in the mean cell volume as the mean value obtained for the CBZ groups is lower than the values given by Oyewale et al. [27], Olayemi et al. [28] and Kelani & Durotoye [29], this condition can be said to be a microcytic anemia which could be due to the reduction in the red blood cell count. It was also observed that tocopherol itself as a supplement can enhance the quality of the blood parameter as seen in the increase in the blood packed cell volume, hemoglobin concentration and the red blood cell count for the Vit E group, but the effect of tocopherol as an effective antioxidant against carbendazim exposed animal is not ascertained in this work as the CBZ/Vit E group and Vit E group seems to have the lowest mean value of the packed cell volume, the hemoglobin concentration and the red blood cell count , this could be attributed either to low concentration of the tocopherol administered in the work which could not effectively inhibit the production of the reactive oxygen species or that the tocopherol concentration used for the work is said to be high therefore suppressing or inhibiting the production of reactive oxygen species needed for the normal functioning of the blood thereby leading to oxidative stress, this reason could also be explained for the packed cell volume and the red blood cell count. This work also disagrees with the report of Scholz & Schultes [30] that carbendazim causes leucopenia in animals as the values obtained for the CBZ group and the control group tend to be similar for almost all the types of white blood cells were higher in the CBZ groups than other groups .

5. CONCLUSION

In conclusion, this study has ascertained the toxic effect of carbendazim to Africa giant rat. Exposure of Africa giant rat to carbendazim, a systemic and acute fungicide inhibits the production of microtubules in the mitotic formation in the cell. It has an effect in the hematological profile of the animal, especially in the packed cell volume, the hemoglobin concentration and the red blood cell count in African giant rat. Furthermore, the antioxidant effect of tocopherol was evident on the erythrocyte of the blood component increasing the packed cell volume, the hemoglobin concentration and the red blood cell count of the of Africa giant rat but has little or no effect on the leucocytes components of the blood. The ameliorative effect of tocopherol on the giant rat exposed to carbendazim was effective probably as suppressing the production of the reactive oxygen species.

As more intensive form of agricultural practices tend to increase, leading to a more mechanized form of farming and the use of chemicals to reduce pest and increase yield, further studies should be carried out on the effect of carbendazim on the hematological parameter at higher concentration especially, the platelet and the leucocytes component and lymphocytes. Also studies should be furthered in the area of the ameliorative effect of tocopherol on Africa giant rat exposed to carbendazim to get a concentration that can give a balance between the production of reactive oxygen species and the body homeostatic balance.

ACKNOWLEDGEMENTS

We appreciate efforts of Ugo Angela .I. during experimentation of this work and B.O. Emikpe for technical support.

CONFLICTING INTERESTS

Authors have declared that no conflicting interests exist.

REFERENCES

1. Pimentel D, Edward CA. Pesticides and ecosystem. Bioscience. 1982;32:595-600.
2. Manske DD, Jhonson RD. Pesticide and other chemical residues in total diet samples (X). Pestic Monit J. 1977;10:134-148.
3. Mittra A, Chatterjee C, Fatik B, Mandal. Synthetic chemical pesticide and their effect on bird. Research Journal of Environmental Toxicology. 2001;5:81-96.
4. Ajayi SS. Wildlife as a source of protein in Nigeria: Some priorities for development. Nigerian Field. 1971;36(3):115-27.
5. Akinpelu AI. Wildlife conservation in Nigeria and the role of zoological garden. Journal of Environtropica. 2007;4(1&2):83-90

6. Kellman M, Tackaberry R. Tropical environment and functioning and management of the tropical ecosystem published by routelagde publisher, new fetter lane, London; 1997.
7. Dipeolu OO, Akinboade OA, Ogunji F. Observation of the African giant rat (*Cricetomys gambianus*). Bulletin of Animal Health and Production Africa. 1981;29:393–397.
8. Cooper RG. Care, Husbandry and Disease of Africa giant rat (*Cricetomys gambianus*). Journal of south Africa Veterinary Asscociation. 2008;79:62-66.
9. Nowak RM, Paradiso JL. Walker mammals of the World. Hopskin University press. London. John. 1991;1.
10. Omonona OA, Kayode IB. Ecotourism; Implications on human and wildlife health. Ibadan University Press Publishing House. Nigeria; 2011. ISBN: 978-978-8414-67-4.
11. Green MB. Energy in pesticide manufacture, distribution and use. In: Z.R. Helsel (editor) Energy in Plant Nutrition and Pest Control. Elsevier, Amsterdam. 1987;165-177.
12. Pfeil R, Dellarco V. Carbendazim addendum. Federal Institute of Risk Assessment. Berlin Germany; 2005.
13. FAO/WHO. Benomyl. In: Pesticide residues in food-1988: Evaluations 1988. Rome: Food and Agriculture Organization of the United Nations. 1988;5–15.
14. Muthuviveganandavel V, Muthuraman P, Muthu S, Srikumar K. Individual and combined biochemical and histological effect of cypermethrin and carbendazim in male albino rats. Journal of Applied Pharmaceutical Science. 2011;01(09): 121-129.
15. Gianessi LP, Anderson JE. Pesticide Use Trends in U.S. Agriculture. NCAFP (National Center for Food and Agricultural Policy) Discussion, Washington, DC; 1993.
16. Hodgson E. A textbook of Modern Toxicology Third Edition. Published by John Wiley & Sons, Inc., Hoboken, New Jersey; 2004.
17. Worden J. Antioxidant and oxidative stress and it molecular biochemical make up. Mc Graw Hill Press; 2014.
18. Durak D, Uzun F, Kalender S, Ogutcu A, Uzunhisarcikli M, Kalender Y. Malathion-induced oxidative stress in human erythrocytes and the protective effect of vitamins C and E *In vitro*. Environmental Toxicology. 2009;24(3):235-242.
19. Eroglu S, Pandir D, Uzun FG, Hatice B. Protective role of vitamins C and E in diclorvos-induced oxidative stress in human erythrocytes *In vitro*. Biol Res. 2013;46:33-38.
20. Kitagaki M, Yamaguchi M, Nakamura M, Sakurada K, Suwa T, Sasa H. Age-related changes in haematology and serumchemistry of Weiser–Maples guineapigs (*Cavia porcellus*). Laboratory Animals. 2005;39:321–330.
21. Lanzarot MP, Victoria BM, San Andre MI, Ferna´ndez-Garcı´ M, Rodrı´guez C. Haematologic, protein electrophoresis, biochemistry and cholinesterase values of free-living black stork nestlings (*Ciconia nigra*). Journal of Wildlife Diseases. 2005;41(2)379–386.
22. Schalm OW, Jain NC, Carrol EJ. Veterinary haematology 3rd edition Philadelphia; 1986.
23. Misra HP, Fridovik I. Role of superoxide radicals in the acute auto oxidation of epinephrine; a simple assay for SOD. J Biol Chem. 1972;247:70-75
24. Sinha AK. Colourimetric assay of catalase. Analytical Biochem. 1972;47:389-394.
25. Scholz J, Weigand W. Carbendazim. Toxicological examination. Frankfurt, Germany, Hoechst AG, Pharmaceuticals Research, Toxicology Section, (Unpublished Report No. A-00935); 1972.
26. Culik R, Sherman H, Zapp JA. Teratogenic study in rats with 2-benzimidazole-carbamic acid, methyl ester (INE-965). Report No. HLR 466-70 from DuPont Haskell Laboratory, Newark, Delaware, USA. Submitted to WHO by E.I. du Pont de Nemours and Company, Wilmington, Delaware, USA; 1970.
27. Oyewale JO, Olayemi FO, Oke AO. Hematology of wild Africa giant rat (*Cricetomys gambianus* Water House). Veterinarski Arhiv. 1998;68(2):91-99
28. Olayemi FO, Oke AO, Oyewale JO, Ogunsanmi AO. The effect of season on the blood profile of Africa giant rat (*Cricetomiys gambianus* Water House). Isreal Journal of Vet Medicine. 2001;56: 147-150.
29. Kelani OL, Durotoye LA. Hematological responses of Africa giant rat (*Cricetomys*

gambianus) to castration and androgen replacement, Vet arkins. 2002;72:39-49.

30. Scholz and Schultes. Report on a subchronic feeding experiment (93 days) with technical active substance HOE 17411 OF. Unpublished Doc. No. A00409 from Hoechst Pharma Fo. To. FRG. Monheim, Germany: Submitted to WHO by Bayer Crop-Science; 1973.

10

Impact of Organo Minerals and Their Integrated Forms on Soil Fertility and Maize Productivity in Southeastern Nigeria

J. C. Nwite[1*], B. E. Eneruvie[2] and S. O. Nwafor[1]

[1]Department of Crop Production Technology, Federal College of Agriculture, Ishiagu, Ebonyi State, Nigeria.
[2]Department of Agricultural Technology, Akanu-Ibiam Federal Polytechnic, Unwana, Afikpo, Ebonyi State, Nigeria.

Authors' contributions

This work was carried out in collaboration between all authors. Author JCN designed the study, wrote the protocol, and wrote the first draft of the manuscript. Authors BEE and SON managed the literature searches, analyses of the study performed the structural equation modeling and discuss the conclusion. All authors read and approved the final manuscript.

Editor(s):
(1) Inz. Krzysztof Skowron, Department of Microbiology, Nicolaus Copernicus University in Toruń, Collegium Medicum of L. Rydygier, Poland.
Reviewers:
(1) Mohamed Abdel-Ghany Aly Bekheta, National Research Centre, Dokki, Egypt.
(2) Vijay Singh Meena, ICAR-Vivekananda Institute of Hill Agriculture Almora, India.
(3) Olawuyi Odunayo, University of Ibadan, Nigeria.

ABSTRACT

Agricultural soil quality is drastically affected by modern human activities. This study evaluates sole and combined effect of three agro-wastes namely: - rice husk Ash (RHA), rice husk dust (RHD) and poultry manure (PD) on soil fertility nutrient content and grain yield of maize on an Ultisol in Southeast Nigeria. Single and combined effects of the materials were investigated at application rates of 10t/ha for sole PD and RHD, 5 t/ha for sole RHA, 5 t/ha PD + 5 t/ha RHD, 5 t/ha PD + 2.5 t/ha RHA and 3.33 t/ha PD + 3.33 t/ha RHD + 1.67 t/ha RHA with 0 t/ha(control) giving seven treatments. The soil studied was low in pH, organic carbon (OC), total nitrogen, available phosphorous, exchangeable potassium and cation exchange capacity. The soil parameters studied include soil pH, OC, Total nitrogen, exchangeable bases, CEC, EA and available phosphorous

**Corresponding author: Email: johnwhite4real_2007@yahoo.com, johnsmallpot@gmail.com*

while the grain yield of maize was also measured after harvest. Results showed that the amendments significantly improved the soil pH, organic carbon, exchangeable Na, K, Ca and Mg, while the total nitrogen was not affected significantly. Exchangeable acidity was drastically reduced by the amendments, while the available phosphorous was significantly increased by the application of treatments. Maize grain yield was also significantly increased by the treatments applications. The integrated application especially at 5 t ha-1 poultry manure and 5 t ha-1 rice husk dust increased maize yield and most soil nutrient elements higher than other combinations in the study.

Keywords: Organo-minerals; amendments; agro-wastes; nutrient leaching; maize grain yield.

1. INTRODUCTION

The need to meet ever increasing nutrition demands of the expanding human populations makes sustainable agriculture and agro-based sectors a front burner of environmental and social development issue in sub-Saharan Africa. Soils are an integral component of agriculture and serve as medium for numerous eco-biological, chemical and physical processes. Over-burdening of the soil as a natural resource capital has always been an issue due to its widely varied applications in the maintenance of human life activities [1]. The need thus to effectively manage soil resources in order to achieve optimum productivity of soils is obvious. The target of soil management is to create a healthy soil environment which may retain balanced nutrient status such that its fertility is maintained over time.

Soils in Africa are typically highly variable in fertility and in how they respond to inputs [2,3].

In Nigeria and other tropical countries, it is the utmost interest of most farmers to produce sustainable high crop yield. However, according to Ayeni et al. [4], decrease in soil fertility after few years of cropping is always a major limitation of these tropical soils. The need to improve soil fertility and crop production to support the rapidly growing population has led to a renew interest in the use of organic sources of nutrients and mineral fertilizers for soil fertility maintenance [4]. The use of chemical (inorganic) fertilizer in crop production has not been sustainable due to its high cost and scarcity, soil acidity, increased soil bulk density, low water infiltration rate and nutrient imbalance [5,6]. Thus, the needs to investigate alternative sources of nutrients that will be less damaging to the soil environment become imperative.

Hence, the recent interest in the use of agricultural wastes by the dominant resource poor farmers [7,8]. Wastes such as crop residues and animal excreta are used directly or composted. However due to high quantity needed, adequate quantity of an organic waste may not be obtained, hence the farmers often apply different wastes combined. Since the wastes are of different quantity and nutrient composition, their combined use is expected to have positive cumulative and complementary effects in nutrient supply and improvement of crop yield [7].

Animal manures such as that of poultry (chicken) has received much research attention in crop nutrition and had been found to be effective as source of nutrient for crops [9,10]. Recently, studies also showed that poultry manure increased soil organic matter and nutrient status [10]. However, the huge amount of poultry manure required for field crop production and its handling problems limits its use to distant farmers when it is to be applied singly.

There is also a renewed interest in the use of burnt organic residues from plants as sources of phosphate and potash fertilizers [5,11,7]. These materials are often considered less likely to have detrimental effect on soil physicochemical properties compared with mineral fertilizers. The use of plant residues are also limited by large quantity required to meet crop nutritional needs due to their low nutrient content and time lag to mineralize [7], hence the need for combination of these plant materials and animal excreta for improved soil and crop productivity.

Studies [12] have shown the superior effect of integrated nutrient supply over sole use of organic source in terms of balanced nutrient supply, improved soil fertility and crop yield [13,9,10].

There is need to compare the effectiveness of the three sources of bio-fertilizer and their integrated forms (animal and plant residues) on soil fertility and productivity. Hence, the objective of this study was to compare the immediate effects of rice husk ash, poultry manure and rice husk dust organo minerals on soil chemical properties and yield of maize in southeastern Nigeria.

2. MATERIALS AND METHODS

2.1 Location of Study Site

The experiment was conducted in 2011at the Research Farm of Federal College of Agriculture, Ishiagu. The area lies within latitude 05°56′N and longitude 07°41′E in the Derived Savannah Zone of Southeastern Nigeria. The mean annual rainfall for the area is 1350 mm, spread from April to October with average temperature of29°C. The underlying geological material is Shale formation with sand intrusions locally classified as the 'Asu River' group. The soil is hydromorphic and belongs to the order Ultisol. It has been classified as Typic Haplustult [14].

2.2 Field Study

The field was cleared, ploughed and harrowed by a tractor and later prepared into seed-beds manually. The field experiment was laid out in randomized complete block design with three replications of 3 x 2.5 m plot size with plant spacing of 50 x 25 cm. A bulk (composite) auger sample was collected within 0- 20 cm soil depth before treatment application for initial soil characteristics. At the harvest, another soil sample was collected from all the plots to determine the changes that occurred due to treatment application.

Seven treatments sourced from poultry dropping (PD), rice husk dust (RHD) and rice husk ash (RHA) in their sole and combined forms including the control were applied at the rates of 10t/ha for sole PD, 10 t/ha for RHD, 5 t/ha for sole RHA, 5 t/ha PD + 5 t/ha RHD, 5 t/ha PD + 2.5 t/ha RHA and 3.33 t/ha PD + 3.33 t/ha RHD + 1.67 t/ha RHA with 0 t/ha (control – CT).

2.3 Collection of Soil Samples and Laboratory Methods

Using a composite sample from the top- (0-20 cm) soil region at the study site, the soil was characterized before land preparation. Auger samples were collected at harvest from all the identified sampling points in all the plots from the top (0–20 cm) soil. The auger topsoil samples were air-dried and sieved with 2 mm sieve. Soil fractions less than 2 mm from individual samples were then analyzed using the following methods.

Particle size distribution of less than 2 mm fine earth fractions was measured by the hydrometer method as described by Gee and Bauder [15].

Soil pH was measured in a 1:2.5 soil: 0.1 M KCl suspensions [16]. The soil organic carbon (OC) was determined by the Walkley and Black method described by Nelson and Sommers [17]. Total nitrogen was determined by semi-micro kjeldahl digestion method using sulphuric acid and $CuSO_4$ and Na_2SO_4 catalyst mixture [18]. Sodium (Na) and potassium (K) were determined from ammonium acetate leachate using the auto-electric flame photometer. Calcium (Ca) and magnesium (Mg) were determined using the complexometer titration method as described by Thomas [19]. The cation exchange capacity (CEC) was determined by the method described by Rhoades [20], while exchangeable acidity (EA) was measured using the method of McLean [16]. Available phosphorus was measured by the Bray II method [21].

2.4 Data Analysis

Statistical analysis of all the data was performed using GENSTAT 3 7.2 Edition. Statistical differences among treatment means were estimated and compared using Least Significant Difference (LSD) and all inferences were made at 5% Level of probability. The correlation - regression was performed using SPSS stat. 17.0.

3. RESULTS AND DISCUSSION

3.1 Initial Physical and Chemical Properties of the Studied Soil

The soil used for the study has sandy loam texture with high proportion of total sand (72%). The high proportion of total sand in this soil could affect the yield of the maize crop because of probable low soil moisture that may occur especially at tasselling stage and associated high leaching of the soil nutrients. Hence, relatively poor yield of maize on soil not treated with manures was expected [8]. The OC, N, P, K, Mg status and its acidic nature (Table 1) noted to be low in the soil studied are expected to improve with the application of the amendments with their mixtures (poultry manure and also the plant wastes).

3.2 Effects of Different Organo Mineral Sources on the Soil pH, Organic and Total Nitrogen

Table 2 indicates that application of rice husk ash and its addition to poultry manure significantly ($p < 0.05$) increased soil pH higher than plots treated with rice husk dust and its integration with poultry dropping. This could be attributed to its higher contents of Ca, Mg and K compared to the RHD. It has been shown that the oxides and hydroxides of Ca, Mg and K contained in ash from plant residues makes its mode of actions similar to that of burnt or hydrated lime [22,23]. The values ranged from 4.70 – 6.50 with low coefficient variation of 1.8%.

Table 1. Physical and chemical properties of the studied soil (0 – 20 cm) depth

Soil properties	Values
Clay (%)	9
Silt (%)	19
Total sand (%)	72
Textural class	Sandy loam
Organic carbon (%)	0.30
Total nitrogen (%)	0.058
pH (H_2O)	5.0
Exchangeable bases (cmolkg^{-1})	
Sodium (Na^+)	0.72
Potassium (K^+)	0.42
Calcium (Ca^{2+})	2.0
Magnesium (Mg^{2+})	0.6
Cation exchange capacity (CEC) cmolkg^{-1}	9.60
Exchangeable acidity (EA) cmolkg^{-1}	2.2
Available phosphorous (mg/kg)	10.4

The result therefore showed that all the treated plots improved pH higher than the control. This confirms the findings of Nwite et al., [24] who stated that the use of plant materials as organic manures increases the pH of the soil and hence the performance of a crop.

However, the strongly acidic nature of the untreated soil could be adduced to the coarse nature of the soil which should have enhanced leaching of exchangeable bases and to the cropping history of the site which was heavily cropped with application of various mineral fertilizers especially NPK (Table 2).

The effects of treatments on soil organic and total nitrogen (Table 2) indicates that addition of rice husk dust (RHD) amendment to poultry manure and its sole increased significantly ($p < 0.05$) soil organic matter Compared with control, total nitrogen was increased with addition of 5 ton/ha RHD amendment to 5 t/ha poultry manure. Soil organic carbon varied from 0.147% - 0.473%, while the total nitrogen varied from 0.63% - 0.152%. The results indicated that there were more consistency in the soil organic carbon pool among the treatments than what were obtained in soil total nitrogen (10.9% and 30.8%), respectively (Table 2).

3.3 Effects of Different Organo Mineral Sources on the Soil Exchangeable Bases

There were significant ($P < 0.005$) effects on the exchangeable bases due to the amendments. The results (Table 3) showed that the highest mean value (0.86 cmol kg^{-1}) for exchangeable sodium was obtained from plots treated with PD+RHD+RHA. The treatments means varied from each other by 9.7%, which shows better consistency. It is discernible from Table 3 that exchangeable potassium in the studied soil differed significantly. The exchangeable potassium content was significantly ($p < 0.05$) higher in PD+RHA (0.77 cmol kg^{-1}) treated plots and the minimum (0.40 cmol kg^{-1}) was recorded in the control treatment. The mean values ranged from 0.40 – 77 cmol kg^{-1}with a coefficient variation of 16.4%. This means that there was higher consistency (Table 3) among the treatments performance on soil exchangeable sodium than as obtained in exchangeable potassium. Exchangeable calcium was statistically higher in RHA treated soil compared to other treatments with very low coefficient variation (7.1%), showing better consistency among the treatments than other exchangeable bases. Generally, the treated soils in this studied increased the exchangeable bases significantly higher than the control. However, the decreased exchangeable bases (sodium, potassium and calcium) content in control treatment from its initial value (0.72 cmol kg^{-1}, 0.42 cmol kg^{-1} and 2.0 cmolKg^{-1}) could be due to the high rainfall pattern and the proportion of total sand which might affect nutrient retention capacity and also aid in high leaching of the soil nutrients in the area. Exchangeable bases were more in integrated treatments as compared to sole treatments.

3.4 Effects of Different Organo Mineral Sources on the Soil Cation Exchange Capacity (CEC), Exchangeable Acidity (EA) and Available Phosphorous (Avai. P)

The results (Table 4) indicated that there was a short-term improvement on the CEC by the use of different organo mineral sources. This means that CEC of the soil gradually responds to different organo mineral sources for crop production. The higher significant improvement on the CEC by poultry dropping (24.27 cmol kg^{-1}) attributed to edge-advantage it has in improving the soil fertility status. The result is in agreement with the report of Adeniyan and Ojeniyi, [10] that poultry manure increased soil organic matter and other nutrient status. The essence of integrating PD with other organo minerals is to reduce the huge amount of poultry manure required for field crop production and its handling problems which do limits its use to distant farmers when it is to be applied sole.

The exchangeable acidity of the soil decreased significantly with the application of amendments (Table 4). The overall drop in the mean EA values in the treated plots was considered to be attributing of the organic manures nature which has successfully increased the soil pH statistically.

It is depicted (Table 4) that available phosphorus in soil in different treatments differed significantly. The available phosphorus content was maximum (26.0 mg/Kg) in PD treated plots and the minimum (8.1 mg/Kg) was recorded in control treatment. Available phosphorus content in control treatment decreased from its initial value (10.4 mg/Kg) whereas the available phosphorus in rest of the treatments increased from its initial levels. The coefficient variation was very high (25.2%), meaning that there were inconsistency among the treatments in their effects on the available phosphorous.

The maize yield (Table 4) showed significant increase among the treatments with integrated forms of plant wastes (RHD and RHA) with the animal manure (poultry dropping) performing significantly higher than their sole application. The result indicates that the highest significant increase in maize grain yield was obtained from plots amended with sole poultry dropping. However, all the treated plots improved the maize grain yield significantly ($p < 0.05$) higher than the control. It was observed that the yield was moderately stable and consistent among the treatments with coefficient variation (CV %) of 10.0% (Table 4). The yield obtained in this study with organo minerals and their integrated forms were higher than yield recorded by Ayeni and Adetunji [8] with integration of poultry manure and mineral fertilizer at different rate of application where the highest yield recorded was 4.50 t/ha.

Table 2. Effects of amendments on soil pH, organic carbon and total nitrogen

Treatments	pH	Organic carbon (%)	Total nitrogen (%)
PD	6.03	0.360	0.125
RHD	5.83	0.473	0.115
RHA	6.50	0.200	0.080
PD+RHA	6.13	0.387	0.112
PD+RHD	5.73	0.433	0.152
PD+RHD+RHA	5.83	0.437	0.132
CT	4.70	0.147	0.063
LSD $_{0.05}$	**0.1908**	**0.06768**	**NS**
CV %	**1.8**	**10.9**	**30.8**

PD = poultry dropping, RHD = rice husk dust, RHA = rice husk ash, CT = control, NS = not significant

Table 3. Effects of amendments on soil exchangeable bases ($cmol_cKg^{-1}$)

Treatments	Exch. Na	Exch. K	Exch. Ca	Exch. Mg
PD	0.84	0.65	4.00	1.47
RHD	0.51	0.54	3.60	1.10
RHA	0.80	0.70	4.07	1.73
PD+RHA	0.82	0.77	3.73	1.90
PD+RHD	0.81	0.66	3.00	1.43
PD+RHD+RHA	0.86	0.59	3.33	1.60
CT	0.33	0.40	1.37	0.87
LSD $_{0.05}$	0.1220	0.1796	0.4193	0.5116
CV %	9.7	16.4	7.1	19.9

3.5 Important Relationships among the Soil Properties and Maize Grain Yield

Table 5 showed the correlation coefficients among the soil properties and maize grain yield. It was obtained that there was no significant correlations between the total sand percent rest of other soil parameters studied including the grain yield of maize. The soil pH indicated positive significant ($p < 0.05$) correlation coefficient r (0.783*, 0.866*, 0.964** and 0.825*) with all the exchangeable bases (exchangeable sodium, potassium, calcium and magnesium), respectively. It was observed that among the exchangeable bases, the pH gave a stronger significant correlation coefficient r (0.964**) with the exchangeable calcium at both 5 and 1 % ($p < 0.05$ and 0.01) probability levels. The positive correlations of the pH with the exchangeable bases is a prove that liming and application of organic manures to tropical soils improves the soil pH and exchangeable bases, agreeing with the submissions of Lickazz, [22] and Nwite et al. [23] that ash from plant residues makes its mode of actions similar to that of burnt or hydrated lime. Nwite et al. [24] stated that the use of plant materials as organic manures increases the pH of the soil and hence the performance of a crop. On the other hand, the soil pH negatively correlated (r = -968**) significantly ($p < 0.05$ and 0.01) with exchangeable acidity in study. This negative correlation implies that as the soil exchangeable acidity increases, the pH decreases. It is therefore important to adopt agricultural practices that will keep the EA at a minimal level as save and sustain our soil resources. It was obtained that there was a positive significant ($p < 0.05$ and $p < 0.01$) correlation coefficient r (0.881**) between soil organic carbon and total nitrogen. An increase in soil organic carbon might lead to an increase in the total nitrogen percent of the soil and vise-versa. Table 5 indicates that soil total nitrogen correlated positively ($p < 0.05$) with the grain yield with a correlation coefficient r (0.832*). The exchangeable sodium shows a significant ($p < 0.05$) correlation coefficient r (0.852*, 0.882**, 0.781*, 0.767* and 0.870*) with exchangeable potassium and magnesium, CEC, available phosphorous and maize grain yield. The study (Table 5) indicated positive significant correlation ($p < 0.05$) with exchangeable calcium and the maize yield (r = 0.797* and 0.771*), respectively. It was shown also that exchangeable K gave a stronger significant ($p <$ 0.05 and 0.01) correlation coefficient r (0.934**) with exchangeable Mg. However, there was a very high negative significant ($p < 0.05$ and 0.01) correlation (r = -0.899**) of the exchangeable K with exchangeable EA. It was obtained that exchangeable calcium and magnesium also strongly significantly ($p < 0.05$ and 0.01) correlated (r = -0.996** and -0.793*) with the EA. Exchangeable Ca gave positive ($p < 0.05$) correlation coefficient r (0.773* and 0.766*) with CEC and available P, respectively. An increase in CEC of the soil has resulted to an increase in the soil available phosphorous and maize grain yield, as it gave very high strong positive significant ($p < 0.05$ and 0.01) correlation coefficient r (0.957** and 0.939**) for avail. P and grain yield of maize, respectively.

Generally, the results showed that maize yield increases with possible increase in the soil total nitrogen, exchangeable sodium and potassium, cation exchange capacity and available phosphorous of the soil. Therefore, any agricultural practice(s) that may lead to improved availability of these above mentioned soil elements, that are environmental friendly and socially acceptable should be encouraged among the rural farmers in their cropping systems to increase soil and crop productivity.

Table 4. Effects of amendments on soil cation exchange capacity – CEC ($cmol_cKg^{-1}$), exchangeable acidity – EA ($cmol_cKg^{-1}$) and available phosphorous – mg/Kg (Avail. P)

Treatments	CEC	Exch. Acidity	Avail. P	Maize yield (t/ha)
PD	24.27	1.47	26.0	5.87
RHD	17.13	1.60	18.1	4.17
RHA	16.73	1.27	18.5	4.07
PD+RHA	17.40	1.27	20.2	5.27
PD+RHD	19.73	1.73	23.5	5.37
PD+RHD+RHA	18.03	1.87	16.6	4.90
CT	9.27	2.87	8.1	2.17
$LSD_{0.05}$	6.136	0.4931	8.39	0.8055
CV %	19.7	16.1	25.2	10.0

Table 5. Matrix of correlation coefficients for total sand percent of the soil and other soil chemical properties including the maize grain yield of the studied site

	TS	pH	OC	TN	Na	K	Ca	Mg	CEC	EA	P	Yield
TS	_											
pH	-.372	_										
OC	-.605	.318	_									
TN	-.496	.318	.881**	_								
Na	-.155	.783*	.424	.632	_							
K	-.338	.866*	.323	.424	.852*	_						
Ca	-.343	.964**	.448	.403	.746	.797*	_					
Mg	-.134	.825*	.226	.319	.882**	.934**	.729	_				
CEC	-.215	.662	.605	.751	.781*	.633	.773*	.504	_			
EA	.465	-.968**	-.447	-.417	-.752	-.899**	-.966**	-.793*	-.719	_		
P	-.368	.695	.569	.734	.767*	.742	.766*	.549	.957**	-.782*	_	
Yield	-.292	.671	.697	.832*	.870*	.771*	.745	.670	.939**	-.753	.935**	_

** Correlation is significant at the 0.05 level (2-tailed). ** Correlation is significant at the 0.01 level (2-tailed).*
TS = Percent total sand, OC = organic carbon, TN = total nitrogen, Na = exchangeable sodium, K = exchangeable potassium, Ca = exchangeable calcium, Mg = exchangeable magnesium, CEC = cation exchange capacity, EA = exchangeable acidity, P = available phosphorous.

4. CONCLUSION

The Study revealed that there was a short-term improvement on the soil fertility status and maize grain yield by the use of different organo mineral sources. It was noted the superiority of sole poultry dropping and its mixtures with the plant materials on the improvement of chemical properties of the soil studied. These improvements have led to a significant increase in the maize grain yield. It is therefore concluded that the integrated application of these materials at reduced levels were more effective in improving the soil nutrient status and maize yield than sole application of any of these plant materials.

COMPETING INTERESTS

Authors have declared that no competing interests exist.

REFERENCES

1. Omotayo OE, Chukwuka KS. Soil fertility restoration techniques in sub-Saharan Africa using organic resources. African Journal of Agricultural Research. 2009;4(3):144-150.
   Available:http://www.academicjournals.org/AJAR ISSN 1991-637X © 2009 Academic Journals.
2. Hossner LR, Juo ASR. Soil nutrient management for sustained food crop production in upland farming systems in the tropics. Juo Soil and Crop Sciences Department College Station Tennessee 77843, USA; 1999.
   Available: http//www.agnet.org
3. AGRA. Alliance for a Green Revolution in Africa: AGRA at work; 2007.
   Available:www.agra-alliance.org/work/ (Retrieved 13th March, 2008)
4. Ayeni LS, Adeleye EO, Oso OP. Residual effect of cocoa pod ash, poultry manure and NPK 20:10:10 fertilizer on soil nutrients, nutrient uptake and yield of maize (*Zea mays*). J. Soil. Nature. 2009;3(2):04-09.
5. Ojeniyi SO. That our soil may not die. 10th Inaugural Lecture of Federal University of Technology, Akure. 1995;5-10.
6. Nottidge DO, Ojeniyi SO, Asawalam DO. Comparative effect of plant residue and NPK fertilizer on nutrient status and yield of maize in a humid ultisol. Nigerian Journal of Soil Science. 2005;15:1–8.
7. Ayeni LS, Adetunji MT, Ojeniyi SO, Ewulo BS, Adeyemo AJ. Comparative and cumulative effect of cocoa pod husk ash and poultry manure on soil and maize nutrient contents and yield. Am.-Eurasian J. Sustain. Agric. 2008;2(1):92-97.
8. Ayeni LS, Adetunji MT. Integrated application of poultry manure and mineral fertilizer on soil chemical properties, nutrient uptake, yield and growth components of maize. Nature and Science. 2010;8(1):60-67. (ISSN: 1545-0740).
9. Adeniyan ON, Ojeniyi SO. Effect of poultry manure, NPK 15 – 15- 15 fertilizer and

combinations of their reduced levels on maize growth and their chemical properties. SSSN. 2005;15:34-41.
10. Adeniyan ON, Ojeniyi SO. Comparative effectiveness of different levels of poultry manure with NPK fertilizer on residual soil fertility, nutrient uptake and yield of maize. Moor Journal of Agricultural Research. 2006;4:194-197.
11. Ayeni LS. Integrated application of cocoa pod ash and NPK fertilizer on soil chemical properties and yield of tomato. American – Eurasian Journal of Sustainable Agriculture. 2008;2(3):333–339.
12. Uyovbisere EO, Elemo KA. Effect of inorganic fertilizer and foliage *Azadriachta* and *Parkia* species on the productivity of early maize. Nigeria Journal of Soil R. 2000;1:17–22.
13. Awodun MA, Ojeniyi SO, Adesoye A, Odedina SA. Effect of oil palm bunch refuse ash on soil and plant nutrient composition and yield of maize. American-Eurasian Journal of Sustainable Agriculture. 2007;1(1):50-54.
14. FDALR. Reconnaissance soil survey of Anambra state-Nigeria, (1:250,000). Federal Department of Agric. and Land Res. (FDALR). Soil report FDALR, Kaduna. In: FAO, (ed.) 1985. Tropical forestry action plan. FAO, Rome; 1985.
15. Gee GW, Bauder JW. Particle-size analysis. In: Klute A, Ed. Methods of Soil Analysis, Part 1, American Society of Agronomy, Madison. 1986;91–100.
16. McLean EO. Soil pH and lime requirement. In: Page AL, Miller RH, Keeny DR, (eds.). Methods of Soil Analysis, Part 2. Am. Soc. Agron., Madison. 1982;199–224.
17. Nelson DW, Sommers LE. Total carbon, total organic carbon and organic matter. In: Sparks DL (ed) Methods of soil analysis, part 3: Chemical methods. Agronomy Monograph No 9. American Society of Agronomy, Madison. 1996;961–1010.
18. Bremner JM, Mulvancy CS. Total nitrogen. In: Page AL, et al. (eds.). Methods of Soil Analysis. No.9; part 2, Amer. Soc. Of Agron. Inc, Madison, Wisconsin, USA. 1982;595–624.
19. Thomas GW. Exchangeable cations. In: Page AL, Miller RH, Keeny DR, (eds.), Methods of Soil Analysis, Part 2. Am. Soc. Agron. Monogr. Madison. 1982;159–165.
20. Rhoades JD. Cation exchange capacity. In: Page AL, Miller RH, Keeney DR, (eds). Methods of Soil analysis. Part 2. Agron A. Madison W.I. P. 1982;149-157.
21. Bray RH, Kurtz LT. Determination of total organic carbon and available forms of phosphorous in soils. Soil Sci. J. 1945;59: 39–43.
22. Lickazz J. Wood ash: An alternative liming material for agricultural soils. Alberta J. Agriculture, Feed and Rural Development; 2002.
Available:www. agric. gov.ab.cal/department/dept.docs.nsf
23. Nwite JC, Obalum SE, Igwe CA, Wakatsuki T. Properties and potential of selected ash sources for improving soil condition and rice yields in a degraded Inland Valley in Southeastern Nigeria. World Journal of Agricultural Sciences. 2011;7(3):304-310.
24. Nwite JC, Essien BA, Anaele MU, Obalum SE, Keke CI, Igwe CA. Supplementary use of poultry droppings and rice-husk waste as organic amendments in Southeastern Nigeria. 1: Soil Chemical Properties and Maize Yield. Libyan Agriculture Research Center Journal International 2012;3(2):90-97. ISSN 2219-4304, © IDOSI Publications, 2012.

11

Glyphosate Influence on the Soil Microorganism Sensibility, Physiological Parameters of the Plant, Isoflavones and Residues in the Seeds and Soil

G. M. B. Bohm[1*], L. Amarante[2], E. M. Bohm[3], C. V. Rombaldi[4] and M. I. Genovese[5]

[1]Departamento de Gestão Ambiental, Instituto Federal Sul-rio-grandense- IFSul, Praça 20 de setembro 455, Pelotas, 96015-360, RS, Brazil.
[2]Departamento de Bioquímica, Faculdade de Agronomia Eliseu Maciel, Universidade Federal de Pelotas - UFPel, Campus Universitário, Capão do Leão, 96010-900, RS, Brazil.
[3]Gestão Ambiental, Instituto Federal Sul-rio-grandense- IFSul, Praça 20 de Setembro 455, Pelotas, 96015-360, RS, Brazil.
[4]Departamento de Ciências e Tecnologia Agroindustrial, Faculdade de Agronomia Eliseu Maciel, Universidade Federal de Pelotas – UFPel, Campus Universitário, Capão do Leão, 96010-900, RS, Brazil.
[5]Departamento de Alimentos e Nutrição Experimental, Faculdade de Ciências Farmacêuticas da USP, São Paulo, Brazil.

Authors' contributions

This work was carried out in collaboration between all authors. Authors GMBB and CVR designed the study, wrote the protocol and wrote the first draft of the manuscript. Authors GMBB, LA and MIG managed the analyses of the study. Authors GMBB, EMB and MIG performed the statistical analysis. Authors CVR and GMBB managed the literature searches, analyses of the study performed and discuss the conclusion. All authors read and approved the final manuscript.

Editor(s):
(1) Chandra Shekhar Kapoor, Department of Environmental Sciences, University College of Science, Mohan Lal Sukhadia University, India.
(2) Anonymous.
Reviewers:
(1) Petigrosso, Lucas Ricardo, Universidad Nacional de Mar del Plata, Argentina.
(2) Addam Kiari Saidou, International Institute of Tropical Agriculture (IITA), Ibadan, Nigeria.
(3) Anonymous, University of California, Davis, USA.

**Corresponding author: Email: bohmgiani@gmail.com*

ABSTRACT

The use of Glyphosate herbicide is the best way to control weed growing when cultivating genetically modified soybean [*Glycine max* (L.) Merr.] resistant to glyphosate (N-phosphono-methyl-glycine) (GR). However, there have been doubts about the safety of this herbicide use concerning its effects on the plant, quality of grains and on the soil cultivated. Hence, the present study aimed at evaluating the effects of weed management two soybean cultivars (GR) and its conventional isogenic genotype with the use of different doses of glyphosate on soil processes and microorganisms, physiology and metabolism of the plant along with possible contamination of soil and seed by herbicide residues. For this purpose, two soybean genotypes, one GR BRS 243 RR and a conventional (isogenic genotype) were grown under different weed control methods: hand weeding and glyphosate herbicide application. The study was performed in Rio Grande do Sul State, during the crop seasons of 2008 to 2009, 2009 to 2010, 2010 to 2011 and 2011 to 2012, with evaluations of soil microbial biomass and respiration, chlorophyll, nitrogen, ureides, nitrates, carotenoid, isoflavone content in seeds, and glyphosate and aminomethylphosphonic acid (AMPA) residues in seeds and soil. The use of glyphosate positively affected the microbial biomass, basal respiration and seeds yield. On the other hand, the chlorophyll, nitrogen, ureides, nitrates, carotenoid, and isoflavone contents in seeds were unaffected by the treatment. Even though we used the recommended application doses, the glyphosate residues in the seeds were above the levels permitted by the Brazilian law. Also, AMPA residues were detected in the soil and the seeds.

Keywords: Residues; isoflavones; chlorophyll; aminomethylphosphonic acid; transgenic soybean.

1. INTRODUCTION

Brazil is the second largest global producer of glyphosate-resistant soybean [*Glycine max* (L.) Merr.] (GR), already occupied over 20 Mha at 2012 [1], approximately 85% of the Brazilian soybean crop area was planted with GR varieties that was accompanied by heavy use of glyphosate [N-(phosphono-methyl) glycine] [2]. Glyphosate works by inhibiting the enzyme 5-enolpyruvylshikimate-3-phosphate synthase (EPSPS; E.C. 2.5.1.19, CP4 EPSPS), which participates in the biosynthesis of aromatic amino acids, having as substrates shikimate-3-phosphate and phosphoenolpyruvate [3]. The inhibition of shikimate pathway by glyphosate results in the accumulation of shikimic acid and/or hydroxybenzoic acids as well as gallic acid in sensitive plant species [4,5]. Thus, both primary and secondary metabolism are affected with accumulation of intermediate compounds in different plant organs [6,7]. In addition, glyphosate is an ionic molecule that easily forms complexes with metals and other soil components, and can be desorbed and/or degraded to aminomethylphosphonic acid (AMPA) [8]. The glyphosate used at recommended doses is not toxic to plants in the soil. In contrast, the AMPA moves to ground water in some soils [9]. Glyphosate resistance in soybean was produced with a transgene encoding GR EPSPS from a microbe that allows normal functioning of the shikimate pathway when the plant's EPSPS is inhibited by glyphosate [10].

Possible impacts of this genetic modification or glyphosate application were reviewed by Cerdeira et al. [11], Cerdeira and Duke [12], Bohm and Rombaldi [13]. According to them there is little evidence that GR soybean affects either soil microorganisms or physical and chemical plant composition (protein, lipid, carbohydrate and mineral contents) including isoflavone levels in soybean seed. Zablotowicz and Reddy [14] showed little sign of inhibition of nodulation or nitrogenase activity by glyphosate application, even when the plants were treated with twice the recommended rate of herbicide on two occasions during growth. Cerdeira et al. [11] reported that early applications of glyphosate can decrease the resistance of soybean to glyphosate GR, causing chlorosis, incidence of infection by Fusarium solani, reductions in biological N_2 fixation (BNF), losses in productivity and biomass reduction. According to Correa et al. [15], glyphosate has broad spectrum of against weeds, however, the exclusive use of this herbicide in areas of GR soybeans, may have limitations. For adequate control some species may require higher rates of glyphosate, applied sequentially or even the addition of another herbicide. For adequate control, some species may require higher doses of glyphosate, applied sequentially or combination with another herbicide.

The ureides, allantoin and allantoic acid, are synthesized in nodules of soybeans exported via xylem main products of BNF, reaching about 80% of the nitrogen transported, being used for synthesis of nitrogenous compounds and proteins, essential for plant growth [16]. The content of ureides, or xylem of leaf area is a reliable indicator of the efficiency of BNF, since high concentrations of ureides in soybeans and other legumes are usually associated with effective nodulation and high rates of BNF [17].

Negative effects in plant nutrition or high residue levels in harvested grains are unlikely to occur if the recommendations for glyphosate use are followed [18-20]. However, in studies carried out in Brazil [2,21], high residual levels of glyphosate in soybean seeds were detected after applying the recommended rate for the crop. Residue levels were above the limits permitted by Brazilian legislation (10 mg kg^{-1}), which is half of that allowed by most countries consuming this commodity [22]. Even though GR soybean has been widely adopted, very few field studies have been carried out to investigate for possible negative impacts of glyphosate on the plant itself and on the environment and harvested seed. Hence, the present study aimed at evaluating the effects of weed management two soybean cultivars (GR and its conventional isogenic genotype) with the use of different doses of glyphosate on soil processes and microorganisms, physiology and metabolism of the plant along with possible contamination of soil and seed by herbicide residues.

2. MATERIALS AND METHODS

The experiment was performed in the experimental area of the Centro Agropecuário da Palma, Pelotas Federal University, located at Capão do Leão, in the state of Rio Grande do Sul, Brazil (27°57'21" S, 51°48'32"W, at 643 m above sea level). The soil is classified as Dystrophic Red Yellow Podzolic (U.S. soil taxonomy) with, pH 5.6, 1.4% organic matter and 16% clay.

Twenty plots 4-m wide by 5-m long were set up in the area where five treatments were employed with 4 replicates. The treatments were a combination of the GR soybean cultivar BRS 243 RR, or the isogenic conventional cultivar, with different weed-control methods, as follows: (i)T1-BRS 243 RR; hand weeding at 28 days after planting (dap); (ii) T2- BRS isogenic; hand weeding at 28 dap; (iii) T3 - BRS 243 RR; one application of glyphosate at 960 g ai ha^{-1} 28 dap; (iv) T4 - BRS 243 RR; one applications of glyphosate at 1920 g ai ha^{-1} at 28 dap; (v) T5 - BRS 243 RR; one application glyphosate at 3840 g ai ha^{-1} at 28 dap.

The experiment was repeated in the crop seasons of 2008 to 2009, 2009 to 2010, 2010 to 2011 and 2011 to 2012, always in a different area of the same soil. Each crop season was considered as a block, even though the statistical analysis did not show a block effect.

The soybean cultivars were supplied by the Wheat Centre of Embrapa (Passo Fundo, Rio Grande do Sul State). Soybean seeds were inoculated with Bradyrhizobium japonicum strain SEMIA 5079 and Bradyrhizobium elkanii SEMIA 5878. A fertilizer rate of 250 kg ha^{-1} of a 0-20-20 NPK was applied in the seedbed just before seeding.

Yield, chlorophyll a and b, carotenoid, total nitrogen, ureides, nitrates and soil microbial biomass and activity were analyzed in samples taken from the experiment along with residues of glyphosate and aminomethylphosphonic acid (AMPA) in soil and seeds and isoflavone content in seeds.

Thirteen days after glyphosate application four plant samples were taken from each plot. The contents of chlorophyll 'a' and 'b' and carotenoids were determined by the method adopted by Zaicovski et al. [23]. Samples of 1 g of leaf disks taken from young leaves disregarding the central ribbed leaflets were macerated with 5 mL of 80% acetone (v/v). The material was centrifuged at 14000 g for 10 minutes and the supernatant was retained. This operation was repeated three times. The final volume was adjusted to 25 mL. The absorbance of the solution was read in a spectrophotometer Ultraspec 2000 ®, the 647 and 663 nm. The contents of chlorophyll and carotenoids of the samples were determined based on the relationships described by Cardoso et al [24]: Chla = [(12.25*Abs663) – (2.79*Abs647)]*[V/(1000*FW)]; Chlb = [(21.50*Abs647)– (5,10*Abs663)]*[V/(1000*FW)]; Car = [((1 0 0 0 * A b s 4 8 0) – (1 . 8 2 * C h l a) –(85.02*Chlb))/198]*[V/(1000*FW)].

Where: Chla, Chlb and Car are, respectively, the contents of chlorophyll 'a', 'b' and carotenoids; Abs 663 Abs 647 Abs 480 and the absorbance in the length of 663 nm, 645 nm and 480 nm; V is the volume used (ml), and FW fresh weight (g). The results were expressed in mg g^{-1}.

The quantification of total ureides, allantoin and allantoic acid, the dry tissue of the plant canopy was performed according to the method adopted by Tajima et al. [25] estimated that the content of ureides taking as a basis a standard curve of allantoin in amounts ranging between 12.5 and 125.0 nmol and absorbance readings at 535 nm and results were expressed in nmol N-ureides/g dry mass.

The method described by Ramos et al. [26] was used for determine the total nitrogen. Samples of 0.3 g of shoot dry along with 2 g of catalyst mixture (K_2SO_4 and $CuSO_4$), 3 glass beads and 5 mL of H_2SO_4 were slowly digested in digestion block until 380°C to obtain a colorless extract, used for the determination of nitrogen total. In the distiller 20 mL of 50% NaOH was reacted with colorless extract allowing the collection of 80 mL in 50 mL conical flask containing 4% of boric acid diluted to 10 drops of the mixed indicator mixture. This was titrated with 0.1N HCl solution.

Determination of nitrate (NO_3) was performed according to the method adopted by Beninni et al. [27]. The extract dried tissue of the plant in a solution of 5 mL of 0.1 M phosphate buffer, pH 7.0, and 2.5 mL of ethanol. Estimates using as a basis a standard curve of nitrate in amounts ranging from 50 to 500.0 nmol and absorbance readings at 410 nm.

Ninety days after soybean planting (R5 growth stage), four soil samples were taken from each plot at a depth of 0 to 20 cm within plant rows. Carbon in the microbial biomass (MBC) was determined by the method described by Bohm et al [28]. According to this method, soil microorganisms are killed by irradiation at 2450 Mhz for four minutes instead of fumigation with chloroform. The MBC was determined by the difference between the irradiated and non-irradiated soil sample after K_2SO_4 extraction and C determination by dichromate oxidation and titration with ferrous ammonium sulphate. The value for MBC was calculated by the following formula: MBC = (Ci - Cni)/Kc, where Ci and Cni are the C content of the irradiated and non-irradiated samples, respectively; and Kc is the correction factor with a value of 0.33 (Bohm et al. 2011). The results were expressed in $\mu g\ CO_2\ g^{-1}$ soil.

Basal respiration (BR), which consists in measuring microbial activity derived from organic carbon decomposition and the quantification of CO_2 released, was determined according to the method described by Bohm et al. [28]. The quantity of CO_2 released from soil of each treatment was trapped in alkali during seven days of incubation and subsequently titrated against HCl. For the calculation of CO_2 efflux the formula BR = (B-S) x M x 4 was used, where B is the volume of HCl to titrate the blank flask; S is the volume of HCl to titrate the remaining NaOH from the soil sample; M is the HCl concentration; and 4 (standard value) is the equivalent gram of respired carbon by soil microorganisms. The results were expressed in mg $C\text{-}CO_2$ 100 g^{-1}.

Soybean yield was estimated by sampling a central area of 1 m^2 from each plot at harvest and seed was dried to 12% moisture. Seed was separated and weighed. Also, soil samples from the layer of 0-20 cm of each plot were taken for the analyses glyphosate and AMPA. The method described by Veiga et al. [29] was used for detection and quantification of glyphosate and AMPA residues in soil and seed. Soil samples were dried at 40℃ and sieved to pass a 2 mm mesh. Glyphosate and AMPA were extracted by a potassium dihydrogenphosphate solution. A 5 g sample was shaken during 15 min in 25 mL 0.1 M KH_2PO_4 then centrifuged for 10 min at 2500 g and passed through a Whatman N° 2 filter. The extraction was repeated twice until 70 to 75 mL liquid extract were obtained from each sample. The extracts were concentrated under low temperatures (freeze-dried) and dissolved in Milli-Q® water until a final volume of 10 ml was reached. These extracts were poured through a Millipore® 0.45 μm membrane and kept at a temperature of -20℃ until analysis by high performance liquid chromatography (HPLC). For derivation, 0.1 mL of extract was mixed to 0.9 mL borate buffer 0.025 M (pH 9.0), 0.9 mL of acetone and 0.1 ml of 9-fluorenylmethyl chloroformate. The sample was shaken for 5 min and left for 20 min, then washed three times with ethyl ether. Glyphosate and AMPA were separated by HPLC composed by 50 μl loop Rheodyne® injector, an anionic exchange column C18 of 150 x 2.0 mm with 5 μm particles (Supelco, Bellefonte, PA, USA) and spectrofluormetric detector (Shimadzu®). The mobile phase consisted of standard water for liquid chromatography (LC) at pH 2.5 (adjusted with formic acid) in phase 1 and an aqueous mixture of 5 mM ammonia acetate (pH 4.8), standard water LC and acetonitrile in phase 2.

Glyphosate and AMPA residue analysis in soybean seeds were similar to those used for soil, except that the sample consisted of 1 g of ground soybean seeds, which was extracted with 15 mL of Milli-Q® water, shaking for 30 min, sonicated for 20 min, then centrifuged at 2,000 g at 20℃ for 20 min. After that, 4 mL of the supernatant was filtered in Millipore® 0.45 μm membrane. A quantity of 5 mL of water was added to the mixture and the sonication, centrifugation and filtration were repeated. Hence, the same procedures of derivation and quantification were followed as described before. The limits for quantification were 0.04 μg kg^{-1} for glyphosate and 0.60 μg kg^{-1} for AMPA, with 90 to 95% recuperation for both molecules.

For the analysis of isoflavone content in seeds, samples were ground in a Janke and Kunkel A-10 (Wilmington, U.S.A.) mill with 0.25 mm sieve. Isoflavones were extracted under mechanical shaker agitation during 2 hours at 4℃ in a 1:20 (m/v) proportion using 80% aqueous methanol [30]. The extracts were then filtered through a Whatman N° 6 filter paper and evaporated under vacuum (Rotavapor-RE 120 – Büchi, Flawil, Switzerland) to a 2 mL final volume. This solution was then completed to 5 mL with methanol (HPLC degree) and filtered with polyethylene filters with PTFE membrane (Millipore Ltd., Bedford, U.S.A.) of 0.22 μm pores for HPLC analysis. The extractions were done in triplicate. Isoflavones were separated in C18 Nova-pak 4 μm column (Waters, Milford, U.S.A.) according to the method proposed by Song et al. [31]. The liquid chromatograph used was a Hewlett Packard (Palo Alto, U.S.A.) model 1100 equipped with a diode detector (DAD) and the ChemStation software. Samples were injected twice. Identification was made based on the spectra and retention time in comparison to known standards, and quantitation was based on external calibration. The 12 isoflavone standards were from LC Laboratories (Woburn, USA). Calibration was performed by injecting the standards three times at five different concentrations ($R2 \geq 0.999$). Total isoflavone contents were expressed as mg of isoflavone aglycone per 100 g of sample (FW.) after normalization of individual isoflavones to account for differences in molecular weight between glycoside derivatives. The mass of each isoflavone form (β-glucosides, malonylglucosides and acetylglucosides) was multiplied by the ratio of its aglycone molecular weight to the molecular weight of the individual form before summing [31].

All the analysis was performed in triplicate and results expressed as mean ± standard deviation. Differences between means were first analyzed by ANOVA test and then Tukey test ($p < 0.05$).

3. RESULTS

At R5 stage of soybean plant development, the amount of microbial biomass C (MBC) did not vary among treatments (Table 1). On the other hand, the basal respiration (BR) was greater where glyphosate applications were performed (Fig. 1). The highest metabolic quotients (qCO_2) were obtained for treatments with glyphosate application. This result is a consequence of greater microbial activity, with greater release of CO_2 per unit of MBC, caused by the presence of easily assimilable substrate for the development of microbial activity. The qCO_2 has been used as a biological indicator of the balance of the soil, since as microbial biomass becomes more efficient, less carbon is released as CO_2 and respiration by a higher proportion of carbon is incorporated into microbial biomass.

The levels of chlorophyll a and b did not vary among treatments (Table 2). In average, the chlorophyll content varied from 1.275 to 1.663 mg/gFW for chlorophyll a and from 1.294 to 1.988 mg/gFW for chlorophyll b. At R5 stage of plant development, the results for carotenoids were not a clear influence of genotype or glyphosate in this parameter, for example, untreated plants had 0.419 μmol g^{-1} and glyphosate treated soybean had 0.460 μmol g^{-1} (Table 2). The total nitrogen did not vary among treatments, reaching averages from 4.575 to 6.025 g/m^2 (Table 2). Concerning ureides, the results were inconsistent, i.e., there was not a clear influence of genotype or glyphosate in this parameter. For example, BRS isogenic hand-weed showed higher ureide level (7.329 nmol/gms), while BRS 243 RR hand-weed or treated with higher glyphosate dose showed the lower level, but did not vary between the other treatments (Table 2).

The higher yield level was obtained by treatments with glyphosate (Table 2). For example, treatment was we used 960 g a.i. ha^{-1}, the yield was 2057.81 kg ha^{-1}, contrasting to 1394.21 kg ha^{-1} in the treatment BRS 243 RR hand-weed. It indicates that glyphosate contributed to increasing in 47.6% the productivity.

Total isoflavone contents ranged from 63 to 78 mg/100 gFW for isogenic BRS and BRS 243RR receiving or not glyphosate applications

(Table 3). These contents were similar to those previously reported for organically grown soybeans (58 to 77 mg 100g^{-1}), meanwhile more than twice these values were found in conventionally grown BRS 258 soybeans, from 161 to 193 mg 100g^{-1} [32]. There was also a significant difference in the content of daidzein conjugates (almost 50% in isogenic vs 30% in RR) and genistein conjugates (45% in isogenic vs almost 60% in RR). The percentage of glycitein conjugates was half that found in RR (Table 4).

Both glyphosate and AMPA were found in soil and GR soybean seed in the tree glyphosate treatments. Applications of glyphosate at 3840 g a.i. ha^{-1} brought about a residue of 5.66 mg kg^{-1} and AMPA residues of 13.8 mg kg^{-1} (Table 5). Glyphosate concentrations mean of 13.21 mg kg^{-1} were detected in the treatments with glyphosate applications (Table 5). AMPA concentration was similar to that of glyphosate, reaching 18.23, 28.57 and 32.25 mg kg^{-1} seed for 960, 1920 and 3840 g a.i. ha^{-1} applications of glyphosate, respectively.

Table 1. Microbial Biomass C (MBC) and Basal Respiration (BR) of soil samples taken from the areas under different soybean cultivars and weed control methods

Soybean cultivar; weed control method	MBC $\mu g\ g^{-1}$	BR[†] $\mu g\ C\text{-}CO_2\ g^{-1}\ h^{-1}$	qCO_2 (10^{-4})
BRS 243 RR hand-weed [‡]	372.23[a]	0.16 [c]	4.58[c]
BRS isogenic hand-weed	369.93[a]	0.18[c]	5.12[c]
BRS 243RR 960 g ai ha^{-1} glyphosate	410.35[a]	0.24 [bc]	6.91[bc]
BRS 243RR 1920 g ai ha^{-1} glyphosate	383.93[a]	0.32 [ab]	9.44[ab]
BRS 243RR 3840 g ai ha^{-1} glyphosate	352.98[a]	0.41 [a]	11.79[a]
Mean	376.59	0.26	7.57
CV (%)	14.13	18.48	19.03

[†] Means followed by the same letter are not statistically different according to the Tukey test at a probability < 0.05; [‡] At 28 days after planting (dap)

Table 2. Chlorophyl, carotenoid, total nitrogen, ureides, nitrates and yield in samples plants soybean taken from the areas with different soybean cultivars and weed control methods

Soybean cultivar; weed control method	chlorophyll a (mg/gFW)	chlorophyll b (mg/gFW)	Carotenoid $\mu mol\ g^{-1}$	Nitrogen g/m^2	Ureides (nmol /g ms)	Nitrates $\mu mol\ g^{-1}$	Yield kg ha^{-1}
BRS 243 RR hand-weed [‡]	1.663 [a]	1.694 [a]	0.419 [ab]	6.019 [a]	4.737 [b]	90.941[a]	1394.21[c]
BRS isogenic hand-weed [‡]	1.498[a]	1.793[a]	0.388[b]	6.025 [a]	7.329 [a]	67.062[b]	1654.52[b]
BRS 243RR 960 g ai ha^{-1} glyphosate	1.275 [a]	1.294 [a]	0.460 [a]	4.799 [a]	5.232 [ab]	96.753[a]	2057.81[a]
BRS 243RR 1920 g ai ha^{-1} glyphosate	1.488 [a]	1.849 [a]	0.465 [a]	4.813 [a]	5.204 [ab]	93.942[a]	2177.60[a]
BRS 243RR 3840 g ai ha^{-1} glyphosate	1.489 [a]	1.988 [a]	0.436[a]	4.575 [a]	4.773 [b]	90.935[a]	2194.41[a]
mean	1.476	1.815	0.434	5.246	5.456	87.434	1895.67
CV (%)	11.42	11.62	7.56	9.40	19.43	10.89	17.41

[‡] At 28 d after planting (dap); [§] Within columns, means followed by the same letter are not statistically different according to the TUKEY test at a probability < 0.05

Table 3. Total content (mg100 g^{-1}) and profile of isoflavones (%) in seeds of GR soybean BRS 243RR and isogenic BRS receiving applications of glyphosate

Soybean cultivar; weed control method	Total isoflavones	%			
		β-glucosides	Malonylglucosides	Acetylglucosides	Aglycones
BR 243RR; hand-weed‡	63.1±11.7a	56.32^{a}	41.08^{b}	n.d.	2.59^{b}
BR isogenic; hand-weed‡	73.6±12.4a	49.20^{b}	46.92^{a}	n.d.	3.88^{a}
BR 243RR; 960 g ai ha^{-1} glyphosate ¶	77.7±6.1a	55.88^{a}	41.93^{b}	n.d.	2.19^{b}
BR 243RR; 1920 g ai ha^{-1} glyphosate ¶	70.3±10.8a	55.21^{a}	41.96^{b}	n.d.	2.83^{b}
BR 243RR; 3840 g ai ha^{-1} glyphosate #	72.3±5.8a	55.57^{a}	41.96^{b}	n.d.	2.46^{b}
Mean	71.4	54.44	42.77	-	2.79
CV (%)	7.5	1.40	1.55	-	16.87

‡ At 28 d after planting (dap); § Within columns, means followed by the same letter are not statistically different according to the TUKEY test at a probability < 0.05

Table 4. Contents of main types of isoflavone in seeds of GR soybean BRS 243RR and isogenic BRS receiving applications of glyphosate

Soybean cultivar; weed control method	Fractions (%)†		
	Daidzein	Glycitein	Genistein
	———————— % ————————		
BR 243RR; hand-weed‡	30.32^{b}	10.74^{a}	58.92^{a}
BR isogenic; hand-weed	49.37^{a}	5.09^{b}	45.52^{b}
BR 243RR; 960 g ai ha^{-1} glyphosate	30.59^{b}	10.51^{a}	58.90^{a}
BR 243RR; 1920 g ai ha^{-1} glyphosate	30.22^{b}	11.30^{a}	58.48^{a}
BR 243RR; 3840 g ai ha^{-1} glyphosate	29.51^{b}	10.56^{a}	59.92^{a}
Mean	34.00	9.64	56.38
CV (%)	5.19	17.96	3.57

† The total percentage of each aglicone represents the sum of free and conjugated isoflavones expressed as aglicones; ‡ At 28 d after planting (dap); § Within columns, means followed by the same letter are not statistically different according to the TUKEY test at a probability < 0.05

Table 5. Concentration of glyphosate and aminometylphosphonic acid (AMPA) in soil and seed of GR soybean BRS 243RR receiving applications of glyphosate

Soybean cultivar; weed control method	glyphosate (mg.kg^{-1})	AMPA (mg.kg^{-1})	glyphosate (mg.kg^{-1})	AMPA (mg.kg^{-1})
	Soil		Seeds	
BR 243RR; hand-weed‡	n.d.	n.d.	n.d.	n.d.
BR isogenic; hand-weed	n.d.	n.d.	n.d.	n.d.
BRS 243RR 960 g ai ha^{-1} glyphosate	3.28^{c}	10.30^{b}	15,37^{b}	18,23^{b}
BRS 243RR 1920 g ai ha^{-1} glyphosate	4.36^{b}	12.11^{ab}	18,07^{a}	28,57^{ab}
BRS 243RR 3840 g ai ha^{-1} glyphosate	5.66^{a}	13.80^{a}	19,63^{a}	32,25^{a}
CV	14.74	15.01	13.21	19.50
P	0.001	0.020	0.006	0.040

‡ At 28 d after planting (dap); n.d. not detected. Means followed by the same letter are not statistically different according to the Tukey test at a probability < 0.05

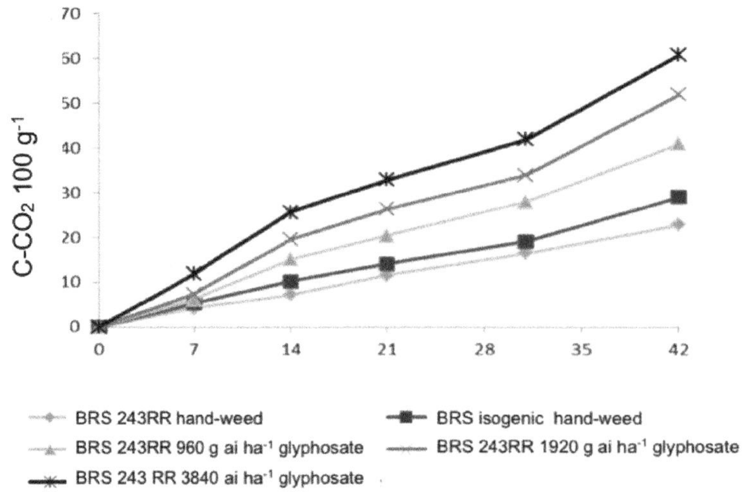

Fig. 1. Basal respiration of soil samples, incubated for 42 days, taken from the areas under different soybean cultivars and weed control methods. The results were expressed in mg C-CO_2 100 g^{-1}

4. DISCUSSION

The putative responses of glyphosate in the GR soybean cultivation on the soil microorganisms activity, grains yield and quality remains barely answered. For example, Haney, Senseman and Hons [33] detected a higher MBC in soils in which there was application of glyphosate herbicide. On the other hand Gomez et al. [34] detected lower levels of MBC in soils with application of glyphosate. Here, in our experiment (four seasons), not changes were observed between treatments. The exact causes of these differences have not been studied, but may be associated with differences in soil and microbial activity of studied environments. Concerning the basal respiration (BR) and metabolic quotients (qCO_2), higher levels were detected in soil were the glyphosate was applied. This change in respiration may be related to the increase in concentration of the easily degradable glyphosate molecule in soil that could stimulate soil N mineralization [33] and decrease the plant reliance on BNF [35]. However, per the authors literature review [7,11,12,20,29,33] glyphosate has generally not been documented to have negative effects on soil. Nitrogen content could be affected by glypohosate application [14]. According Reddy et al. [6] and King et al. [36] the glyphosate applied at the dosage recommended by the manufacturer reduces the number and mass of nodules, the leg-hemoglobin content and the nitrogen content in the plant. In contrast, Bohm et al. [35] in a field experiment with application of glyphosate at 1920 g ha^{-1} did not find effect in the nitrogen contents of the plant, although the treatments with glyphosate in GR soybean reverberated less dependent on plant by BNF (below 70%) when compared to GR soybean under weeding, whose dependence by BNF was estimated at 80%. Here, we observed that Nitrogen did not vary among treatments.

Previous study [14] carried out in a field experiment, showed that glyphosate application had little effect at ureides and total amino acids when applied in the recommended doses (1920 g a.i. ha^{-1}). However, when glyphosate was applied at a dose of 3360 g a.i. ha^{-1} decreases in BNF were observed, as well as the levels of foliar nitrogen and total ureides. The reduction of BNF by glyphosate application was also observed by Bohm et al. [35]. Here, we observed that the ureides contents changes did not showed a co-variation response in function of glyphosate application.

Also, is cited that glyphosate could affect plant chlorophyll stability, if applied at high doses (at 2240 g a.i. ha^{-1}) [6]. However, they observed that, in general, glyphosate application at a dose of 1120 g a.i. ha^{-1} had little or no effect on the chlorophyll content in GR soybean. Here, in this study, doses from 960 to 3840 g a.i. ha^{-1} glyphosate were tested in four seasons and none effect was observed in the leaves chlorophyll content. The main differences between these works are related to model of cultivation: Redddy et al. [6] cultivated the soybean in the greenhouse and we cultivated in the field. In addition, the cultivars were not the same. The carotenoids, that are important photossynthetic protectors [37], increased in the soybean treated with glyphosate. It is possible that the increasing of carotenoid content had a protective effect of chlorophylls, despite of high glyphosate doses. This hypothesis is based in works that shows increasing of carotenoids synthesis in plants submitted to the moderate abiotic stress [38].

As expected the better grain yield was obtained from treatment were weeds were controlled by glyphosate. Cerdeira et al. [11] mentioned that increasing productivity occurs because the application of glyphosate provides excellent weed control favoring less competition, better photosynthesis and nutrient utilization. In contrast, yields of GR soybean crops have been reported to be unaffected by glyphosate application [39,40].

In order to monitoring grains quality, the isoflavone contents were also quantified. The weeds control methods did not affected the accumulation of theses specialized metabolites. Soybean contains three types of isoflavones, in four chemical forms: the aglycones daidzein, genistein, and glycitein; the β-glycosides daidzin, genistin, and glycitin; the acetyl-β-glycosides 6''-O-acetyl-β-daidzin, 6''-O-acetyl-β-genistin, 6''-O-acetyl-β-glycitin; and the malonyl-β-glycosides 6''-O-malonyl-β-daidzin, 6''-O-malonyl-β-genistin, and 6''-O-malonyl-β-glycitin [41]. Aglycones and acetylglycosides are not normally present in seeds and are formed as a result of processing such as drying, defatting, and storage. No significant differences were found among BRS 243 RR receiving or not glyphosate applications in relation to the profile and distribution of isoflavones. However, isogenic BRS presented a higher content of malonylglycosides and a lower content of β-glucosides than 243 RR. There was also a significant difference in the content of daidzein conjugates and genistein conjugates. It is known [41] that malonylglucosides are highly unstable and easily converted to β-glucosides. The sum of malonylglucosides plus β-glucosides is among the expected values, for all the samples. However, regarding the distribution of isoflavones (the total percentage of daidzein and its conjugates, glycitein and its conjugates, and genistein and its conjugates), important differences were detected, with daidzein conjugates being the prevalent forms only in isogenic BRS. Similarly to BRS RR, ten BRS Brazilian varieties previously analyzed presented prevalence of genistein conjugates [41]. This alteration is probably related to an up or down-regulation of their biosynthesis. Daidzein is known to be the precursor of the phytoalexin glyceollin, and the accumulation of glyceollin correlates with resistance to various fungal pathogens [42].

Both glyphosate and AMPA were found in soil and GR soybean seed in the tree glyphosate treatments. The fact that concentration of AMPA in soil was higher than of glyphosate is explained by Ginsing et al. [43], who reported that glyphosate biodegradation happens faster than that of AMPA. This is confirmed in the study of Araujo et al. [44], who found proportionally higher AMPA accumulation in soils with successive glyphosate applications. Even though glyphosate and AMPA levels detected in soil are a matter of concern there are no official parameters establishing a threshold. The glyphosate content in seeds were above the limits permitted by Brazilian Agency for Sanitary Vigilance is 10 mg kg^{-1} [45], with is half of that allowed by USA (20 mg kg^{-1}) and almost all countries of the European community (20 mg kg^{-1}). The reason(s) for the high levels was not explained in this work. However, it is known that the response to treatments, and therefore the prevalence of residues is dependent on edaphoclimatic conditions. The experiment was carried out in an area known to have a significant incidence of abiotic stresses (water stress, temperature changes and intense solar radiation), which are not common (at least concomitantly) in most areas of soybean production.

5. CONCLUSION

The use of glyphosate positively affected the microbiological biomass, basal respiration and seeds yield. On the other hand, the chlorophyll, nitrogen, ureides, nitrates, carotenoid, and isoflavone contents in seeds were unaffected by the treatment. However, even though we used the recommended application doses, the glyphosate residues in the seeds were above the levels permitted by the Brazilian law, but below the international levels allowed. Also, AMPA residues were detected in the soil and the seeds.

ACKNOWLEDGEMENTS

Financial support from the Federal Institute Sul-rio-grandense (IFSul) and Brazilian National

Research Council (CNPq) are greatly appreciated.

COMPETING INTERESTS

Authors have declared that no competing interests exist.

REFERENCES

1. ISAAA- International Service for the Acquisition of Agri-biotech Applications - Global Status of Commercialized Biotech/GM Crops; 2006. Available:<http://www.isaaa.org/Resources/Publications/briefs/35/executivesummary/default.html> (Accessed in 15 ago 2015).
2. Bohm GMB, Rombaldi CV, Genovese MI, Castilhos D, Alves BJR, Rumjanek NG. Glyphosate effects on yield, nitrogen fixation, and seed quality in glyphosate-resistant soybean. Crop Science. 2014; 54(4):1737-1743.
3. Fisher RS, Berry A, Gaines CG, Jensen RA. Comparative action of glyphosate as a trigger of energy drain in Eubacteria. Journal Bacteriology. 1986;168:1147-1154.
4. Duke SO. Glyphosate. In Herbicides: Chemistry, Degradation, and Mode of Action; Kearney PC, Kaufman DD, Eds. Dekker: New York. 1988;3:1-70.
5. Nandula VK, Reddy KN, Rimando AM, Duke SO, Poston DH. Glyphosate-resistant and susceptible soybean (*Glycine max*) and canola (*Brassica napus*) dose response and metabolism relationships with glyphosate. J. Agric. Food Chem. 2007;55(9):3540-3545.
6. Reddy KN, Hoagland RE, Zablotowicz RM. Effect of glyphosate on growth, chlorophyll, and nodulation in glyphosate-resistant and susceptible soybean (*Glycine max*) cultivars. Journal of New Seeds. 2000; 2:37–52.
7. Moldes CA, Camiña JM, Medici LO, Tsai SM, Azevedo RA. Physiological effects of glyphosate over amino acid profile in conventional and transgenic soybean (*Glycine max*). Pesticide Biochemistry and Physiology. 2012;102(2):134-141.
8. Reddy KN, Rimando AM, Duke SO. Aminomethylphosphonic acid, a metabolite of glyphosate, causes injury in glyphosate-treated, glyphosate-resistant soybean. Journal of Agricultural and Food Chemistry. 2004;52:5139-5143.
9. Landry D, Dousset S, Fournier JC, Andreux F. Leaching of glyphosate and AMPA under two soil management practices in Burgundy vineyards (Vosne-Romanée, 21-France). Environmental Pollution. 2005;138(2):191-200.
10. Green JM. Evolution of glyphosate-resistant crop technology. Weed Science. 2009;57:108-117.
11. Cerdeira AL, Gazziero DPL, Duke SO, Matallo MB, Spadotto CA. Review of potential environmental impacts of transgenic glyphosate-resistant soybean in Brazil. Journal of Environmental Science and Health (Part B). 2007;42:539-549.
12. Cerdeira AL, Duke SO. Environmental impacts of transgenic herbicide-resistant crops. CAB Reviews: Perspectives in Agriculture, Veterinary Science, Nutrition and Natural Resources. 2007;2:44-53.
13. Bohm GMB, Rombaldi CV. Effect of genetic transformation and application of glyphosate in soil microbiota, biological nitrogen fixation, quality and safety of genetically modified soybeans. Science Rural. 2010;40:213-221.
14. Zablotowicz RM, Reddy KN. Nitrogenase activity, nitrogen content, and yield responses to glyphosate in glyphosate-resistant soybean. Crop Protection. 2007; 26:370–376.
15. Correia NM, Durigan JC, Spanish M. Weed management in GM soybeans tolerant to glyphosate. Agricultural Research. 2011; 41(2):242-247. DOI: 10.5216/pat.v41i2.8807
16. Ohyama T, Ohtake N, Sueyoshi K, Tewari K, Takahashi Y, Ito S, Nishiwaki T, Nagumo Y, Ishii S, Sato A. Nitrogen fixation and metabolism in soybean plants. In Nitrogen fixation research progress, Editor by Guilhem N. Couto. Nova Science Publishers, Inc. New York. 2008;1-131.
17. Herridge DF, Peoples MB. Ureide assay for measuring nitrogen fixation by nodulated soybean calibrated by 15N methods. Plant Physiology. 1990;93:495-503.
18. Arregui MC, Lenardón A, Sanchez D, Maitre MI, Scotta R, Enrique S. Monitoring glyphosate residues in transgenic glyphosate-resistant soybean. Pest Management Science. 2003;60:163-166.
19. Duke SO, Rimando AM, Pace PF, Reddy KN, Smeda RJ. Isoflavone, glyphosate, and aminomethylphosphonic acid levels in seeds of glyphosate-treated, glyphosate-

resistant soybean. Journal of Agricultural and Food Chemistry. 2003;51:340-344.
20. Duke SO, Reddy KN, Bu K, Cizdziel JV. Effects of glyphosate on the mineral content of glyphosate-resistant soybeans (*Glycine max*). Journal of Agricultural and Food Chemistry. 2012;60(27):6764-6771.
21. Bohm GMB, Genovese MI, Pigosso G, Trichez D, Rombaldi CV. glyphosate residues and aminomethylphosphonic acid, and isoflavones content in soybean BRS 244 RR and BRS 154 grown in Planosol. Journal of Food Science and Technology. 2008;28(supl.):192-197.
22. AGROFIT. Sistema de Agrotóxicos Fitossanitário; 2015. Available:http://www.agrofit.agricultura.gov.br/agrofit_cons/principal_agrofit_cons (Accessed in february 25th, 2015)
23. Zaicovski CB, Pegoraro C, Ferrarezze JP, Dal Cero J, Lund DG, Rombaldi CV. Effect of mechanical damage, temperature reduction and 1-MCP in Legacy broccoli postharvest metabolism. Science and Technology Food. 2008;28(4):840-845.
24. SN Cardoso Oliveira LM, LG Fernandez, Pelacani CR, Souza CLM, ARMF Oliveira. Priming on seed germination, early growth and content Myracrodruon urundeuva fr pigments. Allemão. Journal of Biosciences. 2012;10(4):457-461.
25. Tajima S, Nomura M, Kouchi H. Ureide biosynthesis in legume nodules. Frontiers on Bioscience. 2004;1(9):1374-81.
26. Ramos MG, Villatoro MAA, Urquiaga S, Alves BJR, Boddey RM. Quantification of the contribution of biological nitrogen fixation to tropical green manure crops and the residual benefit to a subsequent maize crop using 15N-isotope techniques. Journal of Biotechnology. 2001;91:105–115.
27. Beninni ERY, Takahashi HW, Neves CSVJ, Fonseca ICB. Nitrate content in lettuce grown in hydroponic and conventional systems. Brazilian horticulture. 2002;20:183-186.
28. Bohm GMB, Schneider L, Castilhos D, Rombaldi CV. Weed control, microbial biomass and soil metabolism due to the application of glyphosate and imazethapyr in soybeans. Semina: Agricultural Sciences. 2011;32(3):919-930.
29. Veiga F, Zapata JM, Fernandez-Marcos ML, Alvarez E. Dynamics of glyphosate and aminomethylphosphonic acid in a forest soil in Galicia, north-west Spain. The Science of the Total Environment. 2001; 271:135-144.
30. Genovese MI, Lajolo FM. Determination of isoflavones from soybeans products. Food Science and Technology. 2001;21:86-93. DOI:org/10.1590/S0101-20612006000400032
31. Song T, Barua K, Buseman G, Murphy PA. Soy isoflavone analysis: Quality control and a new internal standard. American Journal of Clinical Nutrition. 1998;68:1474-1479.
32. Balisteiro DM, Rombaldi CV, Genovese MI. Protein, isoflavones, trypsin inhibitory and in vitro antioxidant capacities: Comparison among conventionally and organically grown soybeans. Food Research International. 2013;51:8-14.
33. Haney RL, Senseman SA, Hons FM. Bioremediation and biodegradation: Effect of roundup ultra on microbial activity and biomass from selected soils. Journal Environmental Quality. 2002;31:730-735.
34. Gomez E, Ferreras L, Lavitti L, Fernandez E. Impact of glyphosate application on microbial biomass and metabolic activity in a vertic argiudoll from Argentina. European Jornal of Soil Biology. 2009;45(2):163-167.
35. Bohm GMB, Alves BJR, Urguiaga S, Boddey RM, Xavier GR, Hax F, Rombaldi CV. Glyphosate and imazethapyr-induced effects on yield, nodule mass and biological nitrogen fixation in field-grown glyphosate-resistant soybean. Soil Biology & Biochemistry. 2009;41(2):420-422.
36. King CA, Purcell LC, Vories E. Plant growth and nitrogenase activity of glyphosate-tolerant soybean in response to foliar glyphosate applications. Agronomy Journal. 2001;93:179–186.
37. Azmat R. An important role of carotenoids in protection of photosynthetic apparatus under VAM inoculation on *Momordica charantia*. Curr Pharm Biotechnol. 2013; 14(9):829-34.
38. Gururani MA, Venkatesh J, Ganesan M, Strasser RJ, Han Y, Kim J-I, Song P-S. *In vivo* assessment of cold tolerance through chlorophyll-a fluorescence in transgenic zoysiagrass expressing mutant phytochrome A. PLoS ONE. 2015;10(5): e0127200. DOI: 10.1371/journal.pone.0127200
39. Elmore RW, Roeth FW, Nelson LA, Shapiro A, Klein RN, Knezevic SZ, Martin A. Glyphosate-resistant soybean cultivar

yields compared with sister lines. Agronomy Journal. 2001;93(2):408-412. DOI: 10.2134/agronj2001.932408x
40. Norsworthy JK. Broadleaved weed control in wide-row soybean (*Glycine max*) using conventional and glyphosate herbicide programmers. Crop Protection. 2004; 23(12):1229-1235.
41. Genovese MI, Hassimotto NMA, Lajolo FM. Isoflavone profile and antioxidant activity of Brazilian soybeans varieties. Food Science and Technology International. 2005;11(3):205-211.
42. Lozovaya VV, Lygin AV, Zernova OV, Ulanov AV, Li SX, Hartman GL, Widholm JM. Modification of phenolic metabolism in soybean hairy roots through down regulation of chalcone synthase or isoflavone synthase. Planta. 2007;225(3): 665-679.
43. Gimsing AL, Borggaord OK, Jacobsen OS, Aamand J, Sorensen J. Chemical and microbiological soil characteristics controlling glyphosate mineralization in Danish surface soils. Applied Soil Ecology. 2004;27:233–242.
44. Araújo ASF, Monteiro RTR, Abarkeli RB, Souza LS. Biodegradation of glyphosate in two Brazilian soils. Pesticides: Journal of Ecotoxicology and Environmental. 2003; 13:157-164.
45. ANVISA. National Health Surveillance Agency; 2007.
Available:<http://www.anvisa.gov.br> (Accessed in July 25th, 2014)

12

Prevalence and Intensity of Infection of Gastrointestinal Parasites in Cavies from Menoua Division-West Region of Cameroon

Vincent K. Payne[1], Ndah Germaine[1], Fusi Ngwa[1], Cedric Yamssi[1*], Leonelle Megwi[1], Fonteh A. Florence[2], Mpoame Mbida[1] and Bilong Bilong C. Felix[3]

[1]Department of Animal Biology, Faculty of Science, University of Dschang, P.O.Box 067 Dschang, Cameroon.
[2]Department of Animal Production, Faculty of Agriculture, University of Dschang, P.O.Box 222 Dschang, Cameroon.
[3]Department of Animal Biology and Physiology, Faculty of Science, University of Yaoundé 1, P.O.Box 812, Yaoundé, Cameroon.

Authors' contributions

This work was carried out in collaboration between all authors. Authors VKP, NG, FN, CY, LM, FAF, MM and BBCF designed the study, wrote the protocol and wrote the first draft of the manuscript. Author NG managed the analyses of the study and the literature searches. All authors read and approved the final manuscript.

Editor(s):
(1) Monica Rosa Loizzo, Department of Pharmacy, Health Sciences and Nutrition of University of Calabria, Italy.
Reviewers:
(1) M. Mohan Varma, Shri Vishnu College of Pharmacy, India.
(2) Ozge Cevik, Cumhuriyet University, Turkey.
(3) Gidi Smolders, Wageningen University, The Netherlands.

ABSTRACT

Background: Domestic cavies, commonly known as guinea pigs are more widely filling a niche in livestock production in West Africa, most especially in Cameroon. Cavies have been shown to play a vital role in the animal protein supply and income generation in rural areas in Cameroon. Despite this, there is a dearth of information concerning diseases and parasites of cavies in Cameroon. The

**Corresponding author: Email: Vk.payne@yahoo.com*

aim of this study was therefore to have an inventory of gastrointestinal parasites of cavies in Menoua Division, West Region of Cameroon; and to characterize the infection rates.

Methods: Three hundred Cavies (171 females and 129 males) aged 1 week to 1 year old were examined from May 2013 to June 2014. A qualitative (double centrifugation flotation technique) and a quantitative (Mc Master Technique) coprologic analyses were used for the egg or oocyst count of gastrointestinal parasites.

Results: Of the total number of cavies examined, 237 showed positive with one or more gastro-intestinal parasites, giving an overall prevalence of 79.0%. Six genera of gastrointestinal nematodes: *Spirocerca* spp. (3.0%), *Trichuris* spp. (3.3%); *Ascaris* spp. (4%); *Trichostrongylus* spp. (10.3%); *Strongyloides* spp. (10.7%); *Paraspidodera* sp. (31.0%) and one species of Protozoa represented by *Eimeria* spp. (48.7%) were identified. There was a significant difference in prevalence (P< 0.05) for *Eimeria* spp, *Strongyloides* spp and *Paraspidodera* sp based on locality while with respect to age group, only *Paraspidodera* spp. and *Eimeria* spp. showed significant difference in prevalence (P< 0.05). As concerns sex of cavies, *Ascaris* spp was significantly more prevalent in females than in males.

Conclusion: The overall high prevalence recorded in this study clearly indicates the need for strategic parasite control in cavies with particular attention focused on middle to old aged cavies which showed significantly higher prevalence and parasitic load for most of the gastrointestinal infections.

Keywords: Prevalence; intensity; helminths; protozoa; cavies.

1. BACKGROUND

Domestic cavies, commonly known as guinea pigs are more widely filling a niche in livestock production in West Africa, and is becoming more widespread in Cameroon [1]. In this country, cavy farming is carried out mainly in the Western High Lands, especially in the West Region. Cavy production in this area is still a secondary activity carried out by many small holder farmers. They are produced in polygamous groups of 10 adult females to 2 adult males (without considering the immature animals). Off-take for home consumption, sales or gifts are usually adult males while females are allowed to reproduce. These cavies are thus produced mainly in the traditional way. This extensive production system shows no defined management practices for most stakeholders. Most feed is provided from kitchen wastes and farm residues, and sometimes, supplemented with vegetables and forages. However, cavies are susceptible to gastrointestinal parasitic infection, as these parasites are known to be widespread in Cameroon and limit livestock production in many areas [2,3]. Studies have shown that helminth and protozoan parasites are by far the most serious causes of production losses in farmed animals and are indisputably the cause of serious production losses in sub-Saharan Africa, and indeed worldwide [4,5]. [6] highlighted the fact that parasitism is generally believed to be the most important disease problem of domestic animals raised in the rural areas. Parasites cause serious economic losses due to emaciation, reduced/low growth rates, anaemia, and death of the infected animals. Cavies have been shown to play a vital role in the animal protein supply and income generation in rural areas in Cameroon [7]. Despite this, the health of these animals have been accorded little scientific efforts. This observation is supported by [1] who reported that one of the major constraints to cavy culture in Cameroon is the health of these animals. They further highlighted the fact that there is a dearth of information concerning diseases and parasites of cavies in Cameroon. The aim of this study therefore was to have an inventory of gastrointestinal parasites of cavies from Menoua Division, West Region of Cameroon, and to characterize the infection rates.

2. METHODS

2.1 Area of Study

This study was carried out between May 2013 and June 2014 in Menoua Division, West Region of Cameroon. Menoua Division covers a surface area of 1380Km^2. It is composed of six Sub-Divisions; the division spans from Santchou at an altitude of 600m through Dschang (1200 m) and topping on the Plateau of Djuttitsa at an altitude of 2200 m. Specifically, the study was carried out in the following localities: Santchou (600 m). Dschang (1200 m), Fokoue (1200 m), Penka - Michel (1500 m) ,Fongo – Tongo (2000 m), and

Nkong-ni (2000 m). Menoua Division was purposively chosen based on its high level of cavy culture; it is a grassland area with above 80% of the inhabitants being farmers [8]. Menoua division is thus one of the major suppliers of vegetables in the country.

2.2 Type of Study

It was a cross-sectional study carried out in 15 villages within Menoua division, West region of Cameroon. The animals were randomly selected using table of random numbers or lottery system. The sample size was estimated according to the methods provided by [9].

2.3 Sample Collection

In order to evaluate the prevalence and intensity of infection of gastrointestinal parasites, a total of 300 cavies were examined from 49 households from 15 villages within the following localities: Santchou (42), Nkong-ni (49), Penka -Michel (51), Fongo–Tongo (52), Dschang (53), Fokoue (53). Mean flock size was about 12 animals; all age groups were represented in all the localities. The number of animals sampled per farm varied, depending on the litter size. Because we encountered only small animal number per household, we decided to sample 50% of the animals from each household. Ages were determined based on weight of the animal, length of the toenails, together with information from the cavy farmers [10,11]. The sampled animals were attributed numbers (ear tattoo), weighed and sex determined, after which they were placed in a ten chambered cage, each per chamber. They were allowed to defaecate within a six-hour period in their various chambers at the different farms. Faeces were put in sealable plastic bags, labeled with the number attributed to a specific animal, age, sex, date and time of collection. The faeces were then dispatched to the laboratory where they were analyzed within 24 hours.

2.4 Farming Methods

Although cavy farming in Cameroon shows basically the same management practices and feed type, variations in husbandry practices were noticed, being associated with cavy housing. The nature of cavy house depended on the purpose of cavy farming and available space. An individual farmer might confine his animals in one corner of the kitchen in order to pile up faeces and decaying forage to be used as manure or organic fertilizer in gardening. Such farmers maintain the same bedding for about six weeks and above. In some cases, cavies shared same bedding with other domestic animals like fowls, ducks goats etc. This was commonly seen in Penka-Michel and Nkong-Ni, which are the major centers for market gardening. Dschang being an urban area and a university town has a high population density. Thus, most cavy farmers confine their animals either in a small kitchen corner or in raised cages (Photo 3). Most farmers in this locality responded that they do a weekly bedding change (usually during weekends). Exceptionally, one farm was surveyed in Dschang with 11 animals reared in raised cages and a more regular bedding change (every day or every 2 days). On the other hand, in Fongo-Tongo, Fokoue and Santchou, most farmers allowed their animals to scavenge freely on kitchen floors. Interspecific animal rearing (rearing of animals of different species in the same house) was very common in Fongo-Tongo because most farmers in that locality were hunters. Therefore cavies were found to be housed with other animals like cane rat, hunting dogs, goats and ducks (Photos 1 and 2). In Santchou, most kitchen floors are linned with bamboos because this locality is swampy, being a lowland area (600 m above sea level, as compared to the other five localities which are found between 1200 and 2200 m above sea level.

2.5 Parasitological Examination

Corprologic analysis was done to have a quantitative and qualitative appreciation of the prevalence and parasitic load of infection of parasites. For the qualitative analysis, faeces were analysed by the double-centrifugal flotation technique using saturated sodium chloride solution (concentration: 6.84 M) as described by [12]. For quantitative analysis, the number of eggs per gram (EPG) in case of helminthic infections or oocyst per gram (OPG) in case of protozoan infection was determined using the Mc Master technique as described by [13,14]. The infection was considered as mild if EPG/OPG were<100, moderate if EPG/OPG were 100–300, and as heavy or severe if EPG/OPG >300 [15].

Photo 1. Inter-specific animal rearing under free range system in Fokoue

Photo 2. Interspecific animal rearing under free range system in Fongo-Tongo

Photo 3. Cavy production under confined system in Dschang

2.6 Statistical Analysis

The collected data was stored in Excel 2007, later transferred to Statistical Packages for Social Science (SPSS version 19.0) for statistical analysis. The prevalence of gastrointestinal parasites was compared using Chi square test. Mann-Whitney test was used to evaluate parasite intensity between sex and Kruskal-Wallis test was used to compare parasite intensity between age and locality. Means were separated using Post Hoc of Turkey. A critical probability of ($P < 0.05$) was adopted throughout as a cut-off point for statistical significance between groups compared. All statistical tables retrieved from analysis with SPSS.

3. RESULTS

3.1 Overall Prevalence of Gastrointestinal Parasites in Cavies

The following six genera of gastrointestinal nematodes were recorded: *Spirocerca* spp. 9 (3.0%), *Trichuris* spp. 10 (3.3%); *Ascaris* spp. 12(4.0%); *Trichostrongylus* spp. 31(10.3%); *Strongyloides* spp. 32 (10.7); *Paraspidodera* sp. 93(31.0%) while one genus of Protozoa represented by *Eimeria* spp. 146 (48.7%) was also identified, giving an overall prevalence of 79.0 for gastrointestinal parasites. Of the 49 farms surveyed, only one farm containing 11 cavies showed negative for gastrointestinal parasitism.

3.1.1 Influence of host age on the Prevalence of gastrointestinal parasites

Eimeria spp., *Paraspidodera* sp., *Ascaris* spp., and *Trichostrongylus* spp. were found in all age groups while *Strongyloides spp., Trichuris* spp. *and Spirocerca* spp. were completely absent in young animals (Table 1). However, there was a significant difference in prevalence ($P<0.05$) based on age groups only for *Spirocerca* sp (between 9-12 weeks and >16 weeks animals) and *Paraspidodera* sp. (between 1-4 weeks and >16 weeks animals).

Table 1. Influence of host age on the prevalence of gastrointestinal parasite [N (%)]

	Age group (weeks)				
	1 - 4	**5 - 8**	**9 - 12**	**13 -16**	**> 16**
N in different age group	65	88	60	50	37
Eimeria spp.	21 (32.3)[a]	41(46.6)[a]	30 (50.0)[ab]	34 (68.0)[ab]	20 (54.1)[ab]
Ascaris spp.	2 (3.1)[ab]	2 (2.3)[b]	2 (3.3)[ab]	1 (2.0)[b]	5 (13.5)[a]
Trichostrongylus spp.	2 (3.1)[a]	7 (8.0)[a]	6 (10.0)[ab]	11(22.0)[ab]	4 (10.8)[ab]
Trichuris sp.	0	3 (3.4)[a]	3 (5.0)[a]	4 (8.0)[a]	0
Strongyloides spp.	0	9 (10.2)[b]	9 (15.0)[b]	8 (16.0)[b]	6 (16.2)[b]
Spirocerca spp.	0	0	2 (3.3)[a]	4 (8.0)[b]	3 (8.1)[b]
Paraspidodera sp.	10 (15.4)[b]	30 (34.1)[ab]	18 (30.0)[ab]	16 (32.0)[ab]	18 (48.7)[a]

Means followed by different letters within the same row differ significantly (P<0.05)

3.1.2 Influence of host sex on the prevalence of gastrointestinal parasite

Both sexes were susceptible to all genera of gastrointestinal parasites recorded. Generally, it was observed that of the seven gastrointestinal parasites recorded, prevalence was not significantly higher in females than in males (Table 2) except for *Ascaris* spp.

Table 2. Prevalence of gastrointestinal parasite by sex of cavies [N (%)]

	Sex	
	Female	**Male**
N per sex	171	129
Eimeria spp.	89 (52.1)	57 (44.2)
Ascaris spp.	10 (5.9)[b]	2 (1.6)[a]
Trichostrongylus spp.	17 (9.9)	14 (10.9)
Trichuris spp.	8 (4.7)	2 (1.6)
Strongyloides spp.	16 (9.4)	16 (12.4)
Spirocerca spp.	5 (2.9)	4 (3.1)
Paraspidodera sp.	56 (32.8)	37 (28.)

Means followed by different letters within the same row differ significantly (P<0.05)

3.1.3 Prevalence of gastrointestinal parasite in cavies per locality

Eimeria spp. and *Paraspidodera* sp. were found in all the six localities of the study. The prevalence of *Eimeria* spp. was highest in cavies from Penka-Michel (76,5%) followed by those from Dschang (62.3%) and Nkong-ni (51.0%). The lowest prevalence for *Eimeria* spp. was recorded in Fongo-Tongo (21.2%). The prevalence of *Paraspidodera* sp. was highest in Dschang (43.4%), followed by those from Nkong-ni (38.8%) and Santchou (38.1%). The other five gastrointestinal parasites were found to be completely absent in some localities. On the other hand, during this study, it was noted that *Spirocerca* spp. (17.3%) was found only in Fongo- Tongo. However, the difference in prevalence was significant (P< 0.05) for *Eimeria* spp, *Strongyloides* spp and *Paraspidodera* sp based on locality (Table 3).

3.1.4 Prevalence of gastrointestinal parasitic associations

Of the 300 cavies examined for gastrointestinal parasites, 150 (50%) had no parasites, 150 (50%) were infected with only one parasite (Fig. 1A). Seventy-five (25%) had double parasitic infections while 12 (4%) had triple parasitic infection. From seventy-five cavies with different combinations of double associations of parasites, 29 (9.7%) had a combination of *Paraspidodera* sp. + *Eimeria* spp., 10 (3.3%) a combination of *Eimeria* spp. + *Trichostrongylus* spp. (Fig. 1B). Twelve animals had 3 different combinations of triple association; a combination of *Eimeria* spp. + *Strongyloides* spp. + *Paraspidodera* sp. (2%) and *Eimeria* spp. + *Trichostrongylus* spp. + *Paraspidodera* sp., (1.3%) of the animals.

3.2 Mean Intensity of Infection

3.2.1 Intensity of infection in relation to host age

The intensity of infection of gastrointestinal parasites did not quite vary with age of the animal, except for three parasites (Table 4). The genera *Ascaris* spp, *Spirocerca* spp and *Paraspidodera* increased with age of the animal, with the highest intensity of infection in the older animals (>16 weeks). These parasites showed significant difference in intensity (P<0.05) based on age group.

Table 3. Prevalence of gastrointestinal parasite in cavies per locality [N (%)]

	Localities						Total (%)
	Dschang	**Fokoue**	**Fong-tongo**	**Nkong-ni**	**Penka michel**	**Santchou**	
Gastro-intestinal parasites / N	53	53	52	49	51	42	300
Eimeria spp.	33 (62.3)^{bc}	20 (37.7)^{b}	11 (21.2)^{a}	25 (51.0)^{abc}	39 (76.5)^{c}	18 (42.9)^{abd}	146 (48.7)
Ascaris spp.	0	8 (15.1)^{a}	0	0	0	4 (9.5)^{a}	12 (4.0)
Trichostrongylus spp.	11 (20.8)^{a}	7 (13.2)^{a}	0	0	10 (19.6)^{a}	3 (7.1)^{a}	31 (10.3)
Trichuris sp.	0	0	0	4 (8.2)^{a}	6 (11.8)^{a}	0	10 (3.3)
Strongyloides spp.	8 (15.1)^{ab}	4 (7.6)^{a}	12 (23.1)^{b}	8 (16.3)^{ab}	0	0	32 (10.7)
Spirocerca spp.	0	0	9 (17.3)^{b}	0	0	0	9 (3.0)
Paraspidodera sp.	23 (43.4)^{b}	15 (28.3)^{ab}	10 (19.2)^{a}	19 (38.8)^{ab}	10 (19.6)^{a}	16 (38.1)^{ab}	93 (31.0)

Means followed by different letters within the same row differ significantly ($P<0.05$)

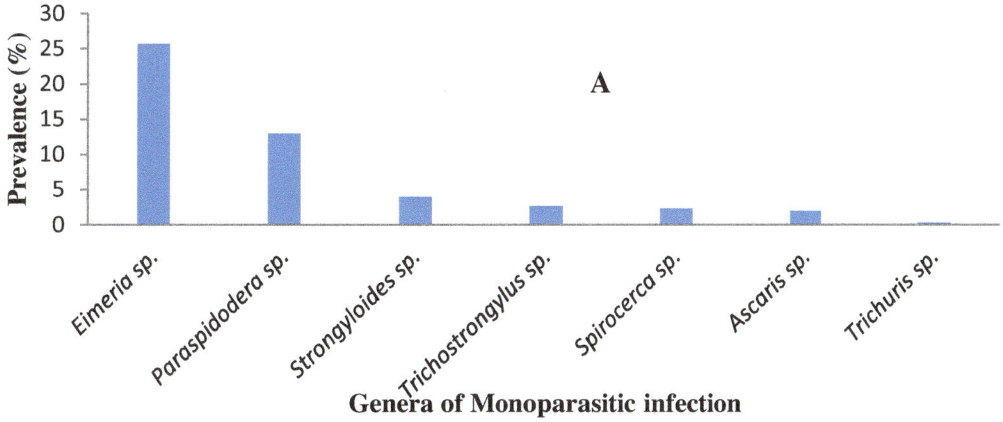

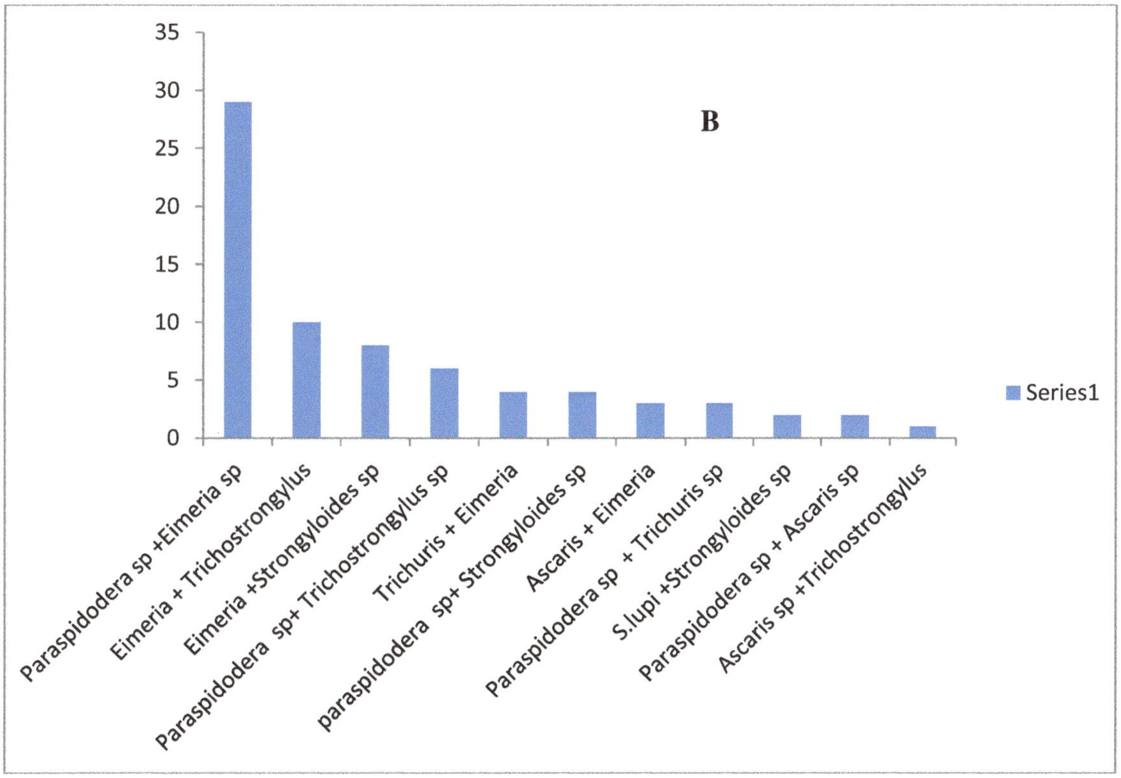

Fig. 1. Parasitic infection and associations

Table 4. Mean intensity of gastro-intestinal parasites (EPG/OPG) with respect to age group

Parasite	Age group (weeks)					Total
	1 - 4	5 - 8	9 - 12	13 - 16	> 16	
Eimeria	219.±280[a]	385±271[a]	481±602[a]	459±600[a]	505±438[a]	430±494
Ascaris	75±35[a]	75±35[a]	250±70[a]	100±0[a]	270±130[b]	255±236
Trichostrongylus	75±35[a]	114±85[a]	183±211[a]	195±165[a]	300±353[a]	187±126
Trichuris	-	66±28[a]	100±0[a]	100±70[a]	-	148±120
Strongyloides	-	144±123[a]	127±112[a]	175±146[a]	150±118[a]	175±186
Spirocerca	-	-	50±0[a]	75±35[ab]	131±55[b]	85±42
Paraspidodera	155±89[a]	121±75[a]	275±271[a]	321±284[a]	458±242[b]	255±236

Means followed by different letters within the same row differ significantly (P<0.05)

3.2.2 Mean intensity of infection in relation to sex

Only *Paraspidera* sp. and *Ascaris* showed significantly higher intensity in females than in males. Generally, most of the gastrointestinal parasites showed no difference in infection intensity between sexes (Table 5).

Table 5. Mean intensity of infection (EPG/OPG) of gastrointestinal parasites in relation to sex

Parasite	Gender	
	Female	Male
Eimeria	452±601	395±505
Ascaris	210±126[b]	75±35[a]
Trichostrongylus	197±209	150±158
Trichuris sp.	81±25	125±106
Strongyloides	156±123	140±121
Spirocerca	50±0	120±48
Paraspidodera	301±265[b]	185±164[a]

Means followed by different letters within the same row differ significantly (P<0.05)

3.2.3 Mean intensity of infection of gastrointestinal parasites in relation to locality

Intensity of infection of gastrointestinal parasites varied with locality. The highest intensity was recorded for the genus *Eimeria* (692.31±761.278OPG) in Penka – Michel, followed by Nkong-ni (434.62±334.60 OPG), then Dschang (427.27±356.87 OPG). The genus *Paraspidodera* equally registered high intensities in all the six localities, with its highest intensity in Nkong-ni (363.16±231.45 EPG) and the lowest in Fongo-Tongo (165.0±88.35 EPG). The genus *Trichostrongylus* registered the highest intensity in Santchou (363.16±231.446 EPG) as compared to the other localities. Statistically, it was noticed that the intensity of *Eimeria* spp, *Paraspidodera* sp and *Trichostrongylus* spp was significantly different with locality (P<0.05).

4. DISCUSSION

Based on faecal examination, six genera of nematodes (*Paraspidodera* sp, *Trichuris* spp, *Trichostrongylus* spp., *Strongyloides* spp., *Ascaris* spp., and *Spirocerca* spp.), were identified, giving a prevalence of 63.3% for helminths, and one genus of Protozoa (*Eimeria*) was equally identified, giving an overall prevalence of 79.0% for gastrointestinal parasites. This high prevalence could be attributed to the fact that in Menoua Division, cavies are reared traditionally, with very little intervention and follow up by the stakeholders. In some cases, cavies are co-reared with other domestic animals, feeding and sleeping in the same area. A sensitive method of corprologic analysis was used (double centrifugation flotation technique) instead of simple flotation technique. [12] revealed that centrifugal flotation is more sensitive than simple flotation because it magnifies gravitational forces, thereby accelerating the downward movement of more dense debris and the upward movement of less dense parasite forms. Furthermore, the genera *Eimeria*, *Trichuris*, *Ascaris*, *Trichostrongylus* and *Strongyloides*, are equally common parasites of domestic animals, and have been recorded in other parasitological studies involving *Cavia sp.* such as that carried out by [15,16] and other ecological associates of cavies such as goats, sheep and cane rat both in Cameroon and Africa at large [2,3,5,17,18,19]. The climatic condition of Menoua is described as humid tropical climate [8] is highly suitable for survival and transmission of infective stages of these parasites.

There was a significant difference ($p < 0.05$) in prevalence and intensity for *Paraspidodera* sp and *Spirocerca* sp in relation to age, being significantly higher in >16 weeks animals. These results corroborate the findings of [3] and [20] who also registered higher intensity of infection with nematode parasites in older animals than the young. However, this observation is not in agreement with the fact that immunity against infections increases with age due to acquired immunity [20]. It was observed that adult animals have a higher potential of scavenging within their kitchen confinement than younger animals. They therefore are more likely to ingest feed contaminated with eggs of nematodes. Furthermore, of the 87 adult cavies surveyed (that is >13 weeks cavies), 72 (82.8%) were females and 15 (28.2%) were males. Literature reveals that reproduction in cavies begin as early as two months, and these animals have a post-partum oestrus of three and a half hours [21]. Therefore, females older than 8 weeks were almost always pregnant and lactating. This placed a stress on the immune system of these pregnant/lactating cavies making older females more susceptible to infection.

The observation of *Spirocerca* eggs in the faeces of domestic cavies is a rare finding. However, this parasite registered low prevalence 9 (17.31) from 3 different farms and intensity (85±42)

Nevertheless, [22,23] reported that rodents are paratenic hosts of this parasite. Although it is uncommon in domestic cavies, it is possible that the infected cavies became contaminated faeco-orally by hunting dogs which were reared in the same household with the cavies in Fongo Tongo. Therefore, the presence of this 'rare' parasite in cavies can be attributed to cross infection from dogs.

As concerns sex of cavies, *Ascaris* spp showed significant difference in prevalence and intensity being higher in females while for *Paraspidodera* sp, difference was significant only for intensity, with females shedding more eggs than males. [24] indicated that the establishment and reproductive potential of these parasites depend on host immunity. Since these animals were reared such that males and females were found in the same house, they had equal exposure to the parasites. The fact that females passed out significantly more eggs than males could be attributed to the triple stress of pregnancy, lactation and care of the young simultaneously. This led to a reduced ability of these females to build up immunity against the parasites.

Regarding locality differences, prevalence was significantly higher for *Eimeria* spp, *Strongyloides* spp and *Paraspidodera* sp, while difference in intensity was significant for *Eimeria* spp, *Paraspidodera* sp and *Trichostrongylus* spp. This could be attributed to the fact that the traditional cavy management system slightly differs in the method of cavy housing within the different localities. It was noted that in those localities where cavies were confined (Dschang, Penka-Michel and Nkong-Ni), the prevalence and intensity of *Eimeria* spp and *Paraspidodera* sp were significantly higher, and lowest where these animals did not have a specific bedding and were allowed freely to scavenge on the floor (Fongo-Tongo and Fokoue). Confining these animals increased the likelihood of transmission of parasites faeco-orally since food becomes easily contaminated with faeces from infected animals. This is in alignment with the reports of [20,25] who both registered 100% prevalence for *Eimeria* sp and *Paraspidodera* sp respectively in confined cavies. Infection of cavies with these two parasites is very common and can reach epidemic proportions [14,24]. Again, these parasites have short life cycle with environmentally resistant eggs or oocysts. Thus confining cavies in a small kitchen corner, especially exacerbated by irregular change of bedding provides the microclimate for easy survival and transmission of extra-host stages. On the other hand, *Strongyloides* spp showed significantly higher prevalence in those localities with free range housing system, where animals are housed in association with other domestic animals like chicken, dogs, ducks and in two cases cane rats. This parasite was completely absent where animals were confined and inter-specific animal housing avoided. Rearing cavies on earth floors, together with other species of domestic animals encourages disease transmission from infected to uninfected animals or from the soil through the skin to achieve transmission to a new host. This is because *Strongyloides* spp is a geohelminth [22] and is easily transmitted through the soil to uninfected animals. Furthermore, Menoua Division has two agro- ecological zones with slightly different temperature and humidity ranges due to a variation in altitude [8]. Temperature is inversely proportional to altitude, and it was noted that higher prevalence of infection with gastrointestinal parasites was in the low land districts (600 m to 1200 m above sea level) while high altitude area like Fongo-Tongo (2000 m above sea level) registered lowest prevalence for these parasites. Again, cavies are frequently exchanged between farmers within the same locality (sub division) than between localities. These exchanges are done to minimize inbreeding (cross between parents and offspring within the same farm). Thus, parasites from one farm are easily transmitted to cavies in another farm within localities than between localities.

It was equally noticed during this study that no gastrointestinal parasite was isolated from cavies reared in raised cages in a household in Dschang. Although only 11 cavies were reared in this manner; none of them was positive for any of the parasites recorded. This is attributed to improved housing since these animals were protected from geohelminths in the soil or from cross infection by other domestic animals. Bedding change for these animals was done every two days, thus minimizing infection and re-infection faeco-orally.

As concerns gastro-intestinal parasitic association, it was observed that the most prevalent gastrointestinal parasitic association was between *Eimeria* spp. and *Paraspidodera* sp. This is because both parasites have similar lifecycle, being transmitted by infective eggs or sporulated oocyst and both have short duration of 2-5 days for eggs or oocycts to become

Table 6. Mean intensity of infection (EPG/OPG) of gastrointestinal parasites in relation to locality

Parasite	Localities					
	Dschang	Fokoue	Fongo Tongo	Nkong-Ni	Penka Michel	Santchou
Eimeria sp.	427±356[b]	155±95[a]	168±128[a]	434±334[b]	692±761[c]	330±285[ab]
Ascaris spp.	-	218±125[b]	-	-	-	125±119[ab]
Trichostrongylus spp.	77±46[a]	307±297[b]	-	-	150±70[ab]	316±275[b]
Trichuris spp.	-	-	-	75±28[a]	100±54[a]	-
Strongyloides spp.	162±130[ab]	125±119[ab]	195±138[ab]	75±26[a]	-	-
Spirocerca spp.	-	-	85±42[a]	-	-	-
Paraspidodera sp.	250±294[ab]	220±168[ab]	165±88[a]	363±231[b]	220±191[ab]	246±277[ab]

Means followed by different letters within the same row differ significantly (P<0.05)

infective. Again, [26] reported that coccidia are intracellular microparasites, which replicate within the epithelial cells of the intestinal mucosa, often resulting in physical damage of the mucosal immune system, while *Paraspidodera* sp is extracellular. Intracellular and extracellular parasites invoke opposite and cross-regulating immune responses. As a result, hosts may have difficulty simultaneously mounting a strong response to co-infection by intracellular and extracellular parasites, leading to increased disease severity in co-infected animals.

5. CONCLUSION

This present study indicated that cavies in Menoua Division are infested with helminthic parasites and *Eimeria* oocyts. The overall high prevalence recorded in this study clearly indicates the need for strategic parasite control in cavies with particular attention focused on adult to old aged cavies which showed significantly higher prevalence and parasitic load for most of the gastrointestinal parasites. Thus, the services of Veterinarians should be employed to ensure appropriate treatment with specific anthelminthics in order to reduce worm loads of the parasitized animals. This will help to improve cavy productivity and enhance smallholder livelihood in this area. The results further underscore the importance of proper sanitation and improved housing of these animals so as to prevent soil and faeco-oral transmission of gastrointestinal parasites.

ACKNOWLEDGEMENT

Great appreciation goes the cavy farmers in Menoua Division, for kindly providing animals for the present study. Special thanks to the Laboratory for Animal Biology and Applied Ecology of the University of Dschang where technical procedures were carried out, Bamou Roland for assisting in the Statistical analysis.

COMPETING INTERESTS

Authors have declared that no competing interests exist.

REFERENCES

1. Niba AT, Meutchieye F, Fon D, Laisin AG, Taboh H, Njakoi H, Bela-Tomo A, Djikeng A, Manjeli Y. Current situation of cavy production in Cameroon: Challenges and Opportunities. Livestock Research for Rural Development. 2012;24(11):241-249.
2. Ndamukong KNJ. Strongyle infestations of sheep and goats at Mankon station Recherches Zootechniques, Mankon Station, Bamenda, Cameroon. Journal of Veterinary. Parasitology. 1985;1(4): 95-101.
3. Ntonifor HN, Shei SJ, Ndaleh NW, Mbunkur GN. Epidemiological studies of gastrointestinal parasitic infections in ruminants in Jakiri, Bui Division, North West Region of Cameroon. Journal of Veterinary Medicine and Animal Health. 2013;5(12):344-352.
4. Food and Agricultural Organization. Distribution and impact of helminth diseases of livestock in developing

countries. FAO Animal Production and Health paper, 96. FAO, Rome; 1992.

Available:http://www.fao.org/docrep/004/t0584e/t0584e00.htm

5. Odoi A, Gathuma JM, Gachuiri C, Omore A. Risk factors of gastrointestinal nematode parasite infections in small ruminants kept in smallholder mixed farms in Kenya. Veterinary. Research Community. 2007;3(6):1746-1186.
6. Rosenio CS, Fe MC, Juliana QS. Prevalence of Ectoparasites and Endoparasites of Carabao in Region I. International Scientific Research Journal. 2013;5(3):94–114.
7. Fonteh AF, Niba AT, Kudi AC, Tchoumboue J, Awah-Ndukum J. Influence of weaning age on the growth performance and survival of weaned guinea pigs. Livestock research for rural Development. 2005;17:12.

   Available:http://www.cipav.org.co/lrrd/lrrd17/12/font17133.htm
8. Seino RA, Ghogomu RT, Manjeli Y, Dongmo TI, Chifon RN. A faunal survey of the Oedipodinae (Acrididae: Orthoptera) of the Menoua Division in Cameroon. International Journal of Fauna and Biological Studies. 2013;1(1):15-19.
9. Thrusfield M. Veterinary epidemiology, 3rd ed. Blackwell Science, Oxford, UK; 2005.
10. Anye NH, Manjeli Y, Ebangi AL. A method of live weight evaluation for local Cavies (*Cavia porcellus* L.) of the western highlands of Cameroon: Synthesis of linear body measurements. Livestock Research for Rural Development. 2010b;22(184): 142-149.
11. Available: www.wikicavy.org/wiki/guineapig Retrieved June 2013.
12. Baker DG. Flynn's parasites of laboratory animals. 2nd ed. American College of Laboratory Animal Medicine. Blackwel publishing. UK; 2007.
13. Thienpont D, Rochette FR, Vanparijs O. Diagnostic de Vermose par examen coprologique. Janssen Research Foundation, Beerse, Belgique; 1979. French.
14. Foreyt W. Veterinary parasitology: Reference manual. 5th edn. Blackwell Publishing, Lowa State University Press; 2001.
15. Gressler TL, Aleksandro SS, Marcos KS, Alexandre AT, Silvia GM. Gastrointestinal parasites of cavy (*Cavia aperea aperea*) in southern Brazil. Research in Veterinary Science. 2010;89:206–208.
16. Dittmar K. Arthropod and Helminth Parasites of the wild cavies, *Cavia aperea*, From the Andes and the cordillera in Peru, South America. Journal of Parasitology. 2002;88(2):409-411.
17. Mpoame M. Gastrointestinal helminths of the cane rat *Thryonomys swinderianus* in Cameroon. Tropical Animal Health. Health Production .1994;26:239-240.
18. Omonona AO. Gastrointestinal parasites of domesticated glasscutters (*Thryonomys swinderianus)* in South-western Nigeria. Journal of Human Ecology. 2011;36(2): 117-120.
19. Meaza M, Tewodros F, Basaznew B. Gastrointestinal Helminthes Parasites in Sheep: Prevalence and Associated Risk Factors, in and Around Gondar Town, Northwest Ethiopia. Advances in Biological Research. 2012;6(5):191-195.
20. Tadesse G, Hailu MS, Alemu, Haileleul N. Prevalence of gastrointestinal parasite of laboratory animals at Ethopian Health Nutrition Research Institute (EHNRI) Addis Ababa. Journal of Infectious diseases and Immunity. 2011;3(1):1-5.
21. Wagner EJ, Manning PJ. The biology of cavies. 1st ed. Academic Press, London; 1976.
22. Anderson RC. Nematode parasites of vertebrates: Their development and transmission. 2nd ed. CABI Publishing; 2000.
23. Cameron J. *Spirocerca lupi* animal diversity Web. 2003; Accessed: March 2014,

    Available:http://animaldiversity.ummz.umich.edu/accounts/Spirocerca_lupi/
24. Soulsby EJL. Helminths, arthropod and protozoa of domesticated animals. 7th ed. Balliere, Tindall and Cassel London; 1982.
25. Tanideh N, Sadjjadi SM, Mohammadzadeh T, Mehrabani D. Helminthic infections of laboratory animals in animal house of shiraz university of medical sciences and the potential risks of zoonotic infections for researchers. Iranian Red Crescent Medical Journal. 2010;12(2):151-157.

26. Gorsich EE, Ezenwa VO, Jolles AE. Nematode-coccidia parasite co-infections in African buffalo: Epidemiology and associations with host condition and pregnancy. International Journal of Parasitology: Parasites of Wildlife. 2014;3(2):124-34.

13

Perceptions, Husbandry and Disease Management Practices amongst Cattle Owners of Mwandi District, Zambia

Ngonda Saasa[1*], Joyce Siwila[2], Ethel M'kandawire[1], Walter Muleya[3], Milton Mwenda[4] and King Shimumbo Nalubamba[2]

[1]*Department of Disease Control, School of Veterinary Medicine, University of Zambia, P.O.Box 32379, Lusaka 10101, Zambia.*
[2]*Department of Clinical Studies, School of Veterinary Medicine, University of Zambia, P.O.Box 32379, Lusaka 10101, Zambia.*
[3]*Department of Biomedical Sciences, University of Zambia, School of Veterinary Medicine, P.O.Box 32379, Lusaka 10101, Zambia.*
[4]*Department of Veterinary Services, Ministry of Agriculture and Livestock, Mwandi 10101, Zambia.*

Authors' contributions

Authors NS, JS and KSN designed the study, drafted the questionnaire and drafted and proof read the manuscript. Authors EM and WM participated in the design of the questionnaire and in drafting the manuscript. Author MM administered the questionnaire. All authors read and approved the final manuscript.

Editor(s):
(1) Mo Salman, Professor of Veterinary Epidemiology, Animal Population Health Institute, College of Veterinary Medicine and Biomedical Sciences, Colorado State University, USA.
Reviewers:
(1) Bergaoui Ridha, Department of Animal Production, National Institute of Agronomy, Tunisia.
(2) Anonymous, Makerere University, Uganda.
(3) Terry Whiting, Department of Agriculture, University Crescent, Canada.

ABSTRACT

A cross-sectional questionnaire survey was conducted in Mwandi district of Western province of Zambia. The objective was to obtain preliminary data on cattle owner perceptions and awareness of livestock husbandry practices and disease management due to lack of information on livestock activities in the area. The study showed that cattle were the main livestock kept (60%), mainly of

Corresponding author: Email: nsaasa@yahoo.co.uk, nsaasa@unza.zm

the indigenous local Barotse breed. The study further showed that public veterinary officials were the main source of advice on animal health (97%, 29/30 respondents vs. 3%, 1/30 for family members), and choice of drugs (100%) for the treatment of sick animals. All the cattle owners (30/30) reported having seen ticks on the cattle. Foot and mouth disease (27/30), contagious bovine pleuropneumonia (26/30) and anaplasmosis (22/30) were reported to be the most important cattle diseases. Surprisingly, none of the cattle owners responded to questions on theileriosis (corridor disease); nor did they report presence of trypanosomosis despite all of them (30/30) using antitrypanocides. The study further showed that oxytetracycline (30/30), Diminazene aceturate (Berenil®) (29/30), penicillin (29/30), and albendazole (28/30) were the most commonly used drugs. In addition to these drugs, local plants/herbs were also used. Inadequate water, stock theft and inaccessible veterinary services were among the challenges. Although farmers were familiar with many factors affecting their cattle, there was apparent lack of agreement between what the farmers knew and their management practices.

Keywords: Perceptions; livestock diseases; Mwandi district; Zambia.

1. INTRODUCTION

Livestock plays an essential role in the economy of developing countries in Africa [1,2]. The livestock sub-sector accounts for 6% of the Zambian economy [3]. In addition, majority of the rural population depends at least in part on livestock as a source of protein in form of milk and meat, income and for other crop-related activities [4,5]. The majority of the cattle are kept under the traditional system as mixed crop-animal systems with subsistence utilization of outputs [6].

In Zambia, livestock diseases account for serious losses through cattle morbidity and mortality. The diseases reported in Mwandi include contagious bovine pleuropneumonia (CBPP) [7], foot and mouth disease (FMD) [8], lumpy skin disease (LSD) and some soil-borne diseases such as anthrax [9] as well as clostridia myositis (Black quarter, BQ) [10]. In addition, Trypanosomosis has also been reported in the area [11]. These developments show that veterinary diseases are of paramount importance in Zambia, particularly in Mwandi district.

Mwandi district lies on the border with Southern province where there have been outbreaks of theileriosis (Corridor disease) due to *Theileria parva lawrencei* transmitted by the hard tick *Rhipicephalus appendiculatus* [12-14]. Wild animals such as the Buffalo *(Syncerus caffer)* have been identified as important natural reservoirs and a source infection for naive cattle [15]. In view of the increasing disease burden and probably their associated vectors in the region [16], coupled with the limited veterinary extension service delivery to livestock farmers, there was need to assess community knowledge and perceptions on theileriosis since the disease has received a lot of medical veterinary attention of recent [17]. As information on husbandry practices and disease management in Mwandi district was lacking, this study sought to obtain baseline data that would facilitate further in-depth investigations.

2. MATERIALS AND METHODS

2.1 Study Area

Mwandi district lies between Sesheke and Kazungula districts in the South-Eastern part of the Western province of Zambia (Fig. 1A). The district has a human population of 23,201 [18], with an estimated 35,000 cattle population. The land is generally not suitable for maize, the staple crop, because of the Kalahari sands, Mopane woodlands with a hot, semi-arid climate. The local community relies significantly on livestock as a source of income and draught power. The vast area of wetland along the Zambezi river flood plains and river banks is important for cattle grazing. Rainfall is generally low and variable, usually inconsistent, with occasional drought spells that affect grazing pastures. As such, cattle owners practice transhumance with their livestock between the higher lands and the low lying flood plains without feed supplementation. During the rainy season (November to April) they move with their animals to the upper lands and return to the low lands during the dry season (May to October).

2.2 Study Design and Data Collection

This cross-sectional study, using a questionnaire, was conducted between July and November

2013. The cattle owners were conveniently included in the study using the available local veterinary office data and snowballing technique due to mobility of cattle owners during the dry period of the year. The survey targeted cattle owners located on the eastern boundary with Southern province bound by the kasaya river (Fig. 1B). A total of thirty (30) cattle owners were identified and interviewed at their residences/ villages. A village/herd comprised cattle belonging to one or more individuals. The interviews were conducted by the local veterinary official who was able to administer the questionnaire in the local language.

The questionnaire was divided into seven sections including: location and interview details; household information; livestock inventory; cattle management; animal health and disease management; ticks and tick-borne disease management, and theileriosis. The range of questions included; the species and number of animals owned, the reasons for keeping cattle, cattle breeds, duration of keeping cattle. Husbandry data collected included information such as who took care of healthy and sick animals, challenges, disease management and preventive practices, extension services, veterinary drug sources, and treatment options. In addition, the final section dealt with knowledge and any experiences with theileriosis.

2.3 Data Analysis

Data was entered into a Microsoft Excel® Spreadsheet for some descriptive analyses. Graphing was performed in R statistical software package (R Core Team, 2014, Vienna Austria). Data are presented as absolute figures or proportions.

3. RESULTS AND DISCUSSION

3.1 Livestock Inventory and Uses

The cattle owners that were included in the study represented 7% (2,338/35,000) of the total cattle population in Mwandi district. As shown in Fig. 2, cattle constituted the highest proportion (60%, 2,338/3,893) of the livestock units followed by chickens, goats, pigs and donkeys. None of the farmers owned sheep. The average cattle herd size per village was 77.9 heads (range 13-250).

The majority (63.3%, 19/30) of the farmers had been keeping cattle for over 50 years, thirty percent (9/30) for at least 10 years and the remaining seven percent (2/30) for less than 10 years. Such relatively long periods probably since settling in the area in the mid seventeenth century after the great migration from the south [19] would suggest that cattle rearing are an important part of the Lozi people of Mwandi district. Rivers and lakes were the main sources of drinking water for the cattle (Data not shown). Fifty percent of the respondents only had the local Barotse breed of cattle while some respondents had introduced some exotic breeds.

Unlike in other parts of the country such as Southern province where sheep are present [20], no one owned any sheep. There was an indication however, of an increase in the number of small ruminants (goats), pigs and donkeys in the district. Two villages (7%) had donkeys for draught power probably due to their resilience to infectious diseases compared with cattle. All the respondents had similar reasons for keeping cattle such as milk source, income and draught power for household chores and tilling the land for mixed crop and livestock production activities. Two respondents (7%) mentioned traditional ceremonies and lobola (dowry payment for marriage).

3.2 Cattle Management Practices

All the respondents used communal grazing lands for the cattle all year round. Their animals spent nights in the kraals during the farming season, except when grazing on the flood plains (during the dry season) when cattle movement was less restricted. It was noted, from the responses, that most (87%, 26/30) of the herding activities were performed by the husband or the male head of the family with the help of the children. Only two villages (7%) reported hiring a herdsman (Table 1). This could indicate the lack of sufficient income to hire a herdsman. The husband was also the main person responsible for taking care of sick animals. Treatment, i.e. choice and actual administration of drugs or herbs to the sick animals was mainly performed by either a family member or public veterinary official (Table 1). From these findings, it can be speculated that having such a large proportion of family members making decisions on disease and treatment options could potentially compromise the efficacy of treatments and contribute to the development of drug resistance as dosages may not be adhered to.

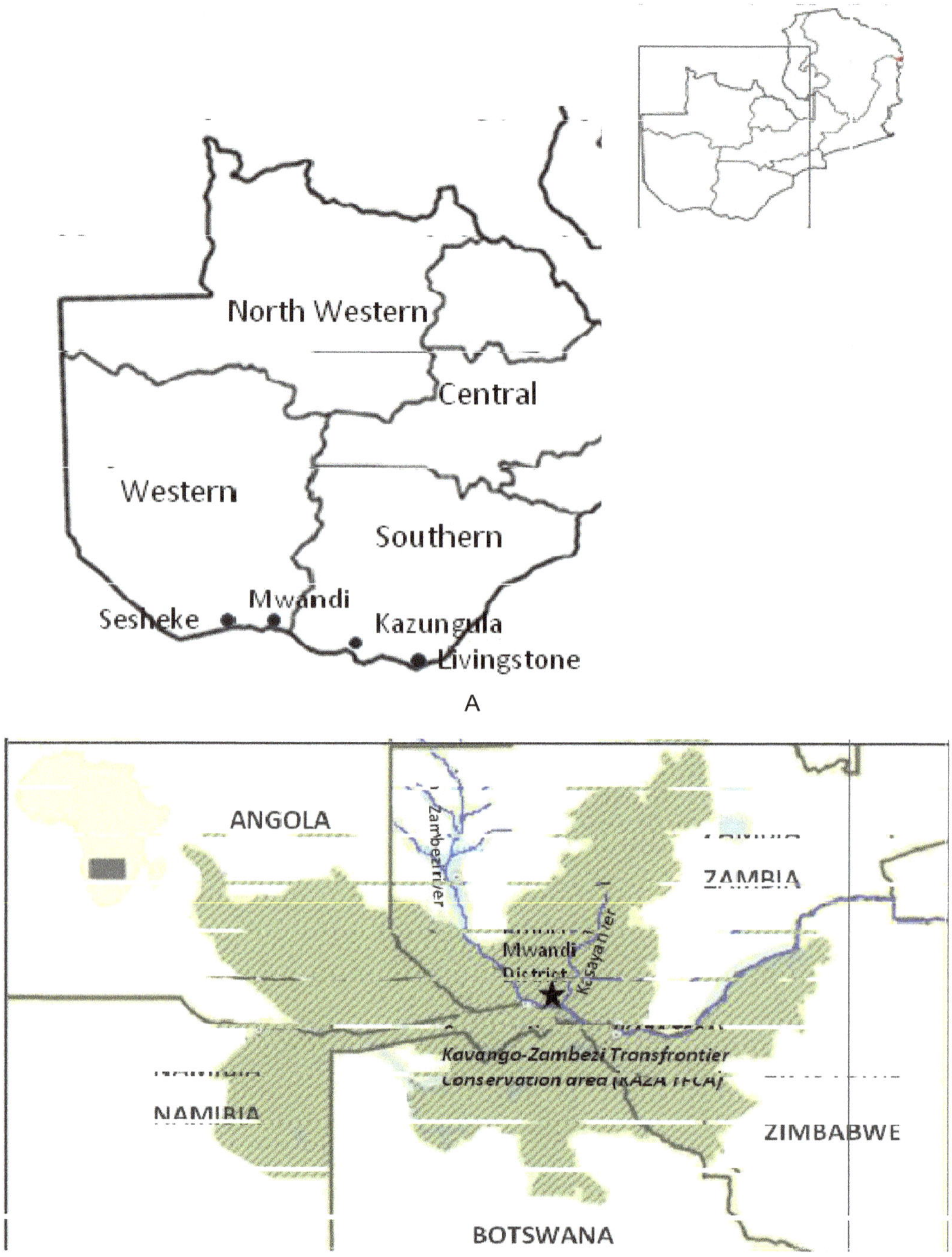

A

B

Fig. 1. A) Map of Zambia showing the location of Mwandi district and adjacent Sesheke district in Western province; Kazungula and Livingstone districts (Southern Province), Northwestern and Central provinces. Full map of Zambia (top right insert). B) Map showing the location of Mwandi district, Zambezi and Kasaya river (boundary of Western and Southern province) lying within the Kavango Zambezi Transfrontier conservation Area (KAZA TFCA) stretching across Angola, Zambia, Zimbabwe Botswana and Namibia. Adapted from www.peaceparks.org

Table 1. Individuals responsible for milking and herding, caring and treating sick animals (*n*=30)

A. Herding/milking	Number	Percentage
Husband/Children	26	87
Husband/Children/Herdsman	2	7
Children/Herdsman	1	3
Herdsman	1	3
B. Care for sick animals		
Husband	15	50
Husband/Children	15	50
C. Drug administration/treatment		
Family Member/Vet. Official	23	77
Veterinary official	7	23

3.3 Tick Control

All the farmers had seen ticks on their livestock at some point in time and the control measures used were spraying (87%, 26/30), tick grease (33%, 10/30), ivermectin (23%, 7/30), scissors (10%, 3/30), kerosine (7%, 2/30) and engine oil (7%, 2/30) (Table 2). Most of the respondents took action only when they saw the ticks on the animals (77%, 23/30) (Table 3). Others controlled ticks at intervals of; twice a week, three times a month, during the rainy season only, three times during the rainy season, once a year, twice a year, and three times a year (Table 3). It was evident that there were various approaches to controlling ticks; however, it was doubtful whether the measures were effective since the activities were not carried out regularly. The lack of consistency in the carrying out tick control activities was similar to reports from other studies on rural communities in the region [21].

It was interesting to note that although the majority of the farmers indicated that they practiced some form of external parasite control on their cattle, the efficacy of these measures in preventing disease transmission was doubtful. This was because the frequency of activities was based on seeing ticks on the animals [22]. We are not aware of any government veterinary department-driven formal policy on dipping of animals in Mwandi district although the farmers reported that they were encouraged to do so. Although the use of engine oil to control ticks has been found to have some acceptable effects [23], it may not be effective, just like the use of scissors to cut off attached ticks from an animal. The use of these practices to control ticks could be attributed to the non-availability of professional advice on external parasite and tick control or owners may not be willing to spend considerable amounts of money on modern conventional acaricides.

3.4 Livestock Diseases

The cattle diseases were given as proportion of the respondents who perceived the diseases as important. Nearly all the farmers thought that FMD (90%, 27/30) was important followed by CBPP (87%, 26/30) and anaplasmosis (73%, 22/30) as shown in Fig. 3. Conditions such as "soil-licking" (pica) were considered a disease and not a trace mineral deficiency. Tick-borne diseases that farmers were aware of included anaplasmosis, babesiosis, and heart water. No farmer reported having knowledge about corridor disease or had previously experienced the disease in their herds, despite the reported outbreaks in the neighbouring Southern province (Fig. 1A). This would probably have been due to farmers' limited knowledge on the clinical presentation of the disease thereby confusing it with other conditions.

3.5 Livestock Diseases and Wildlife

More than half (53%, 16/30) of cattle owners reported their cattle to have had contact with wild animals. Recently, wildlife were introduced in the Simalaha community conservancy by the Peace parks foundation through the Kavango Zambezi Transfrontier Conservation Area (KAZA TFCA) linking protected areas and cultures of Angola, Namibia, Botswana, Zambia and Zimbabwe (Fig. 1B). The study coincided with this re-introduction of the wildlife, which could explain the reported livestock contact with wildlife by the farmers. After this introduction of wildlife in the area, follow-up studies on disease patterns would be necessary as increased contact with wild animals has been reported to predispose domestic livestock to infections [20,24,25].

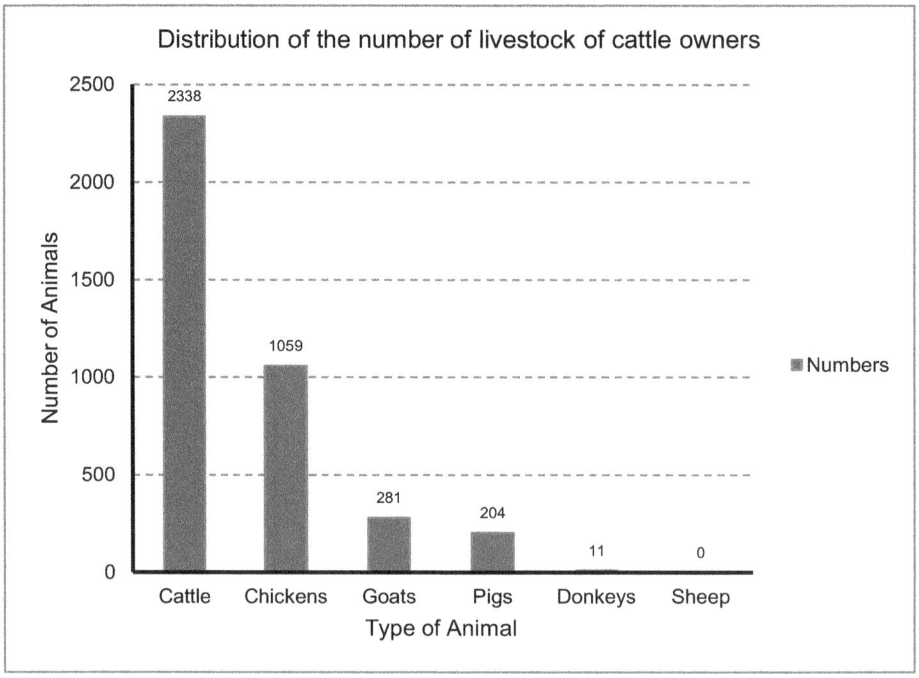

Fig. 2. Number of various animals kept by livestock owners (*n*=30) in Mwandi district

Table 2. Various methods of tick control and the proportion (%) of cattle owners using them

Dipping	Spraying	Tick grease	Ivermectin	Scissors	Kerosine	Engine oil
0/30 (0%)	26/30 (87%)	10/30(33%)	7/30 (23%)	10/30 (3%)	2/30 (3%)	2/30 (3%)

Table 3. Frequency of tick control in cattle and the proportion of respondents for each

When ticks are seen	Twice a week	Twice a year	3 times a year	3 times in rain season	3 times a month	Once a year	Rainy season
23/30 (100%)	1/30(3.3%)	1/30(3.3%)	1/30(3.3%)	1/30(3.3%)	1/30(3.3%)	1/30(3.3%)	1/30(3.3%)

Table 4. Drugs used for treatment of animal conditions and the proportion of cattle owner using them

Oxytetracycline	Penicillin	Diminazene aceturate (Berenil®)	Albendazole	Ivermectin	Isometamidium chloride (Samorin®)	Other Dewormers
30/30 (100%)	29/30 (97%)	29/30 (97%)	28/30(93%)	13/30(43%)	3/30 (10%)	1/30(3%)

3.6 Antibiotics and Other Medicines

As shown in Table 4, nearly all the cattle owners reported the use of oxytetracycline (100%, 30/30), Diminazene aceturate (Berenil®) (97%, 29/30), and penicillin (97%, 29/30), albendazole (93%, 28/30), ivermectin (43%, 13/30) isometamidium chloride (Samorin®) (10%, 3/30) and other dewormers (3%, 1/30). There were three main sources of these drugs; local veterinarian official, chemists and retail shops. The most common source of the drugs was the local veterinary official and shops (87%, 26/30), the chemists accounted for the remaining 13% (4/30). From these findings, it was evident that the farmers used a limited variety of drugs for treatment of sick animals. Some drugs could have been used inappropriately. For instance, despite the use of antitrypanocides, there was no mention by the livestock owners of the presence of trypanosomosis among the important diseases. However, the study did not establish which drug was used to treat which specific diseases. A number of plants/herbs were also used for treatment various conditions (Table 5). This finding was not surprising considering the majority of cattle owners had been keeping cattle for a relatively long time.

3.7 Challenges and Level of Importance

The respondents were asked to rank the challenges (cattle diseases, inadequate water sources, poor grazing pastures, infertility, stock theft and inaccessibility to veterinary services) in their order of importance on a scale of 1-3 (rank 3 as more important and 1 as least) (Table 6). Availability of water and stock theft were among the most frequent responses regarding the challenges the respondents faced (Table 6). This suggests that to the local people availability of water and stock thefts should not be overlooked when designing mitigation measures that would assist farmers in the management of the challenges. The frequent mention of these challenges by the farmers could be an indication that they significantly affected their livelihoods. These challenges faced in this area were common to the region [26].

Table 5. List of specific diseases of cattle in Mwandi district and some local plant herbs or substances named for treatment

Disease		Local plant herbs/substances		
Name	**Local name**	**Local name**	**Name**	**Proportion of users**
Anaplasmosis	Nyooko	Sikubabe	*Dioscorea quartiniana*	3/30
		Katima	*Cassia occidentalis*	5/30
		Sitati	*Aloe vera*	5/30
		Mupumangoma	*Albizia versicolor*	1/30
		Kapusi-pusi	NA	1/30
Anthrax	Lubete	Permanganate	Potassium permanganate	1/30
Diarrhea	Musululo	Mupani	*Colophospermum mopane*	4/30
		Mukulwani	*Borassus aethiopum*	1/30
Diarrhea (bloody)	Mali mwa Musululo	Muhonono	*Terminalia sericea*	4/30
Mange	Sikwekwe	Sikubabe	*Dioscorea quartiniana*	1/30
Senkobo	Sikwekwe	Lungwatanga	*Acanthosicyos naudinianus*	1/30
		Musikili	*Trichlia emetica*	1/30
		Simalembo	*Gnidia daphnifolia*	1/30
		Sifulukunde	NA	1/30
		Malumani	NA	1/30
Eye infection	Meeto	Kupa	Shell (ashes)	1/30
FMD	Taku ni Mulomo	NA	NA	NA
Hemorrhagic Septicemia	Kang'oto	NA	NA	NA
Liver flukes	Minyopi	NA	NA	NA
"Soil licking"/pica	Ku kuma mubu	NA	NA	NA
Swollen joints	Mutumbu	NA	NA	NA
Babesiosis	Mali mwa maswafu	NA	NA	NA

**NA- Not available*

Table 6. Challenges mentioned by the farmers and the frequency for each of the scores 1-3 (3 as more important and 1 as least)

Score	Cattle diseases	Inadequate water sources	Inadequate pastures	Fertility problems	Stock theft	Veterinary services
1	5	0	1	11	1	6
2	0	1	18	11	2	7
3	25	29	11	8	27	17
N	**30**	**30**	**30**	**30**	**30**	**30**

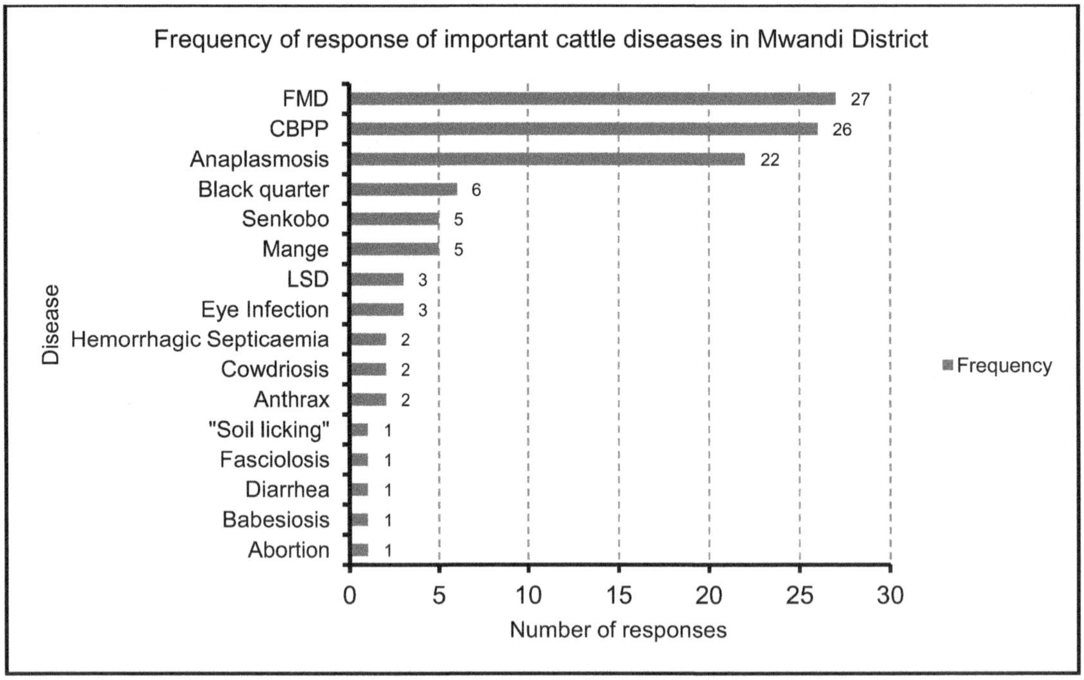

Fig. 3. The proportion of responses of important cattle diseases in Mwandi district (*n*=30). (FMD - Foot and mouth disease; CBPP - Contagious bovine pleuropneumonia; LSD - Lumpy skin disease)

4. CONCLUSION AND RECOMMENDATIONS

This survey has highlighted some key aspects on husbandry practices, farmer awareness of livestock disease management and prevention, antimicrobial use, as well as challenges that farmers faced. It was evident that a significant proportion of the activities of the livestock owners had shortcomings. Overall, cattle owners were aware of the various livestock husbandry practices and cattle diseases. However there were obvious disparities between what the cattle owners knew and the interventional activities they practiced. These preliminary findings showed the need for further in-depth investigation to address and establish the impact of the specific findings from this study.

COMPETING INTERESTS

Authors have declared that no competing interests exist.

REFERENCES

1. Food and Agriculture Organization (FAO). The role of ruminant livestock in food security in developing countries. Committee on World Food Security. Seventeenth Session, Rome, Italy; 1992.
2. Intergovernmental Authority on Development (IGAD). The Contribution of Livestock to the Economies of IGAD MEMBER States: Study Findings, Application of the Methodology in Ethiopia and Recommendations for Further Work. IGAD Livestock Policy Initiative. Addis Ababa. 2010;2-10. Available:http://www.fao.org/fileadmin/user upload/drought/docs/IGAD%20LPI%20WP%2002-10.pdf
3. Daka DE. Livestock sector in Zambia: Opportunities and limitations. In; Development and field evaluation of animal feed supplementation packages. Proceedings of the final review meeting of an IAEA Technical cooperation regional AFRA project. Joint FAO/IAEA Division of Nuclear Techniques in Food and Agriculture, Cairo, Egypt. 2000;141-143. Available:http://www.iaea.org/inis/collection/NCLCollectionStore/_Public/33/032/33032981.pdf
4. Perry BD, Mwanaumo B, Schels HF, Eicher E, Zaman MR. A study of health and productivity of traditionally managed

cattle in Zambia. Preventive Veterinary Medicine. 1984;2(5):633-653.

5. Lubungu M, Chapoto A, Tembo G. Smallholder Farmers participation in livestock markets: The case of Zambian farmers. Indaba Agricultural Policy Research Institute (IAPRI) Working Paper No 66. Lusaka; 2012. Available:http://www.fsg.afre.msu.edu/zambia/WP66.pdf
6. Lungu JCN. Mixed crop-livestock production systems of smallholder farmers in sub-humid and semi-arid areas of Zambia, In: Development and field evaluation of animal feed supplementation packages (Proceedings of the final review meeting of an IAEA Technical cooperation regional AFRA project. Joint FAO/IAEA Division of Nuclear Techniques in Food and Agriculture), Austria. 2002;135-140.
7. Muuka GM, Songolo N, Kabilika S, Fandamu P, Buonavoglia D, Scacchia M. Private sector involvement in the control of contagious bovine pleuropneumonia (CBPP) in the Kazungula district of Zambia benefitted the community and the control strategy. Tropical Animal Health and Production. 2013;45(3):699-703.
8. Hamoonga R, Stevenson MA, Allepuz A, Carpenter TE, Sinkala Y. Risk factors for foot and mouth disease in Zambia, 1981-2012. Preventive Veterinary Medicine. 2014;114(1):64-71.
9. Munang'andu HM, Banda F, Chikampa W, Mutoloki S, Syakalima M, Munyeme M. Risk analysis of an anthrax outbreak in cattle and humans of Sesheke district of Western Zambia. Acta Tropica. 2012; 124(2):162-5.
10. Hang'ombe BM, Isogai E, Lungu J, Mubita C, Nambota A, Kirisawa R, Kimura K, Isogai H. Detection and characterization of Clostridium species in soil of Zambia. Comparative Immunology, Microbiology and Infectious Diseases. 2000;23(4):277-84.
11. Corten JJ, ter Huurne AA, Moorhouse PD, de Rooij RC. Prevalence of trypanosomiasis in cattle in south-west Zambia. Tropical Animal Health and Production. 1988;20(2):78-84.
12. Nambota A, Samui K, Sugimoto C, Kakuta T, Onuma M. Theileriosis in Zambia: etiology, epidemiology and control measures. Japanese Journal of Veterinary Research. 1994;42(1):1-18.
13. Mulumba M, Speybroeck N, Billiouw M, Berkvens DL, Geysen DM, Brandt JR. Transmission of theileriosis in the traditional farming sector in the southern province of Zambia during 1995-1996. Tropical Animal Health and Production. 2000;32(5):303-14.
14. Makala LH, Mangani P, Fujisaki K, Nagasawa H. The current status of major tick borne diseases in Zambia. Veterinary Research. 2003;34(1):27-45.
15. Pascucci I, Capobianco Dondona A, Camma C, Marcacci M, Di Domenico M, Lelli R, et al. Survey of Ixodid ticks and two tick-borne pathogens in African buffaloes, *Syncerus caffer,* from the Caprivi Strip, Namibia. Journal of Zoological Wildlife Medicine. 2011;42(4):634-40.
16. D'Haese L, Penne K, Elyn R. Economics of theileriosis control in Zambia. Tropical Medicine and International Health. 1999; 4(9):49-57.
17. Makungu C, Mwacalimba KK. A quantitative risk assessment of bovine theileriosis entering Luapula Province from Central Province in Zambia via live cattle imports from traditional and commercial production sectors. Preventive Veterinary Medicine. 2014;116(1-2):63-74.
18. Central Statistical Office (CSO). Census of population and housing. 2012;43-46.
19. Mainga Mutumba Bulozi under the Luyana Kings. Political evolution and state formation in pre-colonial Zambia. 2nd Edition; London: Longman; 1973.
20. Muma JB, Samui KL, Siamudaala VM, Oloya J, Matop G, Omer MK, et al. Prevalence of antibodies to Brucella spp. and individual risk factors of infection in traditional cattle, goats and sheep reared in livestock-wildlife interface areas of Zambia. Tropical Animal Health and Production. 2006;38(3):195-206.
21. Hesterberg UW, Bagnall R, Perrett K, Horner R, Gummow B. A questionnaire survey of perceptions and preventive measures related to animal health amongst cattle owners of rural communities in KwaZulu-Natal, South Africa. Journal of the South African Veterinary Association. 2007;78(4):205-8.
22. Cook AJ. Communal farmers and tick control-a field study in Zimbabwe. Tropical Animal Health and Production. 1991; 23(3):161-6.
23. Dreyer K, Fourie LJ, Kok DJ. The efficacy of used engine oil against ticks on cattle.

Onderstepoort Journal of Veterinary Research. 1998;65(4):275-9.
24. Muma JB, Samui KL, Oloya J, Munyeme M, Skjerve E. Risk factors for brucellosis in indigenous cattle reared in livestock-wildlife interface areas of Zambia. Preventive Veterinary Medicine. 2007; 80(4):306-17.
25. Munang'andu HM, Siamudaala V, Matandiko W, Mulumba M, Nambota A, Munyeme M, Mutoloki S, Nonga H. Detection of *Theileria parva* antibodies in the African buffalo *(Syncerus caffer)* in the livestock-wildlife interface areas of Zambia. Veterinary Parasitology. 2009;166(1-2): 163-6.
26. Maass BL, Musale DK, Chiuri WL, Gassner A, Peters M. Challenges and opportunities for smallholder livestock production in post-conflict South Kivu, eastern DR Congo. Tropical Animal Health and Production. 2012;44(6):1221-32.

14

Effects of the Two-early-flowering-time Genes, *Ef1* and *m-Ef1*, Founded in Low Latitude were Reexamined in One of the Origins of Donor Parent under Fluctuating Field Environments as to Earliness Effect and Fitness

Youki Itoh[1*]

[1]*Laboratory of Wildlife Ecology, Department of Environmental Symbiotic Science, College of Agriculture, Food and Environmental Sciences, Rakuno Gakuen University, Japan.*

Author's contribution

The sole author designed, analyzed and interpreted and prepared the manuscript.

Editor(s):
(1) Abdel-Tawab H. Mossa, Environmental Toxicology Research Unit (ETRU), Pesticide Chemistry Department, National Research Centre, Egypt.
Reviewers:
(1) Rezzoug Waffa, Ibn Khaldoun University, Algeria.
(2) Klára Kosová, Crop Research Institute, Prague, Czech Republic.

ABSTRACT

Aims: *Early flowering 1* (*Ef1*) and *modifier of Early flowering 1* (*m-Ef1*) genes were firstly detected in the cultivars fitted to northern margins of rice cultivation. Hence, it has been inferred that the genes might be essential for adaptation to the regions. However, the genes have been identified in low latitudes and the effects on earliness and fitness in the origins have not been tested. To know genetic shift in response to climatic change, the present study was performed as to earliness effect and fitness.

Study Design: The six near-isogenic lines (NILs) associated with *Ef1* and *m-Ef1* genes and the recurrent parent were used. In each genotype, twelve plants were divided into two replicates in a paddy field. In each replicate, genotypes were placed at random.

Place and Duration of Study: The place of this study was National Agricultural Research Center in Sapporo (43 °N) in Japan. Duration of the study was summer season between year 2003 (cool

**Corresponding author: Email: yoshiemetkoneko@hotmail.co.jp*

summer) and 2004 (warm summer).
Methodology: In each plant, flowering time was recorded every day and days to flowering from germination to (DF) were calculated. Earliness effect of a plant of a NIL was defined as subtraction of DF of a NIL from that of the recurrent parent in same ID of same replicate. After harvest, maximum fertility of a panicle (F_{max}), minimum fertility of a panicle (F_{min}), total number of panicles with more than a fertile grain (Effective panicle number: EPN) and total number of panicles (TNP) were recorded in each plant. Fitness in a plant was conventionally estimated from the formulation, $1/2\ (F_{max} + F_{min})\ x\ (EPN\ /\ TNP)$.
Results: Compared with the previous study, which evaluated the effect of *Ef1* and *m-Ef1* genes in low latitude, effect of epistasis between *Ef1* and *m-Ef1* on earliness qualitatively changed between low and high latitudes. Furthermore, it was estimated that one of the epistasis might accelerate flowering time in response to low temperatures. At phenotype level, Individuals with earlier flowering time tended to show higher fitness. However, at genotype level, the tendency was broken in low-temperature year. As a result, correlation of fitness between two years showed no significance.
Conclusion: Epistasis-by-environment interaction might reveal significant role in earliness effects between latitude and between years. In addition, even in the case that earliness effects were preserved among genotypes between years, there is a possibility that fitness might not always be conserved under fluctuating field conditions in the northern limit of rice cultivation.

Keywords: Climatic change; epistasis; fitness; flowering time; genotype-by-environment interaction; rice cultivation; Sapporo.

1. INTRODUCTION

Farmers use phenotypes while plant breeders change genotypes. A genotype often changes the effect on phenotype in response to environments [1,2]. Hence, pursuit of alternations of genotype-phenotype relationship along the ecological context is one of the most important topics in agronomical study. For examining the relationship, flowering time is the preferable trait since flowering time is easily measured adaptive phenotype in higher plant [3,4]. In the most northern regions of rice cultivation, cultivars generally show an early flowering behavior. The two genes, *Ef1* (*Early flowering 1*) and *m-Ef1* (*modifier of Early flowering 1*), were firstly detected in two landraces grown in the northern areas of Japan and China. It has been reported that *Ef1* shortens the duration to floral initiation from germination without changing its photoperiod sensitivity, and *m-Ef1* intensifies the earliness due to *Ef1* in Taiwan (24°N) [5]. *Ef1* and *m-Ef1* showed an epistasis, which contributed to earliness in Taiwan, implying that the epistasis as well as an additive effect might play a significant role for adaptability to the northern margins of their distributions. However, the manner of functional relationship between *Ef1* and *m-Ef1* genes has not been tested in their origins. Then, the first question arises as follows: Can we suppose that the contribution manners of genes to earliness effect might be qualitatively preserved between different locations? This question is also associated with application of quantitative locus analyses, which have recently come into fashions, to plant-breeding program.

In the environment of northern margins of rice cultivation, the two features are remarkable as to temperature change. First, the temperature shows a large difference between years [6,7, see also Fig. 1] Then, we can examine how the differences of year environments change genotype-phenotype relationship in the region [2] (the second question). Second, during cultivation period, temperature drastically increases and decreases in a short span and the difference between maximum and minimum temperatures in the region are equivalent to that between different latitudes. Hence, adequate duration of flowering time is highly limited for obtaining a stable yield [8, see also Fig. 1]. It has been thought that flowering time is agronomically important in rice because this trait is strongly associated with the regional adaptability and fitness [9]. However, fitness is composed of various component traits, which interact each other and change their effects, depending on environmental conditions [2,10,11]. Therefore, to strictly demonstrate the thought, it is the best way that relationship between earliness effect and fitness is examined using flowering-time near-isogenic lines (NILs) in a northern-limit region of rice cultivation where the temperature largely changes in a cultivation period (the third question).

If earliness as phenotype significantly affects fitness, then the last question arises as follows: Can we expect that the flowering genotypes might stably contribute to fitness in the origin of donor parent? When flowering-time NILs are examined as to fitness in the field environments, this question can be easily resolved.

Here, objectives of the present study are answering to the four questions using flowering-time NILs in order to predict genetic shifts in response to climatic changes. That is, first, I examined whether or not contribution manner of *Ef1* and/or *m-Ef1* to flowering time in low latitude is same as that in high latitude in field environment. Second, I examined whether or not *Ef1* and/or *m-Ef1* respond to the difference of year environment in Sapporo. Third, I intended to quantify how degree earliness effect contributes to fitness at phenotype level in Sapporo. Last, I examined whether or not order of flowering-time genotype in the contribution to fitness is preserved under fluctuating field environments if flowering time as a phenotypic trait contributes to fitness across years in Sapporo.

2. MATERIALS AND METHODS

2.1 Genetic Materials

A total of seven strains (six NILs and the recurrent parent) were used in the present study (Table 1). NILs were made by successive backcrossing method and the backcross generation is 20 [5]. The recurrent parent, T65, was derived from Taichung 65 and is a practically photoperiod insensitive strain [5,12]. The NILs were conveniently divided into two groups, T65NILs A and T65NILs B. *Ef1*^{a} and/or *m-Ef1*^{a} involved in T65*E*^{a}, T65*m*^{a} and T65*E*^{a}*m*^{a} which belongs to T65NIL A were originated from Tatong-tailai which is fitted to northern China. While *Ef1*^{b} and/or *m-Ef1*^{b} involved in T65*E*^{b}, T65*m*^{b} and T65*E*^{b}*m*^{b} which belongs to T65NIL B were originated from Bozu 5 which is fitted to northern Japan (Hokkaido islands). Recently, *Ehd1* (*Early heading date-1*), which encodes B-type response regulator, has been proposed to a candidate for *Ef1* since amino acid substitution (Arg to Gly) in exon 4 of *Ehd1* promoted flowering in the genetic background of T65, and the gene was located on chromosome 10 as is *Ef1* [13]. Uwatoko et al. [12] confirmed the proposition based on DNA polymorphism, using T65*E*^{b} and the donor parent, Bozu 5. *m-Ef1* is located on chromosome 7 [14] and the dominant allele clearly showed photoperiod sensitivity when it was combined with *Ef1* although *m-Ef1* has not yet been characterized at molecular level.

2.2 Cultivation and Trait Evaluation

The way of cultivation was summarized in Table 2. Seeds were germinated in petri-dishes in the dark (30°C) in three days and each of the seedlings raised in a hole (1.5 cm diameter) with culture soil of seed-bed. The seedlings were grown in a greenhouse at Hokkaido Agricultural Research Center (HARC) in Sapporo (43°N) until transplanting time. Selected seedlings of each genotype were transplanted in an irrigated paddy field at HARC. In the paddy field, genotypes were placed at random in each of the two replicates in year 2003. In year 2004, genotypes were transplanted in the same area of the same paddy field and the arrangement of genotypes in each of the replicates was same as that of year 2003. It is noted that in year 2003 and 2004 there was cool and very warm summer, respctively (Fig. 1). In year 2003, modern cultivars highly fitted to Hokkaido islands were subjected to cool weather damage. On the other hands, in year 2004, T65 fitted to Taiwan showed a level of fertility. To avoid skewing the estimation of earliness effect on fitness, rice plants were sprayed with an insecticide when panicles emerged. From the same reason, at a month after transplanting, herbicide was used for avoiding competition against weeds.

In each plant, the first date of panicle emergence was observed everyday and the date was regarded as flowering time. Earliness effect of each plant of a NIL was defined as subtraction of flowering time of a NIL from that of the recurrent parent in same ID in same replicate. After harvest, maximum fertility of a panicle (F_{max}), minimum fertility of a panicle (F_{min}), total number of panicles with more than a fertile grain (Effective panicle number: EPN) and total number of panicles (TNP) were recorded in each plant. Fitness in a plant was conventionally estimated from the formulation, $1/2\ (F_{max} + F_{min}) \times (EPN\ /\ TNP)$ since fitness was mostly affected by fertility in the northern margin of rice cultivation [8]. If all the panicles show fertility at 100% level, fitness results in 1.00. If all the panicles do not have fertile grain, fitness results in 0.00. It is noteworthy that fitness does not always reflect yield. All the statistical tests in the present study were carried out using Stat View 5.0 [15].

Table 1. Near-isogenic lines (NIL) of Taichung 65 (T65) used in the present study

Strain	Gene involved[1]		BC[2]	Donor	Days to heading in year	
	Ef1	***m-Ef1***		**parent**	**2003**	**2004**
<T65NILs A>						
T65E^{a}	$Ef1^{a}$	+	20	Tatong-tailai	132.5±1.2	116.5±2.5
T65m^{a}	+	$m\text{-}Ef1^{a}$	20	Tatong-tailai	143.0±1.7	123.3±2.2
T65$E^{a}m^{a}$	$Ef1^{a}$	$m\text{-}Ef1^{a}$	20	Tatong-tailai	121.4±1.7	105.0±1.3
<T65NILs B>						
T65E^{b}	$Ef1^{b}$	+	20	Bozu 5	133.3±1.6	115.9±1.8
T65m^{b}	+	$m\text{-}Ef1^{b}$	20	Bozu 5	142.3±1.8	122.9±2.5
T65$E^{b}m^{b}$	$Ef1^{b}$	$m\text{-}Ef1^{b}$	20	Bozu 5	130.2±1.4	111.0±1.0
<Recurrent parent>						
T65[3]	+	+	-	-	146.1±1.8	126.6±2.8

[1] + indicates wild allele. [2] BC indicates backcross generation. [3] Tatong-tailai and Bozu 5 is a cultivar originated from northern China and Japan (Hokkaido islands), respectively. [4] T65 is a pure line of Taichung 65 (progeny of a self-pollinated single plant)

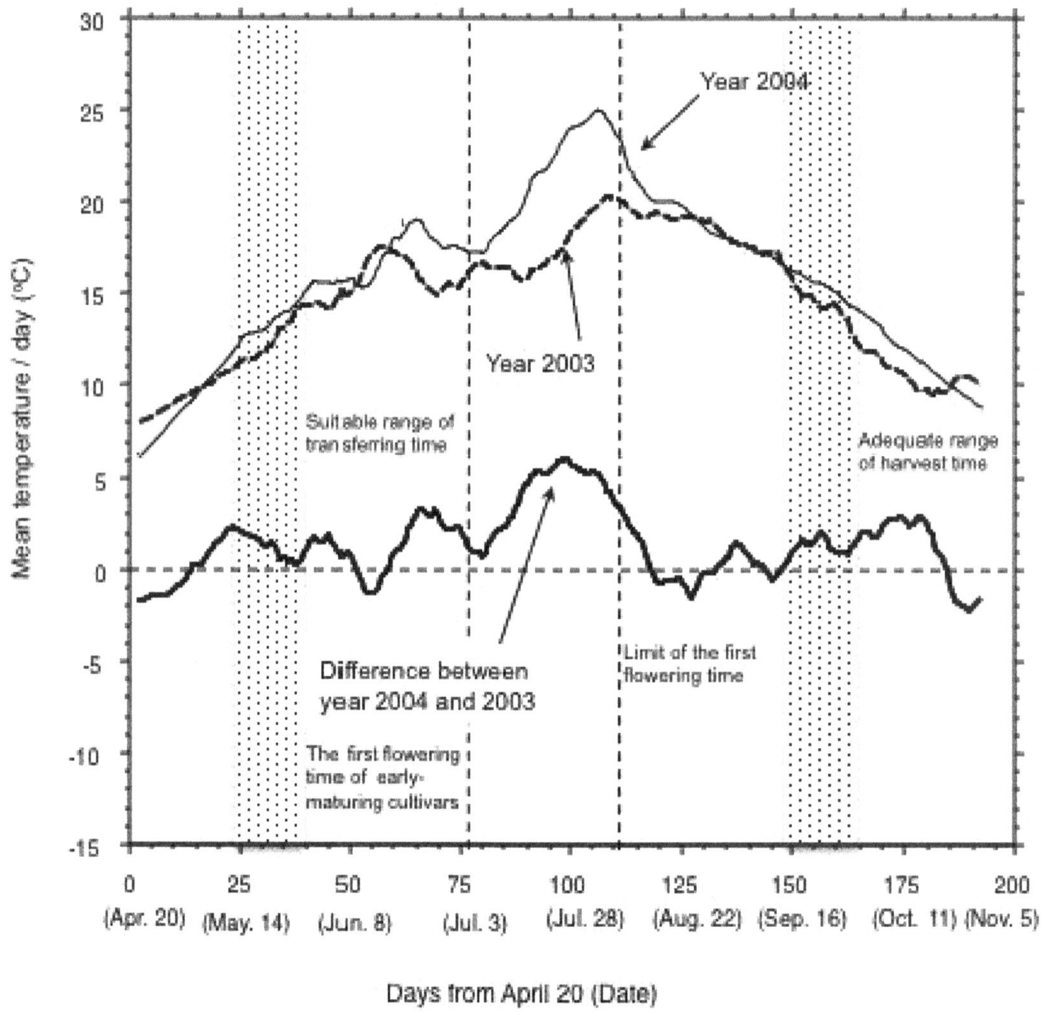

Fig. 1. Changes in daily mean air temperature from sowing to maturation of grain in year 2004 and 2003 and their difference in Sapporo (43°N)

Table 2. Cultivations and year environments in the present study

	Year 2003	Year 2004
Number of plant examined	**Six plants (/ a replicate) x two replicates / a strain**	
Date of germination	April 19	April 21
Date of sowing	April 21	April 24
Date of transfer	May 22	May 19
Density	35-cm x 12.5-cm	35-cm x 12.5-cm
Fertilizer	1kgN-1.2kgP-0.86kgK / a	1kgN-1.2kgP-0.86kgK / a
Limit period of cultivation	October 31	October 7
Year type (See Fig. 1)	Cool summer	Very warm summer

3. RESULTS AND DISCUSSION

3.1 Comparison of Functional Manner of *Ef1* and *m-Ef1* Genes between Low and High Latitudes

All the NILs significantly flowered earlier than the recurrent parent in Sapporo (Table 1; Fig. 2), NILs with *Ef1* genes (i.e. T65E^a and T65E^b) flowered earlier than NILs with *m-Ef1* genes (i.e. T65m^a and T65m^b) and NILs with both of the two genes (i.e. T65$E^a m^a$ and T65$E^b m^b$) flowered earlier than NILs with *Ef1* genes in Sapporo. The order of earliness in Sapporo is nearly consistent with that in Taichung (Fig. 2). However, there are three differences as to functional manner of *Ef1* and *m-Ef1* genes between Taichung and Sapporo. First, *m-Ef1* genes did not show earliness effects in the summer crop season in Taichung, while *m-Ef1* genes stably showed significant earliness effects across years in Sapporo. Considering small range of suitable flowering time in the northern-limit region of rice cultivation [16, see also Fig. 1], the result indicates that *m-Ef1* genes are essential for earliness in the field environments although the effects were small. Second, allelic difference between $Ef1^a$ and $Ef1^b$ was detected in Taichung while the difference was not detected in Sapporo. One may think that allelic difference is a genetic feature which is independent on the environment. It seems to me that such thought may occur due to lack of reaction norm concept (e.g., photoperiodic response curve). The fact that the allelic difference in a low latitude disappeared in a high latitude indicates that each of *Ef1* allele showed genotype-by-environment interaction since these genetic materials are NILs of *Ef1*. Such case is not novel phenomenon especially in regulatory genes [1]. For instance, *photosensitivity 1* (*Se1*) alleles showed different flowering times in a range of photoperiodic conditions but under a photoperiodic condition, the two alleles showed no difference in flowering time [17]. The third difference is associated with epistasis between *Ef1* and *m-Ef1*. Epistasis between $Ef1^a$ and $m\text{-}Ef1^a$ contributed to lateness while that between $Ef1^b$ and $m\text{-}Ef1^b$ contributed to earliness in Taichung [5]. On the other hands, epistasis was detected only between $Ef1^a$ and $m\text{-}Ef1^a$ in Sapporo (Table 3). In addition, epistasis between $Ef1^a$ and $m\text{-}Ef1^a$ contributed to earliness in Sapporo (Fig. 2) Tsai reported that the interaction between $m\text{-}Ef1^b$ and $m\text{-}Ef1^b$ (probably the first detected epistasis in rice) was emphasized, and it has been implied that the effect of epistasis on earliness might be essential for adaptation to the northern margin of rice distribution (or Hokkaido islands) [5]. In addition, Uwatoko et al. [14] also identified the significant effects of epistasis between $m\text{-}Ef1^b$ and $Ef1^b$ on the components of photoperiodic reaction norm in growth-cabinet experiment. However, the epistasis was not detected in both year 2003 and 2004 in the field environments of Sapporo (Table 3), suggesting importance of examination of genic contribution in their origins or regions where plant-breeding programs are performed. The interaction between $Ef1^b$ and $m\text{-}Ef1^b$ might occur under high-temperature condition since Uwatoko et al. adopted so high-temperature condition as the growth resume and the temperatures in Taiwan is higher than that in Sapporo.

Wright emphasized the importance of epistasis in phenotypic variation, which is the material of evolution but it is likely that relationship between epistasis and environments was not kept in his mind [18]. Recently, it has been reported that epistasis associated with fitness-related traits interacted environment in rice and *Arabidopsis* [14,19], suggesting that emergence of variation of epistasis might depend on environmental context. In this sense, epistasis will contribute to enlargement of phenotypic variation, which can easily induce a novel phenotype under a new environment. The present result directly indicates this concept.

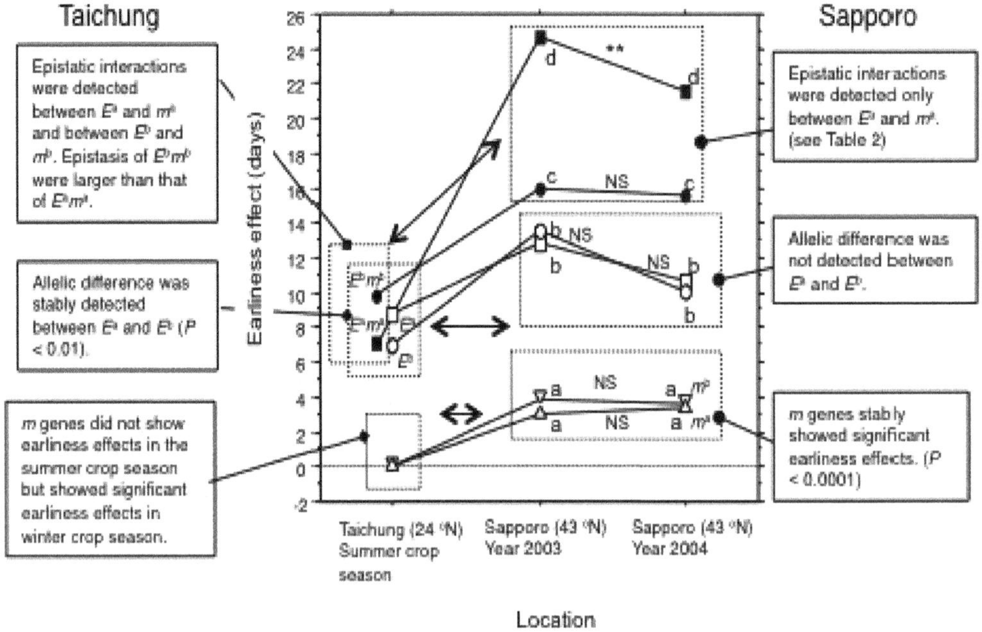

Fig. 2. Comparisons of earliness effect of the six NILs of Taichung 65 (BC20 generation) between Taichung and Sapporo and between years 2003 and 2004 in Sapporo

*Earliness effect of a NIL was calculated by subtracting days to flowering of a NIL from that of the recurrent parent. Data of Taiwan were roughly estimated from Fig. 1 of reference [5]. E and m indicates Ef1 and m-Ef1, respectively. In Sapporo, same latters (i.e. a, b, c and d) indicate that the differences of earliness effects between the strains are not significant in each year (Tukey HSD-test). All the earliness effects of the genes were significant at 0.1% levels. In Sapporo, difference of earliness effect of a NIL between years was tested by t-test. ** and NS indicates significance at 1% level and non-significance (P > 0.05), respectively*

3.2 Response of Function of *Ef1* and *m-Ef1* Genes to Year Environments

There were large differences in temperature between year 2003 and 2004 in Sapporo (Fig. 1). Nevertheless, all the NILs did not show year response as to the earliness effects except for T65$E^a m^a$ (Fig. 2). Both $Ef1^a$ and $m\text{-}Ef1^a$ genes did not show year response. Hence, the significant year response shown by T65$E^a m^a$ might be derived from epistasis-by-year environment interaction. It is noted that T65$E^a m^a$ made earliness effect larger in cool year than in warm year, suggesting the epistasis might increase the earliness effect in response to low temperatures. This suggestion is also supported by the phenomenon that combination of $Ef1^a$ with $m\text{-}Ef1^a$ gene contributed to lateness in Taichung where temperatures are high while that contributed to earliness in Sapporo where temperatures are low (Fig. 2). In the northern-limit region of rice cultivation, suitable range of flowering time is highly limited because cultivation period is mostly covered with low temperatures (Fig. 1). In such situation, two strategies of flowering time are thought. First, when high temperatures are maintained in the early stage of flowering time, it is advantageous for rice plant to accelerate flowering because the rice plants, which cannot have strong photoperiod sensitivity due to the long days in summer time [16], cannot predict the timing when low temperatures occur and there is the capricious timing of occurrence of low temperature in the region [7]. A gene associated with the first strategy was identified in a land race of Hokkaido, and hence, the genetic mechanism for the strategy would exist [20,21]. The epistasis between $Ef1^b$ and $m\text{-}Ef1^b$ might belong to the first strategy. Second, when low temperatures are maintained in the late stage of flowering time, it is advantageous for rice plant to accelerate flowering to avoid cool weather damage. However, genetic mechanism for the second strategy has not been identified. The present result might provide a piece of genetic evidence with the second strategy, that is, epistasis between $m\text{-}Ef1^a$ and $m\text{-}Ef1^a$ might contribute to earliness in response to low temperature.

3.3 Relationship between Earliness Effect and Fitness

Takahashi [8] mentioned that extension of rice cultivation toward the most northern areas has owed to mainly genetic improvements of earliness and cool tolerance of booting stage. However, it has not been quantitatively shown how degree earliness affects fitness since plant breeders have genetically improved both earliness and cool tolerance of booting stage. T65 is a weak strain as to cool tolerance of booting stage [22], and hence, the earliness NILs with the genetic background of T65 are suitable for examining the pure effect of earliness on fitness. At individual level, earliness effects significantly affected fitness in both year 2003 and 2004 in Sapporo (Fig. 3). The present study revealed that earliness effect as phenotype explained at least 51.8% of total variation of fitness (Fig. 3). This is the first depiction of the relationship between earliness effect and fitness as to the quantitative viewpoint. Individual with earlier flowering time tended to show higher fitness, suggesting early-flowering time is essential component of fitness in the northern-limit region of rice cultivation. R^2 value of year 2003 was smaller than that of year 2004, suggesting that low temperatures might disrupt linear relationship between earliness effect and fitness.

The fitness values at genotype level tended to greatly fluctuate depending on year and correlation of the fitness between year 2003 and 2004 was not significant (r = 0.182) (Fig. 4). At genotype level, the fitness values were significantly correlated with the earliness effects in year 2004 (r = 0.886, P < 0.05), however, the tendency disappeared in year 2003 ($r = 0.725^{NS}$) (Fig. 4). Nevertheless, the earliness effect of six genotypes showed similar range between the two years, showing highly significant correlation (r = 0.986, P < 0.0001) (Fig. 4). This seems to be caused by a lower value of fitness in the genotypes with $Ef1^a$ $m\text{-}Ef1^a$ or $Ef1^b$ $m\text{-}Ef1^b$ than those with $Ef1^a$ or $Ef1^b$ alone in year 2003. After time exceeds the limit of suitable range of flowering time, which has been experienced by many rice breeders in the region, environments are averagely covered with low temperatures (Fig. 1) but capricious high temperatures in only several hours in a day can often occur in the northern-limit region of rice cultivation. This may be the cause for the confusion of linear relationship between earliness effects and fitness at genotype level. Whatever the reason, it can be at least said that year environments qualitatively affected the contribution manner of genotype of flowering time (or a fitness component trait) to fitness. This is a very commonplace experience of the rice breeders in the northern margins of rice cultivation, but there has been no scientific depiction that directly shows the experience. Therefore, Fig. 4 provides a foundation with future plant breeding studies and can be a ground for the reason why a rice-breeding program has required many years in the region.

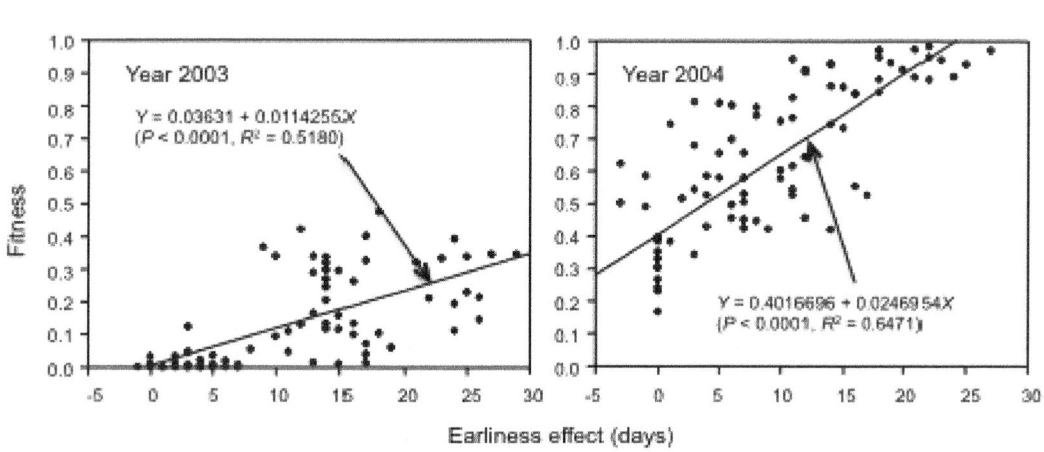

Fig. 3. Relationship between earliness effect and fitness at individual level

Earliness effects of a plant of a NIL were calculated by subtracting days to flowering from germination of a NIL from that of the recurrent parent in same ID of same replicate. Thus, earliness effect of the recurrent parent is zero. Fitness in a plant was estimated from the formulation, 1/2 (Maximum fertility of a panicle + Minimum fertility of a panicle) x (Total number of panicles with more than a fertile grain / Total number of panicles)

Table 3. Mean squares of days to flowering from germination for *Ef1* and *m-Ef1* loci within each of years

Source	df	NILs A[1)]		NILs B	
		Year 2003	Year 2004	Year 2003	Year 2004
Ef-1	1	3383.521****[2)]	2408.333****	1633.333****	1530.021****
m-Ef1	1	475.021****	660.083****	85.333**	221.021****
Ef1 x *m-Ef1*	1	275.521****	200.083****	2.083^{NS}	4.688^{NS}
Residual	44	10.203****	5.231****	10.345*	4.608****

*[1)] In the NILs A, ANOVA was carried out by using the values of T65E^a, T65m^a, T65E^{a}m^a and T65, while in NILs B, ANOVA was carried out by using the values of T65E^b, T65m^b, T65E^{b}m^b and T65. [2)] **, **** and ^{NS} indicates significance at 1%, 0.01% levels and non-significance, respectively*

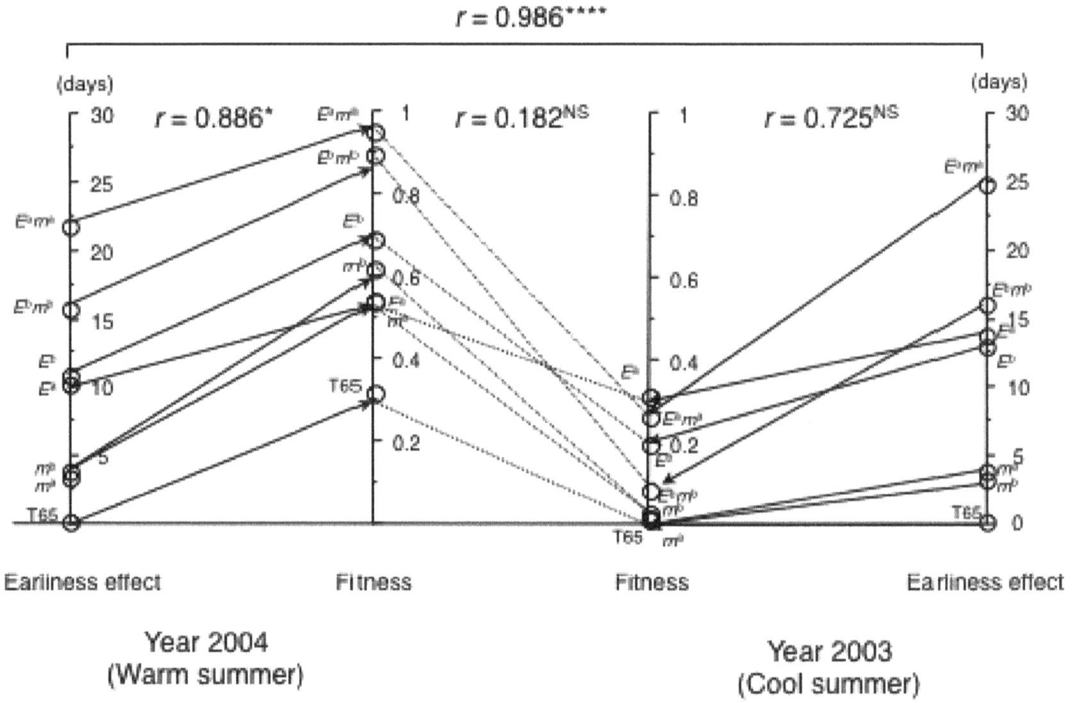

Fig. 4. Relationships between earliness effects and fitness at genotype level in the years 2003 and 2004, and their comparison in Sapporo

*Earliness effects of a NIL were calculated by subtracting mean of days to flowering of a NIL from that of the recurrent parent. Fitness in a plant was estimated from the formulation, 1/2 (Maximum fertility of a panicle + Minimum fertility of a panicle) x (Total number of panicles with more than a fertile grain / Total number of panicles). One-tailed correlation coefficients (r) were calculated by using genotypic mean. *, ****, and ^{NS} indicates significance at 5 and 0.01 % levels and non-significance, respectively*

4. CONCLUSION

Genic effects of *Ef1* and *m-Ef1* on flowering time largely changed between different latitudes. Especially, epistasis between the two genes qualitatively changed, suggesting that the role of epistasis in emergence of novel phenotype in response to environments should not be negligible. In addition, epistasis between *Ef1*^{a} and *m-Ef1*^{a} may respond to low temperatures and accelerate flowering time. This is the first suggested phenomenon in rice. Using flowering-time NILs, it has been firstly demonstrated that earliness effect as phenotype (a phenotypic trait) significantly affects fitness in both cool and warm years in one of the northern-limit regions of rice cultivation. Individual with earlier flowering time tended to show higher fitness. However, this tendency was not always preserved at genotype level. Capricious changes of temperature might disrupt the tendency and this shows the difficulty of plant breeding programs in the region.

ACKNOWLEDGEMENTS

A paddy field was provided for the present study by Mr. Hiroyuki Shimizu (HARC). The use of genetic materials was permitted by professor Yoshio Sano (Research Group of Breeding Science, Division of Applied Bioscience, Research Faculty of Agriculture, Hokkaido University). Y. Itoh thanks Miss Yoshie Sato (GASUTO, Fukuzumi) and Dr. Hiroshi Yorozuya (Brown Bear Research Group of Hokkaido University) for their encouragement and discussions about the present study.

COMPETING INTERESTS

Author has declared that no competing interests exist.

REFERENCES

1. Schlichting CD Pgliucci. Phenotypic evolution a reaction norm perspective. Sunderland, MA: Sinauer Asssiates Inc; 1998.
2. Itoh Y, Sato Y. A genetic basis of reaction norm evolution in density and/or year response of early vegetative plant architecture using recombinant inbred lines derived from the cross between annual-form-wild and cultivated rice under changeable natural climate across years. Int Res J Plant Science. 2013;4(8):248-271.
   DOI: http://www.interesjounals.org/IRJPS
3. Okada K, Shimura Y. Genetic analyses of signaling in flower development using Arabidopsis. Plant Mol Biol. 1994;26:1357-1377.
   DOI: 10.1007/978-94-011-0239-1_8
4. Hayama R, Yokoi S, Tamaki S, Yano M, Shimamoto K. Adaptation of photoperiodic control pathways produces short-day flowering in rice. Nature. 2003;422:719-722.
   DOI: 10.1038/nature01549
5. Tsai KH. Studies on earliness genes in rice, with special reference to analysis of isoalleles at the *E* locus. Jpn J Genet. 1976;51(2):115-128.
6. Sameshima R, Hirota T, Hamasaki T. Temperature trends at the national agricultural research center for Hokkaido region in the 40 years from 1966 to 2005. J Agric Meteorol. 2007;63(2):95-102.
   DOI:http://dx.doi.org/10.2480/agrmet.63.95
7. Nishiyama I. Factors and mechanisms causing cool weather damage. In: Matsuo T, Kumazawa K, Ishii R, Ishihara K, Hirata H, editors. Science of rice plant (Volume 2: Physiology), Tokyo: Food and Agriculture Policy Research Center; 1995.
8. Takahashi M. The history and future of rice cultivation in Hokkaido. Japanese Experience of the UNU Human and Social Development Program series. 1980;22.
   Available:http://d-arch.ide.go.jp/je_archive/english/society/wp_je_unu22.html9
9. Pigliucci, M, Whitton J, Schlichting CD. Reaction norms of *Arabidopsis*. I. Plasticity of characters and correlations across water, nutrient and light gradients. Journal of Evolutionary Biology. 1995;8(4):421-438.
   DOI: 10.1046/j.1420-9101.1995.8040421.x
10. Oka HI. Origin of cultivated rice. Tokyo: Elsevier; 1988.
11. Shimizu H, Itoh Y. Developmental change of vegetative plant architecture of annual-form-wild rice (*Oryza rufipogon* Griff.) elevates competitive ability during the late development under a dense condition. American Journal of Plant Science. 2012; 3(3):670-687.
    DOI: 10.4236/ajps.2012.35081
12. Itoh Y, Sano Y. Phyllochron dynamics under controlled environments in rice (*Oryza sativa* L.). Euphytica. 2006; 150(1-2):87-95.
    DOI: 10.1007/s10681-006-9096-5
13. Doi K, Izawa T, Fuse T, Yamanouchi U, Kubo T, Shimatani Z, Yano M, Yoshimura A. *Edh1*, a B-type response regulator in rice, confers short-day promotion of flowering and controls *FT*-like gene expression independently of *Hd1*. Genes Dev. 2004;18(8):926-936.
    DOI: 10.1101/gad.1189604
14. Uwatoko N, Onishi A, Ikeda Y, Kontani M, Sasaki A, Matsubara K, et al. Epistasis among the three major flowering time genes in rice: coordinate changes of photoperiod sensitivity, basic vegetative growth and optimum photoperiod. Euphytica. 2008(2);163:167-175.
    DOI: 10.1007/s10681-007-9584-2
15. SAS (Statistical Analysis System Institute Inc.) SAS user's guide: Statistics 5th ed. Cary, NC: SAS; 1985.
16. Fujino K. Photoperiod sensitivity gene controlling heading date in rice cultivars in

the northernmost region of Japan. Euphytica. 2003;131(1):97–103. DOI: 10.1023/A:1023088810701

17. Itoh Y, Sano Y. Evolutionary significance of cryptic variation in the reaction norm of photoperiod sensitivity in rice. Euphytica. 2007;155(1-2):153-165. DOI: 10.1007/s10681-006-9318-x
18. Wright S. Evolution in Mendelian populations. Genetics. 1931;16(2):97-159.
19. Weinig C, Dorn LA, Kane NC, German ZM, Halldorsdottir SS, Ungerer MC, et al. Heterogeneous selection at specific loci in natural environments in Arabidopsis thaliana. Genetics. 2003;1:321-329.
20. Sato S, Ogata K, Shinjoh C. Thermo-sensitive action of earliness gene *Ef-x* in rice (*Oryza sativa* L). Jpn J Genet. 1992;67(6):473-482.
21. Itoh Y, Sato S, Sano Y. Developmental changes of phyllochron in near-isogenic lines of rice (*Oryza sativa* L.) with different growth durations. Euphytica. 2001;119(3): 271-278. DOI: 10.1023/A:1017577218630
22. Ishiguro S, Ogasawara K, Fujino K, Sato Y, Kishima Y. Low temperature-responsive changes in the anther transcriptome's repeat sequences are indicative of stress sensitivity and pollen sterility in rice strains. Plant Physiol. 2014;164(2):671–682. DOI: 10.1104/pp.113.230656.

15

Effect of Atrazine on Germination and Growth Performance of Water Yam (*Dioscorea alata*)

O. S. Olabode[1], A. O. Sangodele[1*] and F. A. Akinpelu[2]

[1]Department of Crop Production and Soil Science, Ladoke Akintola University of Technology, Ogbomoso, Oyo State, Nigeria.
[2]Department of Environmental and Crop Protection, Ladoke Akintola University of Technology, Ogbomoso, Oyo State, Nigeria.

Authors' contributions

This work was carried out in collaboration between all authors. Author OSO designed the study and wrote the protocol. Author AOS wrote the first draft of the manuscript, managed the literature searches. Author FAA did the analyses of the data and the conclusion. All authors read and approved the final manuscript.

Editor(s):
(1) Chandra Shekhar Kapoor, Department of Environmental Sciences, University College of Science, Mohan Lal Sukhadia University, India.
Reviewers:
(1) Oluwaranti Abimbola, Obafemi Awolowo University, Nigeria.
(2) Anonymous, Agricultural Research Council, South Africa.

ABSTRACT

Effect of rates of application of atrazine on germination and growth performance of water yam minisetts was investigated at Ladoke Akintola University of Technology, Ogbomoso, Oyo State, Nigeria. The treatments in the field experiment were four (4) rates of atrazine i.e 100%, 50%, 25%, and 0% of the recommended dosage rate of 2.5 kg active ingredient/ha or 5litres/ha. A plot to which no herbicides was applied served as the control (0%). The experiment was arranged in a randomized complete block design with three replications. Result showed that though yam minisetts emergence was not hampered by the rates of application of atrazine, the growth parameters of the yam minisetts were significantly (p= 0.05) affected at early stage. Yam minisetts vine length and vine thickness were significantly (p=0.05) higher at 50% treatments compared to other treatments and control. Rates of application of atrazine however, did not affect the number of leaves of yam minisetts. The rates of application of atrazine did not significantly (p=0.05) affect the

**Corresponding author: Email: deleopeola@gmail.com*

yield parameters of yam minisetts, though the heaviest weight was produced under 25% treatments. Similarly, the rates of application of atrazine were similar in their effect on weed control and were better than the control.

Keywords: Yam minisetts; atrazine; dosage.

1. INTRODUCTION

Yams, (*Dioscorea spp.*), are important tuber crop and staple food for millions of people in many tropical and subtropical countries. It has become a cash crop in countries along the coast of West Africa [1]. They produce edible starchy storage tubers, which are of cultural, economic, and nutritional importance in the tropical and subtropical regions of the world [2]. Generally, yam tubers are boiled, roasted, baked, or fried. However, in some regions, the tubers are boiled and then pounded to glutinous dough called "fufu" (in Yoruba language) Cooked yam tubers or their products are usually eaten in association with protein-rich sauces. It has also gained importance in pharmaceutical industries [3]. Nigeria accounts for about 70% of the world's production of yam, generating a global output of over 33 million metric tons [4]. *D, alata* is popular and prevalent in Abakaliki agro ecological zone of Ebonyi State, Nigeria where it is called "Mbala or Nvula" (Igbo names).

Traditionally, water yams are propagated vegetatively by means of tuber. Whole tuber called "seed" yam from l00 g to 1500 g are used as planting materials. Alternatively, larger tubers are often cut into approximately 200 g pieces and used to establish the new crops [5,6]. A major constraint to yam production however, is the availability of the planting material. Since the tuber for which the crop is grown is equally needed as planting material, it becomes difficult to get enough material for planting since this constitute a competitive use for the tuber.

Furthermore, the conventional seed yam and or yam sett are usually very big thereby consuming a greater percentage of the total yam produced. The practical implication is that the seed represent the highest cost items, accounting for as much as 50 percent of operating expenses [7] while planting material makes up 35 percent and 54 percent of the total cost of production [8].

The National Root Crop Research Institute in Nigeria developed a system termed the "minisett" technique which aims at overcoming the seed yam problem. A "minisett" is defined as set less than one quarter of the minimum size (l00 g) of yam sett often planted [6]. The technique involves cutting the selected "mother seed" yam into pieces 20 g to 25 g each, treating them in a fungicides or insecticides suspension and sprouting them in nursery beds or polyethylene bags. After three to four weeks, the sprouted minisetts are transplanted into the main fields [9].

Generally, weed constitute a major hindrance to crop production especially in the tropics due to factors ranging from weather condition (torrential rain and adequate insolation), farming system and illiteracy which does not allow for maximum employment of chemical weed control [10]. The debilitation effect of weeds is a function of weed type, crop type, season of the year amongst others. In yam production, yam is mostly susceptible to weed infestation during the early growth during which time the canopy is not formed yet thereby allowing for ample space for the weed to develop. Thus, any practice that will allow prompt germination and rapid canopy formation will go a long way to checking the weed menace in yam production. A common practice in the tropics is the use of big size yamsetts or seedyams. This technique relies on the fact that since the young yam plant relies on the stored nourishment in the planting material at the early age, the size of the planting material, therefore is of great significance for a proper and a more competitive development of the yam seedling during the early growth stage through timely and adequate canopy formation which reduce weed seedlings recruitment and development.

From the foregoing, it could be inferred that weeds and weed control is a major constraint to yam minisett production. The tiny nature of the emerging vines due to the size of the planting material delays canopy formation thereby yielding more space for weed emergence, development and competition. Similarly the fragility of the vines makes the use of manual weeding, using hand hoeing, a delicate activity aside labour intensiveness, time constraint and cost of labour.

Thus, a more effective and less cumbersome weeding using the herbicide becomes more appealing. The use of chemical weeding has remained very popular in crop production due effectiveness, timeliness, reduce labour and reduce cost. The practice has also been reputed to be desirable due to reduced erosion arising from reduced soil tillage [11,12]. It is in the light of this that this experiment was designed to evaluate the effect of atrazine, a popular pre-emergence herbicide, on the germination and growth performance of water yam minisett. Atrazine used to prevent pre- and postemergence broadleaf and grassy weeds in crops such as sorghum, maize, sugarcane, lupins, pine, and eucalypt plantations, and triazine-tolerant canola [13].

[14], stated that in the United States as of 2014, atrazine was the second-most widely used herbicide after glyphosate, with 76 million pounds of it applied each year [15,16]. Atrazine continues to be one of the most widely used herbicides in Australian agriculture, [13].

The result of this experiment is expected to enhance the success of rapid yam minisetts production.

2. MATERIALS AND METHODS

Field experiment was carried out at Ladoke Akintola University of Technology, Teaching and Research Farm, Ogbomoso, during the rainy seasons of 2011 and 2012. The raining season, which is between April and October, is characterized by maximum rainfall distribution in southern part, due to the influence of the south westerly monsoon wind on the atmosphere. The dry season covers between November and March and is characterized by hot weather. The soil of the experimental plot was sandy loam and slightly acidic (Table 1).

The common weeds on the experimental plot include: - *Tithonia diversifolia, Argeratum conyzoides, Digitaria horizonntalis,* and *Tridax procumbens.* The predominant weed species were *Tithonia diversifolia and Digitaria horizonntalis* (50% and 35% respectively). The experiments were laid out in a randomized complete block design, with three replicates. The minisetts, 25g each, were treated with wood ash and cut, the surfaces were allowed to dry for 24hrs before planting. The minisetts were planted in situ at 6 plants per bed at a spacing of 0.5 m by 0.5 m. The bed was dimensioned 2 m x 2 m. Grass mulch was applied on planting spots to conserve moisture and prevent excessive sunlight. Four rates of application were tested in the experiment namely; atrazine 100%, atrazine 50%, atrazine 25% and control treatment to which no herbicide was applied. Atrazine was applied to rain moistened soil using a knapsack sprayer model Cp3, which delivers at a rates of 2001/ha. To achieve the 100%, 50%, 25% of dosage rates of herbicides, 250 ml, 125 ml and 62.5 ml of the herbicides were dissolved each in 10 litres of water. Stakes were provided for plant stands and the growing young vines were trained as soon they are long enough using threads. Data were collected on growth and yield parameters of water yam. Weed population density was estimated from 3 randomly placed 0.25 m^2 quadrat, while the wet weight of weeds harvested from the quadrats were measured on a sensitive scale. Mean data collected over the two years were subjected to analysis of variance (ANOVA) while the means were separated by Duncan's Multiple Range Test (DMRT) at 5% probability level.

Table 1. Pre-cropping soil analysis of the experimental site

	Year 2011	Year 2012
pH (H_2O)	6.0	6.3
Organic C (%)	1.9	1.6
Total N	0.26	0.18
Available P (ppm)	4.93	5.98
Exchangeable K (Meg/100 g)	0.42	0.38
Sand (%)	87	88
Silt (%)	9	10
Clay (%)	4	2
Textural class	Sandy loam	Sandy loam

3. RESULTS AND DISCUSSION

Effect of rates of application of atrazine on germination of yam minisetts is presented in the Table 2. There were no significant (p=0.05) differences among the treatments and the control at 3 weeks after planting (WAP). Atrazine did not have any effect on the germination of yam minisetts as both the atrazine treated and the untreated minisetts germinated at the same rate. At 6WAP, there were no significant differences on germination rate between atrazine 50%, Atrazine 25% and control. These were however significantly (p=0.05) higher than those of Atrazine 100%.

Table 2. Effect of rates of application of atrazine on germination of yam minisetts

Treatment	% germination	
	3 WAP	6 WAP
Atrazine l00%	16.7a	55.6b
Atrazine 50%	27.7a	77.8a
Atrazine 25%	33.3a	72.2a
Control	33.4a	77.8a

Mean followed by the same superscript along the column are not significantly different (p=0.05) by Duncan's Multiple Range Test (DMRT).
WAP: Weeks after planting

Effect of rates of application of atrazine on growth parameters of yam minisetts is presented in Table 3. There were no significant (p=0.05) effect of rates of application of atrazine on yam minisetts growth parameters across the period of data collection.

Similarly, application rates of atrazine did not have significant (p=0.05) effect on tuber yield of yam minisetts (Table 4). However, the heaviest tuber (1.3 kg) was produced under 25% treatment (1.30 kg) followed by the control (1.0 kg). The smallest tuber (0.87 kg) was produced by the 100% treatment.

The weed control effectiveness of the rates of application as measured by weed density and biomass is presented in Table 5. All the rates were superior to the control in effectiveness (p= 0.05). There were no significant difference (p= 0.05) between the rates in effectiveness. This result confirms the efficiency of the herbicide as indicated by the producers. Atrazine had been indicated for the control of broad leaves and some grass weeds when applied pre-emergently on at early weed emergence.

The result from this experiment indicated that atrazine can safely be used for weed control in yam minisett production. The experimented rates did not have any significant (p= 0.05) negative effects on both the growth and tuber yield of water yam compared to the control (Table 7), which confirmed yam's tolerance to timely application of the herbicides. Thus, it could be safely recommended that the lowest rates (25%) be adopted for weed control in yam minisett production more so, when there were no significant (p= 0.05) difference between the weed control efficiencies of the rates (table 6). As such, it is implied that only 25% of the recommended dose will be required. Similar finding had been reported by [17]. In addition, since atrazine had been listed as one of the most persistent herbicides [18], an effective reduced dosage rate will help mitigate the problem of persistence of the herbicides in addition to added economic gains. Furthermore, since the use of herbicides does not involve using equipment like hoes and cutlasses that can easily injure or destroy the tender and fragile yam minisetts vines, the choice of this method of weed control aside reducing labour, drudgery and cost will guarantee safety of the yam minisett vines.

Table 3. Effect of rates of application of atrazine on vine length of yam minisetts

Treatment	Vine length (cm)			
	6 WAP	8 WAP	10 WAP	12 WAP
Atrazine 100%	33. 1b	86.0a	122.6ab	147.0a
Atrazine 50%	58.1a.	98.0a	l41.4a	144.4a
Atrazine 25%	54.4a	71.4a	80.7b	135.9a
Control	46.8a	96.2a	124.6ab	126.4a

Mean followed by the same superscript along the column are not significantly different (p=0.05) by DMRT

Table 4. Effect of rates of application of atrazine on vine thickness of yam minisetts

Treatment	Vine thickness (cm)			
	6 WAP	8 WAP	10 WAP	12 WAP
Atrazine 100%	0.5a	0.5a	0.5a	0.6ab
Atrazine 50%	0.6a	0.5a	0.6a	0.7a
Atrazine 25%	0.5a	0.5a	0.5a	0.6ab
Control	0.5a	0.5a	0.5a	0.5b

Mean followed by the same superscript along the column are not significantly different (p= 0.05) by DMRT

Table 5. Effect of rates of application of atrazine on number of leaves of yam minisetts

Treatment	Number of leaves			
	6 WAP	8 WAP	10 WAP	12 WAP
Atrazine 100%	8.4a	24.0a	33.8a	49.0a
Atrazine 50%	11.8a	32.2a	37.5a	49,0a
Atrazine 25%	11. 8a	27.7a	32.7a	48.8a
Control	10.0a	24.7a	30.2a	52.2a

Mean followed by the same superscript along the column are not significantly different (p=0.05) by DMRT

Table 6. Effect of rates of application of atrazine on weed density and weed biomass of yam minisetts

Treatment	Weed density unit/25 m^2	Weed biomass g/25 cm^2
	8 WAP	8 WAP
Atrazine 100%	10.3b	28.3b
Atrazine 50%	14.4b	33.3b
Atrazine 25%	15.0a	32.3b
Control	28.9a	79.7a

Mean followed by the same superscript along the column are not significantly different (p=0.05) by DMRT.
WAP: Weeks after planting

Table 7. Effect of rates of application of Atrazine on the yield parameters of yam minisetts

Treatment	Yield parameters		
	Tuber weight (kg)	Tuber length (cm)	Tuber diameter (cm)
Atrazine 100%	0.87a	21.20a	9.50a
Atrazine 50%	0.99a	24.90a	8.90a
Atrazine 25%	1.30a	24.90a	10.30a
Control	1.00a	25.30a	9.70a

Mean followed by the same superscript along the column are not significantly (p=0.05) different by DMRT

4. CONCLUSION

This experiment has shown that atrazine at the rates used in this experiment is adequate for use in the rapid multiplication of water yam tubers through minisetts. Though yam performance under the herbicides was not significantly different from those produced under hand hoe weeding, however considering the fragility of the minisett vines, the relative ease of spraying, timeliness coupled with the difficulty of the experienced in securing labour for farm operations especially hoeing, this result is a positive way to enhancing yam production through the use of minisett technology. Furthermore, for economic and environmental concern, it is recommended that 25% of the recommended dosage rate of atrazine should be used for weed control in yam minisetts production.

COMPETING INTERESTS

Authors have declared that no competing interests exist.

REFERENCES

1. Egesi CN, Ogunyemi S, Asiedu R. Genotype x environment interaction in water yam (*Dioscorea alata*) yield trials in five yam-growing states in Nigeria; 2005.
2. Coursey DG. Yams. Longmans, London. 1967;230.
3. Hahn SK. Yams: *Dioscorea* spp. (Dioscoreaceae). In: Smartt J, Simmonds NW (Eds). Evolution of crop plants. Longman Scientific & Technical. UK. 1995; 112-120.
4. FAO 2000. FAOSTAT agricultural data. Food and Agriculture Organisation of the United Nations.
   Available: http//www.fao.org
5. Onwueme IC. The tropical tuber crops, yams, cassava, sweet potato and cocoyam. 1978;14:234.
6. Okoli O, Acha LA, Shiwachi H, Asiedu R, Akoroda MO. Effect of auxin on root development in yam (*Dioscorea rotundatd*) vine. Tropical Science. 1982;44:80-84.

7. Gray A, Crentsil D, Gallat S, Gogoe S. Survey of yam trading practices and loss assessment in Techiman market. Project A0497; R6505. Chatham: Natural Resources Institute; 1997.
8. Osei C. The yield of seed yams grown from mmisette as influenced by mulch in Rudat, Improving farming systems in the interior savanna zone of Ghana: proceedings of the 2nd workshop on improving farming systems; Nyankpala; 1990. Tamale: CRI/NAES; 1993.
9. Vernier P, Ericas N, Aegbo E, Hounhouigan J, N'Kpenu K, Orkwor G. From biological control to a systems approach in post- harvest. Proceedings of an IITA/GTZ meeting on integrated control of insect pests in rural maize stores, with particular reference to the Larger Grain Borer and the future development of the post-harvest sector in sub-Saharan Africa; 1999.
10. Akobundu IO. Weed control strategies for multiple cropping system of the humid and sub-humid tropics. IITA proceeding series 3. 1978;80-95.
11. Singh B, Malhi SS. Response of soil physical properties to tillage and residue management on two soils in a cool temperate environment. Soil & Tillage Res. 2006;85:143-153.
12. Nakamoto T, Yamagishi J, Miura F. Effect of reduced tillage on weeds and soil organisms in winter wheat and summer maize cropping on Humic Andosols in Central Japan; 2006.
13. Australian Pesticides and Veterinary Medicines Authority. Chemical Review: Atrazine. Retrieved 2015.
14. Rachel Aviv. A valuable reputation: Tyrone hayes said that a chemical was harmful, its maker pursued him. The New Yorker; 2014.
15. Walsh Edward. EPA Stops Short of Banning Herbicide. Washington Post. 2003;A14. Retrieved 2007.
16. EPA. Atrazine Updates: Amphibians; 2010.
17. Olabode OS, Adesina GO, Babajide PA. Weed control efficiency of reduced atrazine doses and its effect on soil organisms in maize *(Zea mays* L) on field of South Western Nigeria. Journals of Tropical Agriculture. 2010;48(1-2):52-54.
18. Akobundu IO. Weed science in the tropics. John Wiley and Sons, New York. 1987;522.

16

Prediction of Unsaturated Hydraulic Conductivity of Agricultural Soils Using Artificial Neural Network and c#

M. A. Al-Sulaiman[1] and A. M. Aboukarima[1,2*]

[1]Community College, Huraimla, Shaqra University, P.O.Box 300, Huraimla 11962, Saudi Arabia.
[2]Agricultural Engineering Research Institute, Agricultural Research Centre, Egypt.

Authors' contributions

This work was carried out in collaboration between the two authors. Author MAAS managed the required instrumentations, supervised on the experimental work, contributed in writing the first draft of the manuscript and discussed the conclusion. Author AMA wrote the protocol, managed the literature searches, analyzed data, carried out the analysis of artificial neural network model and wrote the first draft of the manuscript. Both authors read and approved the final manuscript.

Editor(s):
(1) Chandra Shekhar Kapoor, Department of Environmental Sciences, University College of Science, Mohan Lal Sukhadia University, India.
(2) Anonymous.
Reviewers:
(1) Milad Fatehnia, Florida State University, USA.
(2) Edward Ching-Ruey, LUo, National Chi-Nan University, Taiwan.

ABSTRACT

Aims: The objective of this study was to develop an artificial neural network model and interactive application using C# application to predict unsaturated hydraulic conductivity of soil.
Study Design: The actual measurements of unsaturated hydraulic conductivity of soil were obtained using the Mini Disk Infiltrometer (Decagon Devices, Inc.).
Place and Duration of Study: The study was conducted in laboratory located in Community College, Huraimla, Shaqra University, Saudi Arabia during March-April 2015.
Methodology: The experiments were conducted using water having electric conductivity of 2.26 dS/m and sodium adsorption ratio of 4.8. Unsaturated hydraulic conductivity of different soil textures (sand, sandy loam, loam and loamy sand) was determined at suction of -6 cm using Mini Disk Infiltrometer. The soil samples were taken from depth of 0-20 cm and repacked in a plastic 1000 cm^3 container.

**Corresponding author: Email: aboukarima@su.edu.sa*

Results: The predicted unsaturated hydraulic conductivity of soils compared favorably with the actual measurements in testing stage, however, mean relative error was 4.184% and coefficient of determination (R^2) was 0.9979. In general, artificial neural network model gave considerable results but more data is still necessary. The main equations for C# application were obtained from the trained artificial neural network model.
Conclusion: It could be concluded that the developed interactive application is recommended for estimating unsaturated hydraulic conductivity of agricultural soils within the range of the studied variables to provide data for water management in Saudi Arabia.

Keywords: Artificial neural network; C#; mini disk infiltrometer; unsaturated hydraulic conductivity of soil.

1. INTRODUCTION

In the field of irrigation agricultural, water management is required to improve the efficiency and sustainability of agricultural systems when water is scarce [1]. One of the tasks of water management is to quantify, to predict and in the end to control water and solute transport into soil. Consequentially, these processes are dependent on soil hydraulic conductivity which needs to be determined in the field or in the laboratory. On the other hand, direct measurement of hydraulic conductivity of soil is difficult, tedious, relatively costly, labour intensive and time-consuming [2-5]. Thus, indirect methods using predictive approaches have been developed for estimation of hydraulic properties of a soil from easily measurable soil properties [6-8]. However, predictive approaches of the soil hydraulic conductivity have gained considerable attention and efforts have been made by researchers to improve the power of predictability [9]. The predictive approaches of the soil hydraulic conductivity may be developed by different methods such as multiple linear regression, adaptive neuro-fuzzy inference systems and artificial neural network models. These methods have been recommended in the field of hydraulic conductivity as encourage results of different research papers [10-11] were reported.

Hydraulic conductivity is defined as "the meters per day of water seeping into the soil under the pull of gravity or under a unit hydraulic gradient" [12]. Additionally, hydraulic properties of a soil and their applications in the soil water flow models play an important role on solving many water management issues [13].

Hydraulic conductivity is useful to soil and water scientists, land managers, and growers, in knowing how quickly water will infiltrate when applied to a given field or soil type [14]. The selection and determination of soil hydraulic parameters are the basis of predicting water movement and solute transfer in soils [15]. The measurements show that the value of unsaturated hydraulic conductivity varies considerably from soil to soil with different water content [16]. Soil hydraulic conductivity includes two items, the first is saturated hydraulic conductivity and the second is unsaturated hydraulic conductivity [17]. Of all hydraulic properties, the unsaturated hydraulic conductivity is the most difficult to measure. It is considered to be the most important parameter for modelling flow process in unsaturated soil. Furthermore, unsaturated soil hydraulic conductivity is essential to many agriculture and environmental applications [18]. It is considered the most important variable in soil water flow models [13]. It is a function of soil moisture content [19]. It is also one of hydraulic characteristics needed for numerical modelling of water flow and solute transport [20]. It is also often necessary for solving unsaturated flow problems [21].

Measurement of unsaturated soil hydraulic conductivity is a challenging task and requires costly and skilled experimentation [22]. Therefore, the use of indirect methods has become common to estimate the unsaturated hydraulic conductivity from more easily measured soil properties [23]. Additionally, unsaturated soil hydraulic conductivity could be estimated from air permeability [24], from soil electrical conductivity [25] and from moisture retention data [26].

Vereecken et al. [7] measured hydraulic conductivity (saturated and unsaturated) on 127 soil cores, which were taken in different horizons of a wide variety of Belgian soil series. The textural composition in nine fractions, the organic carbon content and the dry bulk density were determined for each of the sample horizons as well. Four different empirical models were evaluated on their performance in describing the measured hydraulic conductivity.

Zhuang et al. [17] estimated the unsaturated hydraulic conductivity of soil based on some physical properties of the soil which were soil texture, the hydraulic conductivity of soil, soil water properties and amounts of gypsum and lime precent, actual and apparent the distribution of particle sizes.

Neshat and Farhad [19] carried out an experiment and made calculations for estimating the unsaturated hydraulic conductivity of a soil to get suitable relationship between the unsaturated hydraulic conductivity of the soil and the soil physical properties.

Malaya and Sreedeep [22] investigated the correlation between grain size distribution curve of a soil and unsaturated hydraulic conductivity of soil. They reported that unsaturated hydraulic conductivity of soil could be estimated from grain size distribution curve of a soil.

Amer et al. [27] proposed an equation to predict unsaturated hydraulic conductivity based on water viscosity, acceleration due to gravity, and density of water, ratio of total volume pores and radius of equivalent cylindrical pore size.

Neyshabouri et al. [28] proposed a simplify method to predict unsaturated soil hydraulic conductivity using readily accessible soil. Tests with 51 soils with estimated electrical properties were confirmed good agreement of the developed model for coarse–medium-textured soils (<35–40% clay).

Direct measurement of soil properties is easier and much rapid than unsaturated hydraulic conductivity of soil [29]. There are several studies that have attempted to use artificial neural networks (ANN) for estimating hydraulic conductivity of soil from its physical parameters due to ANN is becoming a common tool for modeling complex input–output dependencies [30]. Ghanbarian-Alavijeh et al. [31] developed ANN model to predict saturated hydraulic conductivity of soil using calculated fractal dimensions, air entry values, bulk density and effective porosity. In the training and testing steps of ANN, 114 and 28 measured soil samples were used, respectively. Coefficient of determination and mean squared error were 0.76 and 0.0028, respectively. Erzin et al. [32] developed ANN model for determining hydraulic conductivity of fine-grained soils. The ANN model exhibits higher prediction performance based on their performance indices. It has been demonstrated that the developed ANN model could be employed for determining hydraulic conductivity of compacted fine-grained soils quite efficiently. Ghanbarian-Alavijeh et al. [33] developed an ANN model with back propagation algorithm to estimate saturated hydraulic conductivity from available parameters such as sand and clay contents, bulk density, van Genuchten retention model parameters as well as effective porosity functions. The results showed that ANN model estimated saturated hydraulic conductivity in a good way. Moosavi and Sepaskhah [34] used sand, silt, clay, bulk density, soil organic matter and initial and saturated volumetric water content to predict the unsaturated hydraulic conductivity by the help of ANN. They showed that silt, clay, sand, bulk density and soil organic matter were the most basic influential input variables for prediction of unsaturated hydraulic conductivity.

In semi-arid regions like Saudi Arabia, water management is vital goal to improve the efficiency and sustainability of agricultural systems as water is scarce. So, the objective of this study was to use artificial neural network technique to create a model for unsaturated soil hydraulic conductivity predictions based on sand, silt, clay contents, soil electric conductivity, soil adsorption ratio, soil organic matter, initial soil moisture content and soil bulk density to be used as a tool for creating a database collection about soil hydraulic conductivity for agricultural soils in Saudi Arabia.

2. MATERIALS AND METHODS

2.1 Soil and Water Samples Characteristics

Experiments were conducted in laboratory located in Community College, Huraimla, Shaqra University, Saudi Arabia during March-April 2015. Three samples of each soil were taken from the top 20 cm of the selected sites. Textural analyses showed that the soils were classified as sand, sandy loam, loam and loamy sand. The initial soil water content (dry base) of the samples was measured by the help of electric oven for 24 hr at 105°C. The soil was carefully repacked in a plastic container with capacity of 1000 cm^3 with almost volume of 600 cm^3.

Sodium adsorption ratio (SAR) is a measure of the sodicity of soil, as determined from analysis of water extracted from the soil. The formula for calculating sodium adsorption ratio is as follows [35]:

$$SAR = \frac{Na^{+}}{\sqrt{\frac{1}{2}\left(Ca^{++} + Mg^{++}\right)}} \tag{1}$$

Where Na^+, Ca^{++}, and Mg^{++} represent concentrations of sodium, calcium and magnesium, respectively expressed in milliequivalents per litre (meq/L). The characteristics of water used in the laboratory experiments are shown in Table 1.

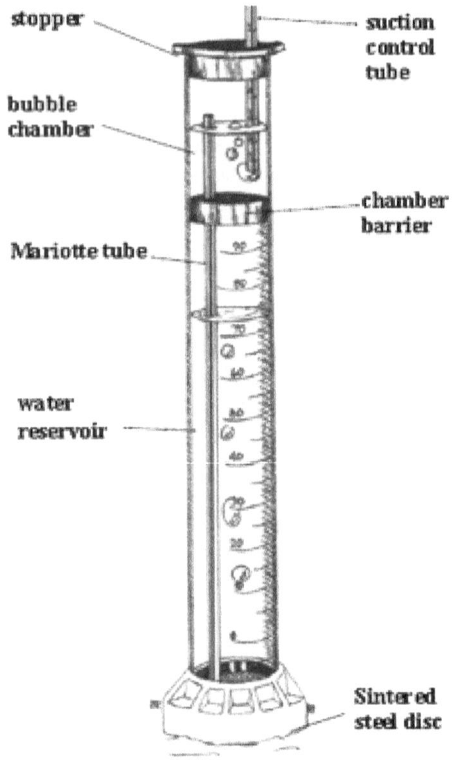

Fig. 1. Mini disk infiltrometer (Decagon devices, Inc.)

2.2 Mini Disk Infiltrometer

Mini disc infiltrometer (Decagon Devices, Inc.) measures soil hydraulic conductivity quickly and easily in any soil type. The infiltrometer (Fig. 1, above) is constructed of a polycarbonate tube with a semi-permeable stainless steel sintered disk on the bottom. An adjustable steel tube above the sample chamber regulates the discharge rate. The top and bottom chambers are both filled with water. The top chamber controls the infiltrometer's suction. The lower chamber contains the water that will infiltrate into the soil and is marked like a graduated. It can measure the hydraulic conductivity in the unsaturated medium (close to near saturation) for adjustable suction ranging from -0.5 cm to -7 cm. At time zero, the infiltrometer is placed on the soil surface. The volume of water that infiltrate into the ground has been recorded as a function of time, based on which infiltration and permeability characteristics is determined.

2.3 Calculating Hydraulic Conductivity

A number of methods are available for determining soil hydraulic conductivity [36]. The method proposed by Zhang [37] is quite simple and works well for measurements of infiltration into dry soil from the recorded data by mini disk infiltrometer. The method requires measuring cumulative infiltration vs. time and fitting the results with the function:

$$I = C_1 t + C_2 \sqrt{t} \tag{2}$$

Where I is the cumulative infiltration (cm), t is the time (sec), and C_1 (cm/sec) and C_2 (cm/sec$^{-0.5}$) are parameters. C_1 is related to hydraulic conductivity and C_2 is related to soil sorptivity. The hydraulic conductivity (*Ki*) of the soil is then computed from

$$Ki = \frac{C_1}{A} \tag{3}$$

Where C_1 is the slope of the curve of the cumulative infiltration vs. the square root of time (Fig. 2) and (A) is a value relating the van Genuchten parameters for a given soil type to the suction rate and radius of the infiltrometer disk. (A) is computed from:

$$A = \frac{11.65\left(n^{0.1} - 1\right)\exp\left[2.92(n - 1.9)\alpha h_0\right]}{\left(\alpha r_0\right)^{0.91}} \qquad n \geq 1.9 \tag{4}$$

$$A = \frac{11.65\left(n^{0.1}-1\right)\exp\left[7.5(n-1.9)\alpha h_0\right]}{\left(\alpha r_0\right)^{0.91}} \qquad n < 1.9 \tag{5}$$

Where n and α are the van Genuchten parameters for the soil, r_0 is the disk radius and h_0 is the suction at the disk surface. The van Genuchten parameters for the 12 texture classes were obtained from Carsel and Parrish [38].

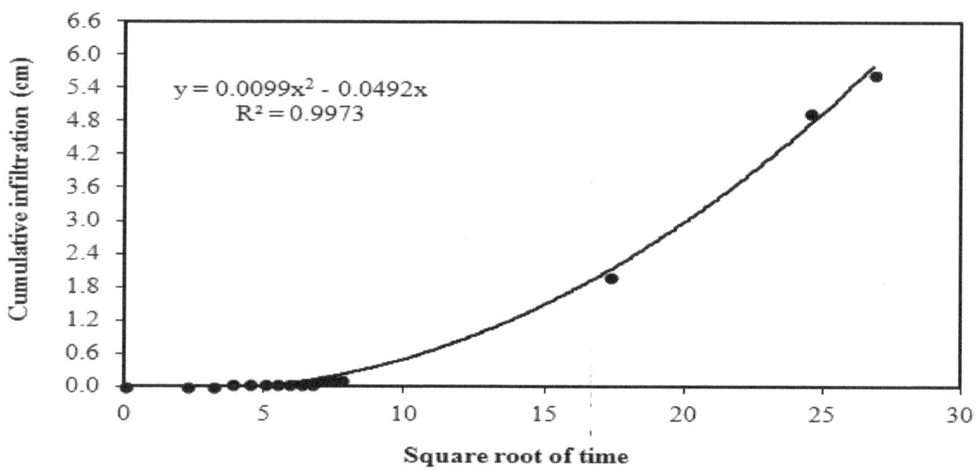

Fig. 2. Cumulative iinfiltration versus square root of time for a soil

Table 1. The characteristics of water used in the laboratory experiments

pH	EC_{water}	SAR_{water}	HCO_3	SO_4	Cl	Na	K	Ca	Mg
(---)	(dS/m)					(meq/L)			
7.7	2.26	4.81	8	5.8	22	15.3	0.3	15	5.2

2.4 Measurement of Soil Hydraulic Conductivity

The unsaturated hydraulic conductivity was measured using a mini disk infiltrometer (Decagon Devices Inc.). The mini disk infiltrometer consists of two chambers (water reservoir and bubble chamber), which are connected via a Mariotte tube to provide a constant water pressure head (suction) of -0.5 to -7 cm as depicted in Fig. (1). The bottom of the mini disk infiltrometer contains a porous sintered steel disk. The water filled tube is placed upon the soil surface (Fig. 3) resulting in water infiltrating into the soil, with the volume of water and speed of infiltration depending on the sorptivity and hydraulic conductivity of the soil. A pressure head (suction) of -6 cm was chosen in this study as the recommended suction for sandy soil [14]. All measurements within one sample test were taken on the same day. The mini disk infiltrometer measurements were taken three times for every soil sample. During the measurement, the volume of the water in the reservoir chamber was documented in regular intervals.

2.5 Artificial Neural Networks

For predicting soil properties, artificial neural networks (ANN) models have become an alternative method as described in different studies [39-40]. The ANN tool is a computational approach that simulates biological neuron function. The performance of an ANN model is sensitive to its architecture, such as the number of input nodes, hidden layer nodes and output nodes. The appropriate an ANN model architecture is, therefore, highly problem dependent [41]. The ANN model is trained to reproduce the input–output relation used to find the optimal weights. The training process consists of calculating the output variables from the input variables; comparing the measured output with the calculated output form ANN model; and then adjusting the weights and bias for each node to minimize the difference between the measured and calculated values.

Fig. 3. Mini disk infiltrometer used for laboratory infiltration measurements

Commercially available QNET 2000 was employed in this study [42]. This software is a Windows-based package, which supports standard back-propagation algorithm for training purposes. QNET 2000 operates via a graphical user interface (GUI) that enables the user to load the data of training and testing sets, design the network architecture and feed values for the training parameters. The ANN type used in this study was a standard back-propagation neural network. The neurons in the ANN layers are connected by weights. The weights connecting input neuron i to hidden neuron j are denoted by w_{ji}^{h} , while the weights connecting hidden neuron j to output neuron are denoted by w_j^o . The input of each neuron is the weighted sum of the network inputs, and the output of the neuron is a transfer function value based on its inputs. More specially, for the jth hidden neuron [43].

$$\begin{cases} net_j^h = \sum_{i=1}^{P} w_{ji}^h x_{t-1} + b_j \quad , \\ y_j = f(net_j^h) \end{cases} \tag{6}$$

While for the output neuron

$$\begin{cases} net^o = \sum_{j=1}^{m} w_j^o y_j + c \quad , \\ \tilde{x}_t = f(net^o) \end{cases} \tag{7}$$

Where b_j and c are thresholds (bias), this network has P neurons in the input layer and m neurons in the hidden layer, f is typically taken to be an transfer function and in this study, it was changed to be sigmoid function as shown in equation (8) or hyperbolic tangent (tanh) as shown in equation (9).

$$f(x) = \frac{1}{1+e^{-x}} \tag{8}$$

$$f(x) = \frac{e^x - e^{-x}}{e^x + e^{-x}} \tag{9}$$

It must be noted that because the variables (input or output) presented were of different orders of magnitude, all of the original inputs or output variables were normalized between 0.15 and 0.85 before entering into the network structure using the following equation:

$$T = \frac{(t - t_{\min})}{(t_{\max} - t_{\min})} \times (0.85 - 0.15) + 0.15 \tag{10}$$

Where t is the original values of input and output parameters, T is the normalized value; t_{max} and t_{min} are the maximum and minimum values of the input and the output parameters in the training data set, respectively. The training data set was used to compute the network parameters. The testing data set was used to ensure robustness of the network parameters. Table 2 illustrates minimum and maximum values of input and output variables in training date set for ANN model.

In the study, trial and error approach [44] was used to determine the optimum neurons in the hidden layers of the network. Transfer function was also varied; however, they were sigmoid and hyperbolic tangent (tanh) in the hidden layers. The training data set was consisted of 9 laboratory-measured patterns. Electric conductivity of soil, soil sodium adsorption ratio, initial soil water contents, sand, silt and clay contents, soil bulk density and soil organic matter were measured. However, these variables were considered as input variables. The output variable in this study was consisted of 9 laboratory measured unsaturated hydraulic conductivity of soil at suction of -6 cm. Preliminary trails indicated that two hidden layers network performed better results to predict unsaturated hydraulic conductivity of soil from the studied parameters. To determine the optimal number of the neurons in the hidden layers, the training network was performed using 8-n1-n2-1 architectures. The number of neurons in the first hidden layer (n1) was varied from 1 – 12. The number of neurons in the second hidden layer (n2) was varied from 2 – 20. The results showed that among the various structures, the best training performance to predict unsaturated hydraulic conductivity of soil was belong to the 8-8-14-1 structure. However, the best ANN model was elected based on the highest correlation coefficient and the lowest training error. Fig. (4) illustrates the best ANN structure in the study. Meanwhile, final training error after 10000 iterations was 0.001277 as shown in Fig. (5) and the details of network definitions to predict unsaturated hydraulic conductivity of agricultural soils is depicted in Table 3. Additionally, Table 4 illustrates network statistics after training process.

2.6 Evaluation of ANN Model Predictability

In order to perform a supervised training, a way in which the ANN output error between the actual and the predicted output could be evaluated is therefore required. A popular measure is the mean absolute error (MAE), root mean square error (RMSE) and mean relative error (MRE) as follows:

$$MAE = \frac{1}{N} \times \sum_{i=1}^{i=N} \left| K_{viobs} - K_{vipre} \right| \tag{11}$$

$$RMSE = \sqrt{\frac{\sum_{i=1}^{i=N} \left(K_{vi\,obs} - K_{vi\,pre} \right)^2}{N}} \tag{12}$$

$$MRE = \frac{100}{N} \times \sum_{i=1}^{i=N} \left(\frac{K_{vipre} - K_{viobs}}{K_{iobs}} \right) \tag{13}$$

Where $K_{v\,iobs}$ and $K_{v\,ipre}$ are experimental and predicted unsaturated hydraulic conductivity of soil, N is number of observations. In addition, the coefficient of determination (R^2) was selected to measure the linear correlation between the calculated and the predicted values. However, R^2 reflects the degree of fit for the mathematical model [45]. The closer the R^2 value is to 1, the better the model fits to the actual data [46].

2.7 C# Application

C# (C-Sharp) programming language available under.NET programming environment has been used for developing an application to predict unsaturated hydraulic conductivity of agricultural soils. This application has been developed keeping in view of its user friendliness and easily operable. Moreover, C# was chosen because it is a generic portable language and will run on other operating systems [47]. In addition, C# is also a clean, well-designed and implemented object-oriented language. Additionally, Sharma and Lal [48] reported that by using C# language, a fast development of MS-Windows based applications could be implemented for easy analysis of agricultural data. The current C# application was developed based on the weights obtained from the developed ANN model during training stage. These weights were formulated into equations in C# for easy prediction of unsaturated hydraulic conductivity of agricultural soils. The developed C# application was validated with experimental data to ascertain its suitability for unsaturated hydraulic conductivity of agricultural soils predictions. However, Fig. (6) depicts the starting screen of the developed C# application.

Table 2. Statistical values of input and output variables in training data set for ANN model

	Input variables								Output variable
	Sand	Silt	Clay	EC	SAR	Organic matter	Initial soil water content	Soil bulk density	Unsaturated hydraulic conductivity
	(%)	(%)	(%)	(dS/m)	(---)	(%)	(%,db)	(g/cm^3)	(cm/sec)
Minimum	52	2	1	1.06	2.00	0.26	3.60	1.30	0.00017
Maximum	95	28	39.	91.10	57.60	2.62	7.54	1.54	0.02037
Average	76.22	11.11	12.67	13.93	11.42	1.68	5.65	1.46	0.00624
Standard deviation	15.18	9.57	13.71	29.09	17.46	0.82	1.16	0.07	0.00813
Coefficient of variation (%)	5.02	1.16	0.92	0.48	0.65	2.05	4.86	19.49	0.7678

SAR= soil sodium adsorption ratio, EC =soil electric conductivity

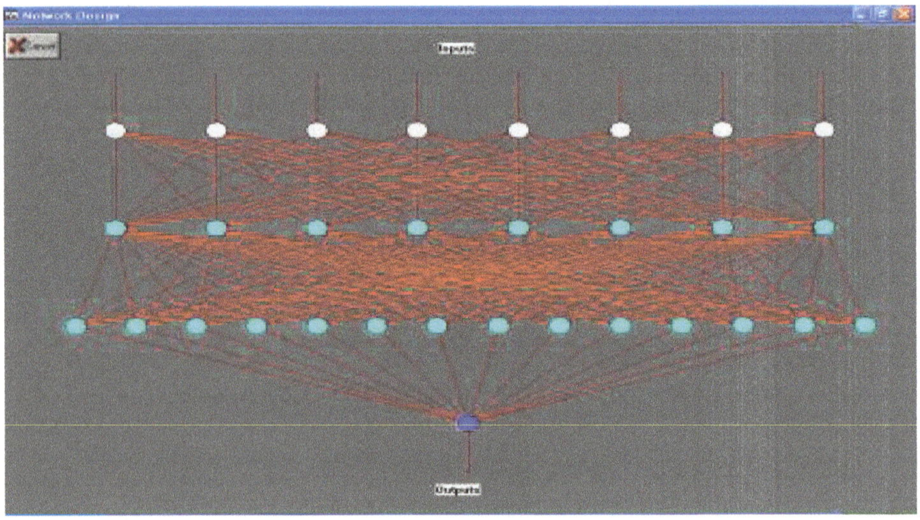

Fig. 4. Structure of the best ANN used in the study

Table 3. Network definitions to predict unsaturated hydraulic conductivity of agricultural soils

Network definitions	Value
Number of layers	4
Input layer nodes	8
Hidden Layer 1 Nodes	8
Transfer function	Sigmoid
Hidden Layer 2 nodes	14
Transfer function	Sigmoid
Iterations	10000
training error	0.001277
learn rate	0.095193
momentum factor	0.8
training patterns	9

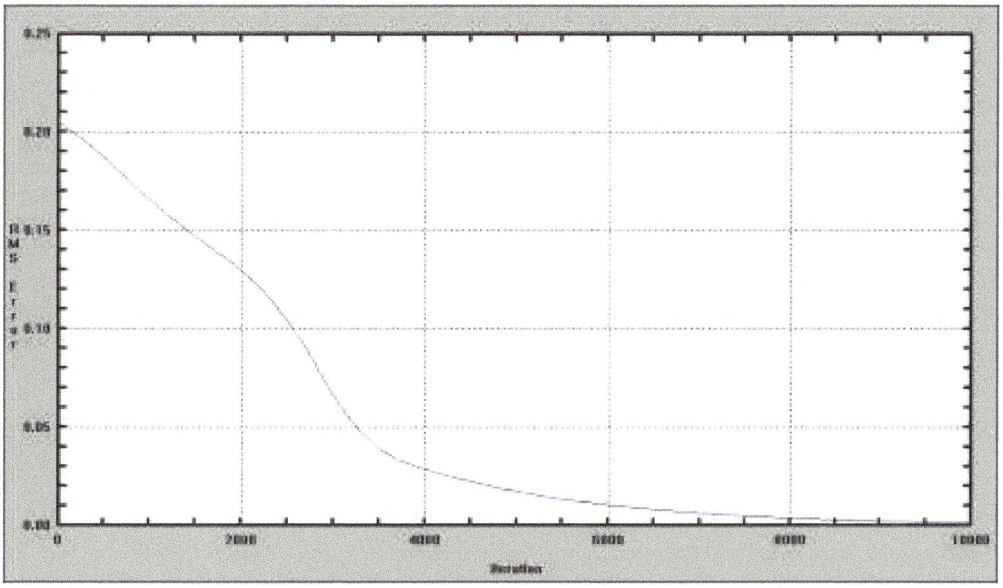

Fig. 5. Error curve during training proces

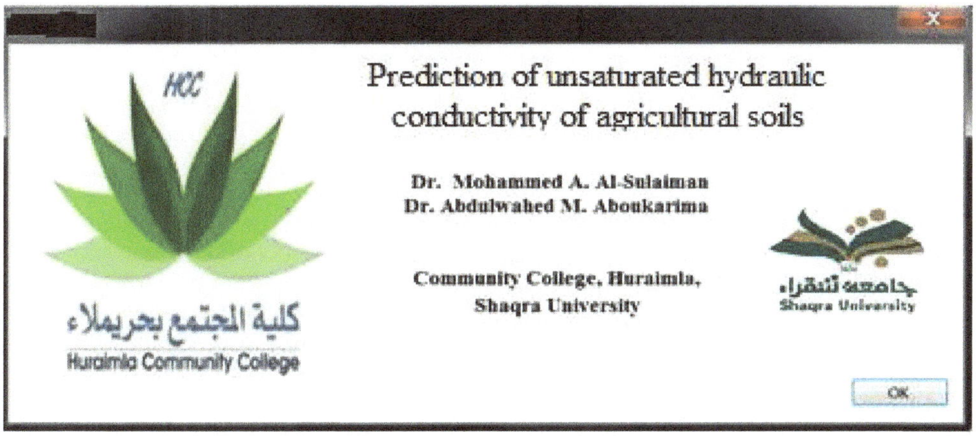

Fig. 6. The starting screen of the developed C# application

Table 4. Network statistics from Qnet software after training process to predict unsaturated hydraulic conductivity of agricultural soils

Standard deviation (cm/s)	Bias (cm/s)	Maximum error (cm/s)	Correlation coefficient
0.0000369	-0.000000395	0.0000728	0.99999

3. RESULTS AND DISCUSSION

3.1 Analysis of Unsaturated Hydraulic Conductivity Data

From Table 2, it is clear that the lowest and the highest unsaturated hydraulic conductivity of the soils were 0.00017 and 0.02037 cm/sec, respectively with coefficient of variation of 0.7678 % which indicates that no big variations between measured unsaturated hydraulic conductivity of soil. However, Gulser and Candemir [49] reported that hydraulic conductivity gave significant positive correlations with silt, sand contents and bulk density values of soils. Besides, hydraulic conductivity gave negative correlations with clay content and soil moisture constant. Schaap et al. [40] reported that if clay content increases so hydraulic conductivity of soil decreases. Additionally, Hudson [50] stated that increased soil organic matter generally produced a soil with increased water holding capacity and conductivity, largely as a result of its influence on soil aggregation and associated pore space distribution. Furthermore, Olourunfemi and Fasinmirin [51] indicated that hydraulic conductivity of soil increases with increase in soil organic matter content. Neshat and Farhad [19] showed that increases in volume soil moisture content and apparent density results increase in unsaturated hydraulic conductivity of soil.

3.2 Performance of ANN Model

In this study, ANN technique was employed to predict unsaturated hydraulic conductivity of agricultural soils based on physical and chemical properties of soil. The results showed that structure of ANN model with two hidden layers gave best result compared to other structures; however, Moosavi and Sepaskhah [34] also employed ANN with two hidden layers for estimating unsaturated hydraulic conductivity from soil properties.

After training process, the developed ANN model was tested by 3 patterns of data which were not use in training process. The error criteria during testing process of ANN model for prediction of unsaturated hydraulic conductivity of agricultural soils were shown in Table 5. In general, the RMSE, MRE and MAE (0.000333251 cm/sec, 4.184% and 0.000284699 cm/sec) measure showed a small error between the measured the predicted values for unsaturated hydraulic conductivity of agricultural soils (Table 5), suggesting that the employed ANN model was very accurate in predicting the values of it. The results suggest that the ANN model could be used as a reliable tool that can be employed for estimating unsaturated hydraulic conductivity of agricultural soils at any values of independent variables, falling within the range of values in the study.

Table 5. Error criteria during testing process of ANN model for prediction of unsaturated hydraulic conductivity of agricultural soils

Error criteria	Unit	Value
RMSE	(cm/sec)	0.000333251
MAE	(cm/sec)	0.000284699
MRE	(%)	4.184
R^2	(---)	0.9979

Fig. (7) shows the relationship between the measured and the predicted unsaturated hydraulic conductivity of soil during testing phase using the developed ANN model. The figure clearly shows that the points, during the testing process, are uniformly scattered around the regression line with high correlation represented by values of coefficients of determination (R^2) that were 0.9979.

3.3 Contribution of Inputs on Predicted Unsaturated Hydraulic Conductivity of Agricultural Soils

The Qnet provides a contribution calculation for how the change in each input, changes the output prediction. The contribution percentage of the eight input variables to the output (unsaturated hydraulic conductivity of agricultural soils) was calculated using the developed ANN model and the results are illustrated in Table 6. It can be deduced from Table 6 that the major contribution to the unsaturated hydraulic conductivity of agricultural soils was attributed to the clay content in the soil with a contribution percentage of 22.2%. Hence, clay content in a soil should be obtained accurately as it greatly influence the resulting predicted the unsaturated hydraulic conductivity of agricultural soils within the boundaries of the training data set used in the study. This result is in agreement with the finding by Chiu and Shackelford [52] who reported that the major factor to predict the unsaturated hydraulic conductivity of the sand-clay mixture was the amount of clay in the mixture. In addition, Al-Shayea [53] reported that the hydraulic conductivity sharply decreases with increasing clay content up to 40%.

The other variables, i.e initial soil water content and soil bulk density were also found to have high influence on the predicted the unsaturated hydraulic conductivity of agricultural soils with a contribution percentage of 20.63 and 19.12%, respectively. Meanwhile, silt content, soil organic matter were also found to have moderate influence on the predicted the unsaturated hydraulic conductivity of agricultural soils with a contribution percentage of 13.6 and 12.62%, respectively. Finally, soil electric conductivity, soil sodium adsorption ratio and sand content in a soil were found to have lower influence on the predicted unsaturated hydraulic conductivity of agricultural soils with a contribution percentage of 6.88, 3.17 and 2.5%, respectively. However, McNeal and Coleman [54] reported that saturated hydraulic conductivity of soil decreased with decreasing total electrolyte concentration and increasing SAR values.

Table 6. Contribution percentage of eight independent variables used in the 8-8-14-1 ANN model for prediction of unsaturated hydraulic conductivity of agricultural soils

Input variables	Contribution (%)
Sand	2.5
Silt	13.06
Clay	22.02
EC	6.88
SAR	3.17
Organic matter	12.62
Initial soil water content	20.63
Soil bulk density	19.12

3.4 Performance of C# Application

To valid the C# application to predict unsaturated hydraulic conductivity of agricultural soils, an interface for input required data was developed as shown in Fig. (8). When the inputs were 82.9%, 13.08%, 4.02%, 4.6 dS/m, 2.08, 0.98%, 4.71% db and 1.36 g/cm^3 for sand, silt, clay, electric conductivity of soil, soil sodium adsorption ratio, soil organic matter, initial soil water content and soil bulk density, respectively, the unsaturated hydraulic conductivity of soil obtained by the C# application was 0.001815 cm/sec (Fig. 9), meanwhile, the measured unsaturated hydraulic conductivity of soil was 0.001756 cm/sec. Also when the inputs were 86.92%, 6.04%, 7.04%, 2.9 dS/m, 2.15,0.62%,

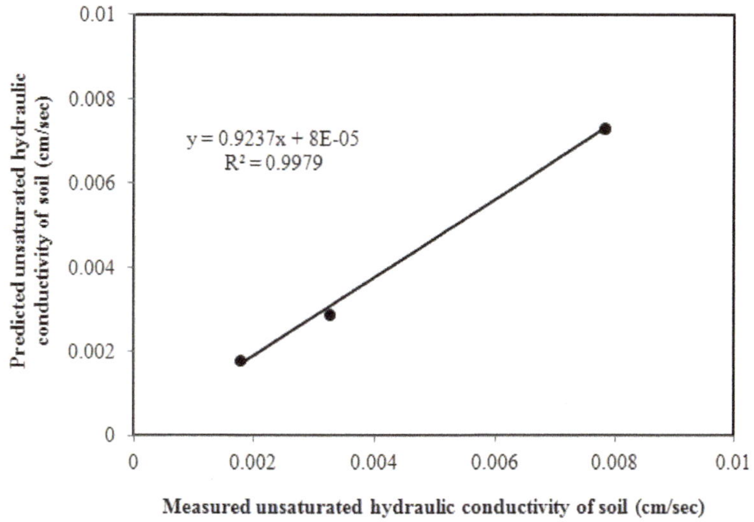

Fig. 7. Relationship between the measured and the predicted unsaturated hydraulic conductivity of soil during testing phase using the developed ANN model

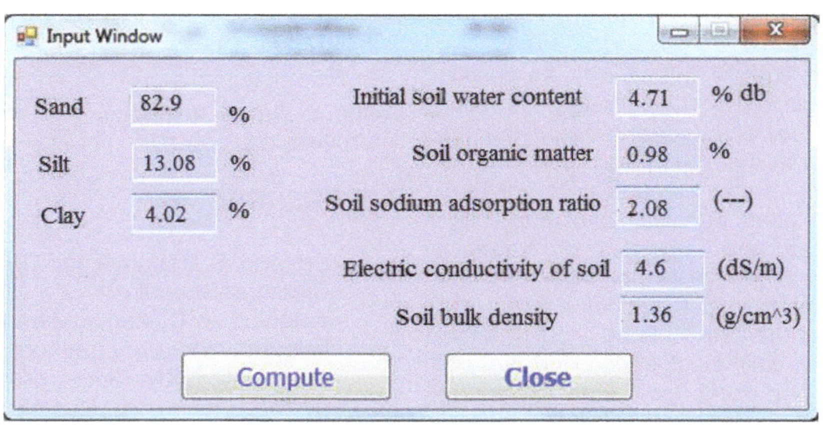

Fig. 8. Screenshot of interface for input required data for unsaturated hydraulic conductivity of soil prediction

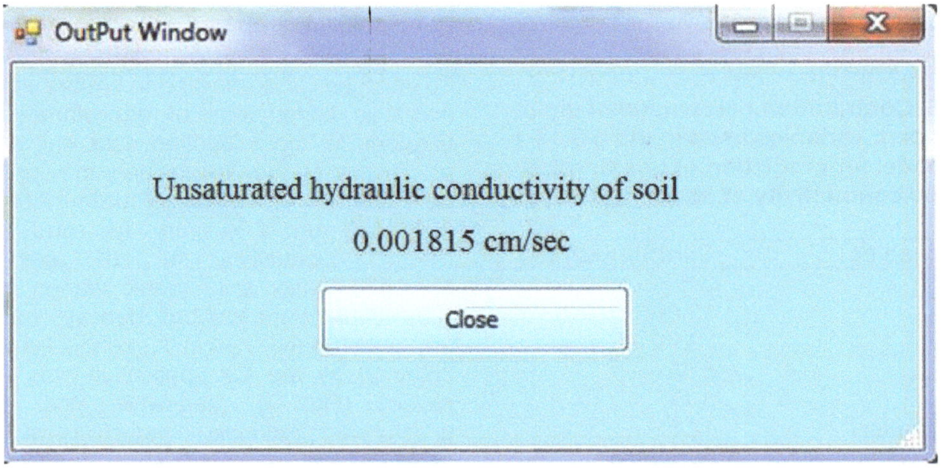

Fig. 9. Screenshot of computer interface for predicting unsaturated hydraulic conductivity of soil at specific input data

6.26% db and 1.41 g/cm^3 for sand, silt, clay, electric conductivity of soil, soil sodium adsorption ratio, soil organic matter, initial soil water content and soil bulk density, respectively, the unsaturated hydraulic conductivity of soil obtained by the C# application was 0.002913 cm/sec, meanwhile, the measured unsaturated hydraulic conductivity of soil was 0.003228 cm/sec. Additionally, When the inputs were 90.95%, 6.04%, 3.02%, 0.69 dS/m, 3.73,1.14%, 4.87%db and 1.54 g/cm^3 for sand, silt, clay, electric conductivity of soil, soil sodium adsorption ratio, soil organic matter, initial soil water content and soil bulk density, respectively, the unsaturated hydraulic conductivity of soil obtained by the C# application was 0.007336 cm/sec, meanwhile, the measured unsaturated hydraulic conductivity of soil was 0.007816 cm/sec.

4. CONCLUSION

As Saudi Arabia is semi-arid country, an applied model for prediction of unsaturated hydraulic conductivity of soil is needed for studying the moving of water in the unsaturated soil in order to prevent of water waste. The higher performance, higher prediction accuracy and ability to re-learn are important to create a powerful model. The results obtained from the study show that the ANN model learned the relationship between the eight input factors (sand, silt, clay, soil electric conductivity, soil sodium adsorption ratio, soil organic matter, initial soil water content and soil bulk density) and unsaturated hydraulic conductivity of soil. The root mean squared error (0.000333251 cm/sec), mean relative error (4.184%) and mean absolute error (0.000284699 cm/sec) for testing data set were showed a small error between the actual and the predicted values of unsaturated hydraulic conductivity of agricultural soils, suggesting that the employed ANN model was very accurate in estimating the values of the unsaturated hydraulic conductivity of agricultural soils.

ACKNOWLEDGEMENTS

The authors are grateful to King Abdulaziz City for Science and Technology (KACST) for funding of the research project (Grant No. ARP-33-51). Also, the authors are grateful to ElBahloul Abdussalem, Lecture of programing, Shaqra University, Community College, Huraimla, Saudi Arabia for his valuable work in building C# application.

COMPETING INTERESTS

Authors have declared that no competing interests exist.

REFERENCES

1. Siltecho S, Hammecker C, Sriboonlue V, Clermont-Dauphin C, Trelo-ges V, Antonino ACD, Angulo-Jaramillo R. Use of field and laboratory methods for estimating unsaturated hydraulic properties under different land-use. Hydrology and Earth System Sciences Discussions. 2014; 11(6):6099-6137.

2. Lakzian A, Aval MMB, Gorbanzadeh N. Comparison of pattern recognition, artificial neural network and pedotransfer functions for estimation of soil water parameters. Notulae Scientia Biologicae. 2010; 2(3):114-120.
3. Emami H, Shorafa M, Neyshabouri MR. Evaluation of hydraulic conductivity at inflection point of soil moisture characteristic curve as a matching point for some soil unsaturated hydraulic conductivity models. JWSS - Isfahan University of Technology. 2012; 16(59):169-182.
4. Kalkhajeh YK, Arshad RR, Amerikhah H, Sami M. Multiple linear regression, artificial neural network (MLP, RBF) and ANFIS models for modeling the saturated hydraulic conductivity of tropical region soils (a case study: Khuzestan province: southwest Iran. International Journal of Agriculture: Research and Review. 2012; 2(3):255-265.
5. Fereshte HF. Evaluation of artificial neural network and regression PTFS in estimating some soil hydraulic parameters. Pro-Environment. 2014;7:10–20.
6. Schuh WM, Bauder JW. Effect of soil properties on hydraulic conductivity moisture relationships. Soil Science Society of America Journal. 1986;50:848-855.
7. Vereecken H, Maes J, Feyen J. Estimating unsaturated hydraulic conductivity from easily measured soil properties. Soil Science.1990;149:1-12.
8. Vereecken H. Estimating the unsaturated hydraulic conductivity from theoretical models using simple soil properties. Geoderma. 1995;65:81-92.
9. Skaggs TH, Arya LM, Shouse PJ, Mohanty BP. Estimating particle-size distribution from limited soil texture data. Soil Science Society of America Journal. 2001;65:1038-1044.
10. Tamari S, Wosten HM, Ruiz-Suarez JC. Testing an artificial neural network for predicting soil hydraulic conductivity. Soil Science Society of America Journal. 1996;60:1732.
11. Nosrati KF, Movahedi NSAR, Hezarjaribi A, Roshani A, Dehghani AA. Using artificial neural networks to estimate saturated hydraulic conductivity from easily available soil properties. Electronic Journal of Soil Management and Sustainable Production. 2012;2(1):95-110.
12. Kirkham MB. Principles of Soil and Plant Water Relations. Elsevier Academic Press: Burlington, MA. 2005;145-172.
13. Torabi FN, Ghahraman B, Kashkouli HA, Mahmoud HS. Measurement of unsaturated soil hydraulic conductivity and comparison with four estimation models. Iranian Journal of Soil and Waters Sciences. 2006;20(2):296-308.
14. Decagon Devices. Minidisk infiltrometer. User's manual. 2012;26. Available:www.decagon.com
15. Lu D., Shao M. Models and determining methods of unsaturated soil hydraulic parameters. Ying Yong Sheng Tai Xue Bao. 2004;15(1):163-166.
16. Khaleel R, Relyea JF. Evaluation of van Genuchten–Mualem relationships to estimate unsaturated hydraulic conductivity at low water contents. Water Resources Research. 1995;31:2659–1668.
17. Zhuang J, Nakayama K, Yu GR, Miyazaki T. Predicting unsaturated hydraulic conductivity of soil based on some basic soil properties. Soil & Tillage Research. 2001;59:143–154.
18. Kadhim AJ. Models to predict unsaturated hydraulic conductivity with the use of retc code. Iraqi Journal of Agricultural Sciences. 2011;42(Special Issue):105-163.
19. Neshat A, Farhad M. A presentation of an experimental model for unsaturated hydraulic conductivity under affection of physical properties of Soil: A case-by-case study of Baghin Plain in Kerman, Iran. Academic Journal of Plant Sciences. 2012;5(3):70-75.
20. Klipa V, Sacha J, Snehota M, Dohnal M, Zumr D., Tacheci P. Automated multi-point mini-disk infiltrometer measurements of unsaturated hydraulic conductivity. GU General Assembly. 2014, held 27 April-2 May, 2014 in Vienna, Austria, id.7230.
21. Abrisqueta JM, Plana V, Ruiz-Canales A, Ruiz-Sanchez MC. Unsaturated hydraulic conductivity of disturbed and undisturbed loam soil. Spanish Journal of Agricultural Research. 2006;4(1):91-96.
22. Malaya C, Sreedeep S. Correlation between grain size distribution curve and unsaturated hydraulic conductivity curve of soils. Proceedings of Indian Geotechnical Conference December 22-24, 2013, Roorkee.

23. Van Genuchten M, Leij FJ. On estimating the hydraulic properties of unsaturated soils. In: van Genuchten M.Th, Leij FJ, Lund LJ, editors. Proceedings of the International Workshop, Indirect Methods for Estimating the Hydraulic Properties of Unsaturated Soils, University of California, Riverside. 1992;1-14.
24. Neyshabouri MR, Rafiee Alavi SR, Rezaei H, Nazemi AH. Estimating unsaturated hydraulic conductivity from air permeability. In '19th World Congress of Soil Science Soil Solutions for a Changing World'. 1–6 August 2010, Brisbane, Qld. (IUSS); 2010.
25. Doussan C, Ruy S. Prediction of unsaturated soil hydraulic conductivity with electric conductivity. Water Resources Research. 2009;45:1-12.
26. Campbell GS. A simple method for determining unsaturated conductivity from moisture retention data. Soil Science. 1974;117:311–314.
27. Amer AM, Logsdn SD, Davis D). Prediction of hydraulic conductivity in unsaturated soils. Soil Sci. 2009;174(9):508-515.
28. Neyshabouri MR, Rahmati M, Doussan C, Behroozinezhad B. Simplified estimation of unsaturated soil hydraulic conductivity using bulk electrical conductivity and particle size distribution. Soil Research. 2012;51(1):23-33.
29. Maren AJ, Harston CT, Pap RM. Handbook of neural computing applications. San Diego (CA): Academic Press.1990
30. Minasny B, Hopmans JW, Harter T, Eching SO, Tuli A, Denton MA. Neural networks prediction of soil hydraulic functions for alluvial soils using multistep outflow data. Soil Sci. Soc. Am. J. 2004;68:417–429.
31. Ghanbarian-Alavijeh B, Liaghat AM, Sohrabi S. Application of artificial neural networks in prediction of saturated hydraulic conductivity using soil physical parameters. Journal of Agricultural Engineering Research. 2009;10(1):97-112.
32. Erzin Y, Gumaste SD, Gupta AK, Singh DN. Artificial neural network (ANN) models for determining hydraulic conductivity of compacted fine-grained soils. Canadian Geotechnical Journal. 2009;46(8):955-968.
33. Ghanbarian-Alavijeh B, Liaghat AM, Sohrabi S. Estimating saturated hydraulic conductivity from soil physical properties using neural networks model. International Scholarly and Scientific Research & Innovation. 2010;4(2):108-113.
34. Moosavi A A, Sepaskhah A. Artificial neural networks for predicting unsaturated soil hydraulic characteristics at different applied tensions. Archives of Agronomy and Soil Science. 2012;58(2):125-153.
35. Suarez DL, Wood JD, Lesch SM. Infiltration into cropped soils: Effect of rain and sodium adsorption ratio–impacted irrigation water. Journal of Environmental Quality. 2008;37(5) Supplement:169-179.
36. Fatehnia M, Tawfiq K, Abichou T. Comparison of the methods of hydraulic conductivity estimation from mini disk infiltrometer. EJGE. Bund. E. 2014;19: 1047-1063.
37. Zhang R. Determination of soil sorptivity and hydraulic conductivity from the disk infiltrometer. Soil Science Society of America Journal. 1997;61:1024-1030.
38. Carsel RF, Parrish RS. Developing joint probability distribution of soil water retention characteristics. Water Resources Research. 1988;24:755-769.
39. Pachepsky YA, Timlin DJ, Varallyay G. Artificial neural networks to estimate soil water retention from easily measurable data. Soil Sci Soc Am J. 1996;60:727–773.
40. Schaap MG, Leij FL, Van Genuchten M. Neural network analysis for hierarchical prediction of soil hydraulic properties. Soil Sci Soc Am J. 1998;62:847–855.
41. Rumelhart DE, Hinton GE, Williams RJ. Learning representations by back propagating errors. Nature. 1986;323:533–536.
42. Vesta Services. Vesta Services, Inc. Qnet2000 Shareware, Vesta Services, Inc., 1001 Green Bay Rd, STE 196, Winnetka, IL 60093; 2000.
43. Zhang D, Jiang Q, Li X. Application of neural networks in financial data mining. International Journal of Computational Intelligence. 2005;1:106-109.
44. Ahmed FE. Artificial neural networks for diagnosis and survival prediction in colon cancer. Molecular Cancer. 2005;4:29.
45. Nath A, Chattopadhyay PK. Optimization of oven toasting for improving crispness and other quality attributes of ready to eat potato-soy snack using response surface methodology. Journal of Food Engineering. 2007;80(4):1282-1292.
46. Sin HN, Yusof S, Shikh Abdulhamid N, Abdulrahman R. Optimization of enzymatic clarification of sapodilla juice using

response surface methodology. Journal of Food Engineering. 2006;73(4):313-319.

47. Post E. Designing and implementing a grid application for cumulative agrichemical residue tracking using third-party data sources and software components. Journal of Research and Practice in Information Technology. 2007;39(2):125-136.
48. Sharma A, Lal SB. SSDA Analysis - A class library for analysis of sample survey data. International Journal of Modern Engineering Research. 2012;2(1):242-246.
49. Gulser C, Candemir F. Using soil moisture constants and physical properties to predict saturated hydraulic conductivity. Eurasian Journal of Soil Science. 2014;3:77-81.
50. Hudson B. Soil organic matter and available water capacity. J Soil Water Conservation. 1994;49:189–193.
51. Olourunfemi IE, Fasinmirin JT. Hydraulic conductivity and infiltration of soils of tropical rain forest climate of Nigeria. Proceedings of the Environmental Management Conference, Federal University of Agriculture, Abeokuta, Nigeria. 2011;397-413.
52. Chiu TF, Shackelford CD. Unsaturated hydraulic conductivity of compacted sand-kaolin mixture. Journal of geotechnical and Geo-environmental Engineering.1998;124(2):160-170.
53. Al-Shayea NA. The combined effect of clay and moisture content on the behavior of remolded unsaturated soils. Engineering Geologey. 2001;62:319-342.
54. McNeal BL, Coleman NT. Effect of solution composition on soil hydraulic conductivity. Soil Sci. Soc. Am. Proc. 1966;30:308-312.

17

Macro Fungi Biodiversity at the Central and Northern Biosphere Reserved Areas of Tropical Moist Deciduous Forest Region of Bangladesh

M. I. Rumainul[1] and F. M. Aminuzzaman[1*]

[1]Department of Plant Pathology, Faculty of Agriculture, Sher-e-Bangla Agricultural University, Sher-e-Bangla Nagar, Dhaka-1207, Bangladesh.

Authors' contributions

This work was carried out in collaboration between both authors. Author MIR wrote the protocol, carried out the research and wrote the first draft of the manuscript. Author FMA designed and supervised the study also edited the manuscript. Both authors read and approved the final manuscript.

Editor(s):
(1) Anonymous.
Reviewers:
(1) Anonymous, Mexico.
(2) Onguene Awana Nérée,University of Dschang, Cameroon an University of Yaoundé II, Cameroon.
(3) P. N. Krishnan, Tropical Botanic Garden and Research Institute, Palode, Thiruvanathpuram, Kerala, India.
(4) Rajesh Kumar, Rain Forest Research Institute, Jorhat, Assam, India.

ABSTRACT

Mushrooms are considered as macro fungi as well as seasonal fungi, which occupy diverse niches in natural forest ecosystem. It appears predominantly during rainy season, if temperature and moisture become favorable in that ecosystem. Investigation of diversity of macro fungi is gaining importance because of the disappearance due to the threat of habitat destruction. This investigation deals with the biodiversity of macro fungi at Dhaka, Gazipur, Bogra, Rajshahi, Pabna, Jaipurhat and Dinajpur district of central and northern region of Bangladesh, where 50 samples were collected, morphologically characterized, photographed and preserved. They were identified to 8 genera and 9 species. The predominant genera were *Trametes*, *Daedaleopsis*, *Collybia* and *Armillaria*. This preliminary investigation suggested that, the central and northern region of Bangladesh under tropical moist deciduous forest region is enriched with diversity of wild mushrooms.

**Corresponding author: Email: aminsaupp@yahoo.com*

Keywords: Macro fungi; biodiversity; tropical moist deciduous forest; Bangladesh.

1. INTRODUCTION

Mushroom is typically above ground producing fleshy spore bearing fruiting body of a fungus. Some of them are edible along with high nutritive value. Wild types of edible mushroom have been consumed by human being from thousands of years. Though its nutritive value is still untapped, edible species of mushroom are associated with people living 13,000 years ago in Chile, according to the geological records [1]. Nowadays, some mushrooms are also considered as a source of diet in human health and fetch income in Bangladesh, due to the increasing demand of consumption of mushroom as food and medicine.

Bangladesh is a South Asian country with very low proportion of forests. The total 10.2% of the land area in Bangladesh are covered by forest, which amounts approximately 1.3 million hectares [2]. The government forests area are situated within 12 districts in eastern and south-eastern part of the country, which covers more than 90% [3]. Only 4.65% of forested area are situated in the central and north-western part of the country, which covering 0.12 million ha [4]. These forests area are classified as tropical moist deciduous forest region of Bangladesh [5]. The texture of this forest's soil is generally sandy loam as well as silty loam. The surface soil of this forest area are consisting of organic matters, which ranges from low (<1.5% under grassland) to moderate (2-5%under forest cover) and acidic in nature (pH5.2 -5.5), as well [6]. In the northern part, the mean annual rainfall increases more than 2,300 mm, whereas the mean annual temperature is 26.3°C. The average maximum and minimum temperatures are approximately 27.5°C and 18.5°C, accordingly [7].

Mushrooms are typically found in waste lands, where generally absence of artificial activities of any human-being such as, forest lands, field lands or, on the moist branches of trees. Throughout this investigation, we tried to represent the existing biodiversity of mushroom in this forest region. Bangladesh has basic four types of forest, among them this investigation was conducted in Dhaka, Gazipur, Bogra, Rajshahi, Pabna, Jaipurhat, and Dinajpur, which are situated in the tropical moist deciduous forest region of Bangladesh. Due to the wide variation in climates, soils and plant communities, the tropical moist deciduous forest region of Bangladesh has within its boundaries is one of the greatest diversity of mushrooms. Several types of mushroom are grown abundantly in this forest region during and after the rainy season. Therefore, the aim of this initial investigation is to promote knowledge about the available common type of wild mushroom of the central and northern forest area of Bangladesh belongs to tropical moist deciduous forest region. This article represents as part of continuing investigations describing some newly reported mushrooms from the central and northern biosphere reserved areas of tropical moist deciduous forest region of Bangladesh.

2. MATERIALS AND METHODS

2.1 Survey Area

The survey area was Dhaka, Gazipur, Bogra, Rajshahi, Pabna, Jaipurhat and Dinajpur of central and northern biosphere reserved areas of tropical moist deciduous forest region of Bangladesh (Fig. 1). Survey was conducted during the end of rainy season starting from July to October, 2013 and 2014 respectively. According to the Bangladesh Meteorological Department, Dhaka is located at 23°42′37″ N (Latitude), 90°24′26″ E (Longitude) and it has an average elevation of 4 meters (13.12 ft.). Gazipur is located at 24.00°N (Latitude), 90.43°E (Longitude) and it has an average elevation of 10 meters (32.8 ft.). Bogra is located at 24°51′00″ N (Latitude), 89°22′00″ E (Longitude) and it has an average elevation of 24 meters (78 ft.). Rajshahi is located at 24°22′00″ N (Latitude), 88°36′00″ E (Longitude) and it has an average elevation of 31 meters (101 ft.). Pabna is located at 24°00′00″ N (Latitude), 89°15′00″ E (Longitude) and it has an average elevation of 8 meters (26 ft.). Jaipurhat is located at 25.1° N (Latitude), 89.02° E (Longitude) and it has an average elevation of 29 meters (94.25 ft.). Dinajpur is located at 25°37′37″ N (Latitude), 88°38′19″ E (Longitude) and it has an average elevation of 34 meters (111 ft.). University campus, residential area, botanical garden, parks, forest and nearby villages of those above mentioned areas of tropical moist deciduous forest region were selected as collection area.

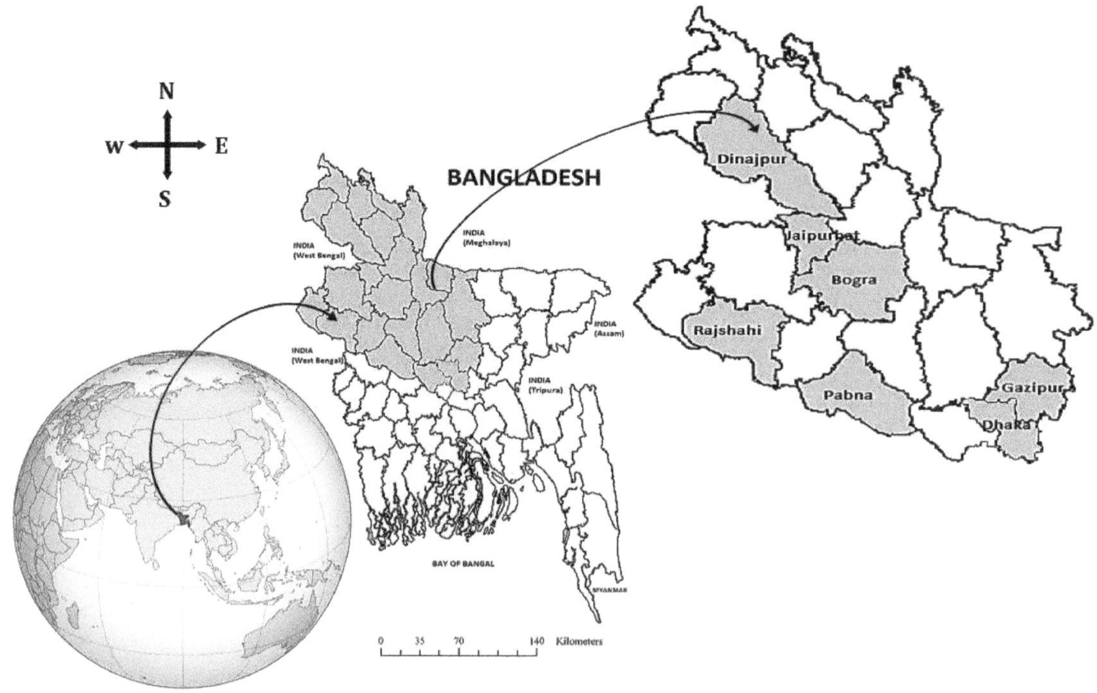

Fig. 1. Central and northern biosphere reserved areas of tropical moist deciduous forest region of Bangladesh

2.2 Collection of Mushrooms

Collection was conducted following the method of Hailing [8]. Survey of different locations was done systematically at different area in the central and northern part of tropical moist deciduous forest region of Bangladesh during July to October, 2013 and 2014, respectively. Required equipment was carried out while collection, such as isolation kit, slants, isolation chamber, data recording sheet, digital camera for photograph, digging tools, heat convector card board. During day time, collection of specimens were usually done, while a field characteristics of mushrooms were recorded in the data sheet, which was also prepared by following Molina et al. [9]. After that, the specimens were dried in hot air oven at 40℃ to 50℃ and stored in air tight containers with desiccators like, silica gel for future microscopic studies. A detailed study was conducted into the laboratory. Photographs were taken through using Sony Cyber-Shot Camera with the magnification of 14.2 megapixels.

2.3 Identification of Collected Mushrooms

The collected mushroom samples were identified on their morphological, macroscopic, physiological and ecological features according to the previously published documents by Arora, Singer [10-12]. The morphological parameters used for identification of mushrooms specimens were cap color, cap surface, cap margin, cap diameter, stipe length, gill attachment, gill spacing and spore diameter [1]. The spore diameter and the photograph of spores were calculated using the Motic microscope (Motic Images Plus 2.0) with the magnification of 40x. One mature and one immature (growing) mushroom were sampled for each collection to record accurate information. Through comparing recorded characteristics the mushrooms were identified following Dickinson and John, Jorden, Pegler and Spooner [13-15].

3. RESULTS

Through the investigation, 8 genera and 9 species of mushroom were identified [Figs. 2 & 3]. The identified genera were *Trametes*, *Daedaleopsis*, *Pycnoporus*, *Boletus*, *Tuber*, *Collybia*, *Panaeolus* and *Armillaria*. The details about the identified mushrooms through this investigation on that particular time in the tropical moist deciduous forest region (central and northern biosphere reserved areas) of Bangladesh were described below:

3.1 Genus: *Trametes*

3.1.1 *Trametes versicolor*

Family: Polyporaceae
Common name: Polypore or Bracket mushroom
Color: Young: Dark violate along with white border; Mature: Dark violate along with white border
Length: Mature: 8.2 cm; Young: 4.8 cm
Width: Mature: 5.4 cm; Young: 3.8 cm
Spore bearing surface under cap: Pores on hymenium
Spore diameter (Average): Length: 8.62 μm; Width: 5.6 μm
Pileus: Color: Creamy; Surface character: Dry in nature; Margin: Regular in shape; Cuticle: Not peeling
Texture of the fruiting body: Brittle and woody
Flesh odor: Fragrant and disagreeable
Pores color: White
Ecological features: Locality found: Dinajpur; Nature: Saprotrophic; Habitat: On the bark wood of the tree, in closely associated with Royal siris (*Albizziaprocera*) tree; Habit: Solitary and somewhat are caespitose cluster; Type of association: Stem associated; Forest type: Leveled; Constancy of occurrence in specific habitat: Abundant; Type of soil: Loamy soil in nature; Factors affecting their distribution: Moist weather around the habitat.

3.2 Genus: *Daedaleopsis*

3.2.1 *Daedaleopsis confragosa*

Family: Polyporaceae
Common name: Blushing bracket
Color: Young: Whitish to creamy in circle; Mature: Whitish to creamy in circle and somewhat pinkish circle around the border
Length: Mature: 4.1 cm to 5.4 cm; Young: 1.4 cm to 1.5 cm;
Width: Mature: 6.4 cm to 8 cm; Young: 3 cm to 3.8 cm
Spore bearing surface under cap: Teeth in very young stage
Spore diameter (Average): Length: 6.88 μm; Width: 6.44 μm
Pileus: Color: Pinkish; Surface character: Leathery and dry in nature; Margin: Regular in shape; Cuticle: Not peeling
Texture of the fruiting body: Brittle, tough, leathery and woody
Flesh odor: Fragrant
Firmness: Very hard
Ecological features: Locality found: Botanical garden, Mirpur, Dhaka; Habitat: On bark wood of the tree, in an association near the Ipil-Ipil (*Leucaenaleucocephala*) tree; Habit: Solitary and somewhat are caespitose cluster; Type of association: Trunk associated; Forest type: Leveled; Constancy of occurrence in specific habitat: Abundant; Type of soil: Loamy soil in nature; Factors affecting their distribution: Literally more moist weather around the habitat.

3.2.2 *Daedaleopsis confragosa* var. *tricolor*

Family: Polyporaceae
Common name: Blushing Bracket
Color: Young: Dark red; Mature: Dark red
Length (Mature): 8.5 cm
Width (Mature): 12.3 cm
Spore bearing surface under cap: Teeth (beginning stage)
Spore diameter (Average): Length: 4.84 μm; Width: 4.06 μm
Cap of the carpophore: Size: 8.5 cm x 12.3 cm; Shape: Umbilicate
Pileus: Color: Pinkish; Surface character: Leathery and moist in nature; Margin: Regular in shape; Cuticle: Not peeling
Texture of the fruiting body: Brittle, tough, leathery and woody
Flesh odor: Fragrant
Firmness: Solid and very hard
Ecological features: Locality found: Botanical garden, Mirpur, Dhaka; Nature: Saprotrophic; Habitat: On bark wood of the tree, in an association near the Ipil-Ipil (*Leucaena leucocephala*) and Golden shower (*Acacia auriculiformis*); Habit: Caespitose cluster; Type of association: Root associated; Forest type: Mixed; Constancy of occurrence in specific habitat: Abundant; Type of soil: Loamy soil in nature; Factors affecting their distribution: More moist weather around the habitat.

3.3 Genus: *Pycnoporus*

3.3.1 *Pycnoporus cinnabarinus*

Family: Polyporaceae
Common name: Cinnibar Red Polypore
Color: Young: Orange-Red; Mature: Orange-Red
Length: Mature: 4.2 cm; Young: 3.1 cm
Breadth: Mature: 2.5 cm; Young: 1.8 cm
Width: Mature: 0.8 cm to 9 cm
Spore bearing surface under cap: Pores
Spore diameter (Average): Length: 3.75 μm; Width: 2.48 μm

Cap of the carpophore: Shape: Flat
Pileus: Color: Orange-red; Surface character: Smooth, Silky, somewhat dry in nature; Margin: Incurved in shape; Cuticle: Not peeling
Texture of the fruiting body: Woody
Flesh odor: Fragrant
Firmness: Solid
Ecological features: Locality found: Bogra; Nature: Parasitic; Habitat: On humus, in an association with the stem of Bamboo (*Bambuseae*) tree; Habit: Scattered; Type of association: Closely associated; Forest type: Leveled; Constancy of occurrence in specific habitat: Un-abundant; Factors affecting their distribution: Less moist or moderately dry weather around the habitat

3.4 Genus: *Boletus*

3.4.1 *Boletus amygdalinus*

Family: Boletaceae
Common name: Summer bolete
Color: Young: Yellowish with brown shades; Mature: Yellow
Length (Mature): 9.8 cm
Width (Mature): 7.1 cm
Spore bearing surface under cap: Pores
Spore diameter (Average): Length: 12.1 μm; Width: 5.6 μm
Cap of the carpophore: Size: 7.1 cm × 2.5 cm; Shape: Convex
Pileus: Color: Greyish and brownish; Surface character: Smooth, silky, leathery and dry in nature; Margin: Regular; Cuticle: Half peeling
Texture of the fruiting body: Spongy and some extent woody
Flesh odor: Fragrant
Lamellae: Present; Pores attachment: Adnate; Pores color: Pale yellow; Gill shape and width: Broad; Pores spacing: Crowed; Lamellulae: Present; Forking pattern: Unbranched
Stipe: Present; Size: 7.3 cm; Shape: Bare; Position: Central; Surface characteristics: Dry and polished; Color: Yellowish and some are greyish to light brown (change color upon bruising); Firmness: Solid
Ecological features: Locality found: Botanical garden, Mirpur, Dhaka; Habitat: On soil; near the mango (*Magifera indica*) and Ipil-Ipil (*Leucaenaleucocephala*) trees; Habit: Scattered; Type of association: Scattered; Forest type: Leveled; Constancy of occurrence in specific habitat: Un-abundant; Type of soil: Loom; pH of soil: 6.1; Moisture: 6.8%, Temperature: 35℃; Factors affecting their distribution: Less moist weather

3.5 Genus: *Tuber*

3.5.1 *Tuber aestivum*

Family: Tuberaceae
Common name: Truffle
Color: Young: White; Mature: White
Length (Mature): 1.1 cm to 1.5 cm
Width (Mature): 1.1 cm to 1.3 cm
Spore bearing surface under cap: Glebal
Spore diameter (Average): Length: 12.9 μm; Width: 7.94 μm
Cap of the carpophore: Size: 1.1 cm to 1.3 cm; Shape: Flat
Pileus: Color: White; Surface character: Glabrous and dry in nature; Margin: Irregular; Cuticle: Not peeling
Texture of the fruiting body: Spongy
Flesh odor: Fragrant
Ecological features: Locality found: Botanical garden, Mirpur, Dhaka; Habitat: On bark wood of the plant; in an association with the Dahlia plant (*Dahlia* sp.); Habit: Scattered and some are caespitose cluster; Type of association: Close; Forest type: Leveled; Constancy of occurrence in specific habitat: Un-abundant; Type of soil: Loom; Factors affecting their distribution: Less moist weather.

3.6 Genus: *Collybia*

3.6.1 *Megacollybia platyphylla*

Family: Marasmiaceae
Common name: Broad gilled collybia
Color: Young: White; Mature: White
Length: 1.3 cm
Width: 1.4 cm
Spore bearing surface under cap: Ridges or ribs
Spore diameter (Average): Length: 4.63 μm; Width: 4.13 μm
Cap of the carpophore: Size: 1.4 cm
Pileus: Color: Whitish; Surface character: Smooth and dry in nature; Margin: Incurved
Texture of the fruiting body: Soft and spongy
Flesh odor: Fragrant
Lamellae: Present; Gill attachment: Subdecurrent; Gill color: White; Gill shape and width: Narrow; Gill spacing: Crowed; Lamellulae: Present; Forking pattern: Branched
Stipe: Present; Size: 0.5 cm; Shape: Radicating; Position: Excentric; Surface characteristics: Dry and glabrous; Color: White to creamy; Firmness: Solid
Ecological features: Locality found: Ramna park, Ramna, Dhaka; Habitat: On the bark wood

of the tree; in an association with the Mahogany (*Swietenia mahogani*) tree; Habit: Solitary; Type of association: Closed; Forest type: Leveled; Constancy of occurrence in specific habitat: Un-abundant; Type of soil: Loom; Factors affecting their distribution: More moist weather

3.7 Genus: *Panaeolus*

3.7.1 *Panaeolus foenisecii*

Family: Bolbitiaceae
Common name: Brown hay cap
Color: Young: Pink and purple; Mature: Pink and purple
Length: 4.9 cm
Width: 1.0 cm
Spore bearing surface under cap: Gills
Spore diameter (Average): Length: 11.2 μm; Width: 7.4 μm
Cap of the carpophore: Size: 1.0 cm; Shape: Convex
Pileus: Color: Brownish; Surface character: Smooth; Margin: Regular; Cuticle: Fully peeling
Texture of the fruiting body: Soft and spongy
Flesh odor: Fragrant
Lamellae: Present; Gill attachment: Adnaxed; Gill color: Pale reddish; Gill shape and width: Narrow; Gill spacing: Close; Lamellulae: Present; Forking pattern: Branched
Stipe: Present; Size: 4.5 cm to 4.9 cm; Shape: Bare; Position: Central; Surface characteristics: Dry and glabrous; Color: White to light brown; Firmness: Solid
Ecological features: Locality found: Ramna park, Ramna, Dhaka; Habitat: On humus; Habit: Scattered; Forest type: Leveled; Constancy of occurrence in specific habitat: Un-abundant; Type of soil: Loom; Factors affecting their distribution: Moist weather

3.8 Genus: *Armillaria*

3.8.1 *Armillaria mellea*

Family: Physalacriaceae
Common name: Honey fungus
Color: Young: Brown; Mature: Brown with dark brown spots on cap
Length: 4.8 cm
Width: 1.4 cm
Spore bearing surface under cap: Gills
Spore diameter (Average): Length: 7.15 μm; Width: 3.88 μm
Spore shape: Single walled, smooth and ellipsoidal
Cap of the carpophore: Size: 1.2 cm to 1.4 cm; Shape: Convex or flat
Pileus: Color: Brownish; Surface character and Zonation: Smooth and silky; Margin: Regular
Texture of the fruiting body: Soft and spongy
Flesh odor: Pleasant
Lamellae: Present; Gill attachment: Adnate; Gill color: Pale brownish; Gill shape: Narrow; Gill spacing: Crowed; Lamellulae: Present; Forking pattern: Branched
Stipe: Present; Size: 4.8 cm; Shape: Equal; Position: Central; Surface characteristics: Dry and polished; Color and color changes: Grey to light brown; Firmness: Tubular
Annulus (position): Present and single
Scale: Present and brown
Ecological features: Locality found: Botanical garden, Mirpur, Dhaka; Habitat: On soil, in an association with the Mahogany (*Swieteniamahogani*) and Ipil-Ipil (*Leucaenaleucocephala*) tree; Habit: Solitary; Type of association: Close; Forest type: Leveled; Constancy of occurrence in specific habitat: Abundant; Type of soil: Sandy; Factors affecting their distribution: Moist weather

4. DISCUSSION

Mushroom becomes the object of curiosity and speculation of modern research as it is an important component of the ecosystem, since the beginning of nature. It also becomes more interesting, while the scientist discover its various kinds of properties as well. Nowadays, mushrooms become economically more important, due to its edibility, psychotropic properties, mycorrhizal symbiosis as well as parasitic association with the host trees. The association with such type of similar substrate like wood, litter and soil are also express a role for them in these micro habitats [16]. These kind of fleshy macro fungi generally grown in an alternation of specific season but the most productive seasons are rainy seasons, just starting after summer. Therefore, an investigation was conducted just after the rainy season at the central and northern part of Bangladesh under tropical moist deciduous forest region to identify the preliminary species, those are being available each year into that moist environment on some specific area. As a result among 50 samples, 8 genera and 9 species were able to identify through this present investigation, including 2 species of *Daedaleopsis* and 1 species each of the genera *Trametes, Pycnoporus, Boletus*, *Tuber*, *Collybia*, *Panaeolus* and *Armillaria*. On the other hand, a similar type of investigation was also conducted by Rumainul et al. into these same biosphere reserved areas of tropical moist deciduous forest region of Bangladesh, where 14 genera and 24 species were reported [17].

The genus *Trametes*was found in Dinajpur district of tropical moist deciduous forest region at the northern part of Bangladesh. It was first described by Elias Magnus Fries as a saprobicon the deadwood of hardwoods [18]. The genus has

Fig. 2. Fruiting body of collected mushrooms: a. *Trametes versicolor;* b. *Daedaleopsis confragosa*; c. *Daedaleopsis confragosa* var. *tricolor*; d. *Pycnoporus cinnabarinus;* e. *Boletus amygdalinus;* f. *Tuber aestivum;* g. *Megacollybia platyphylla* (growing phase); h. *Panaeolus foenisecii;* i. *Armillaria mellea*

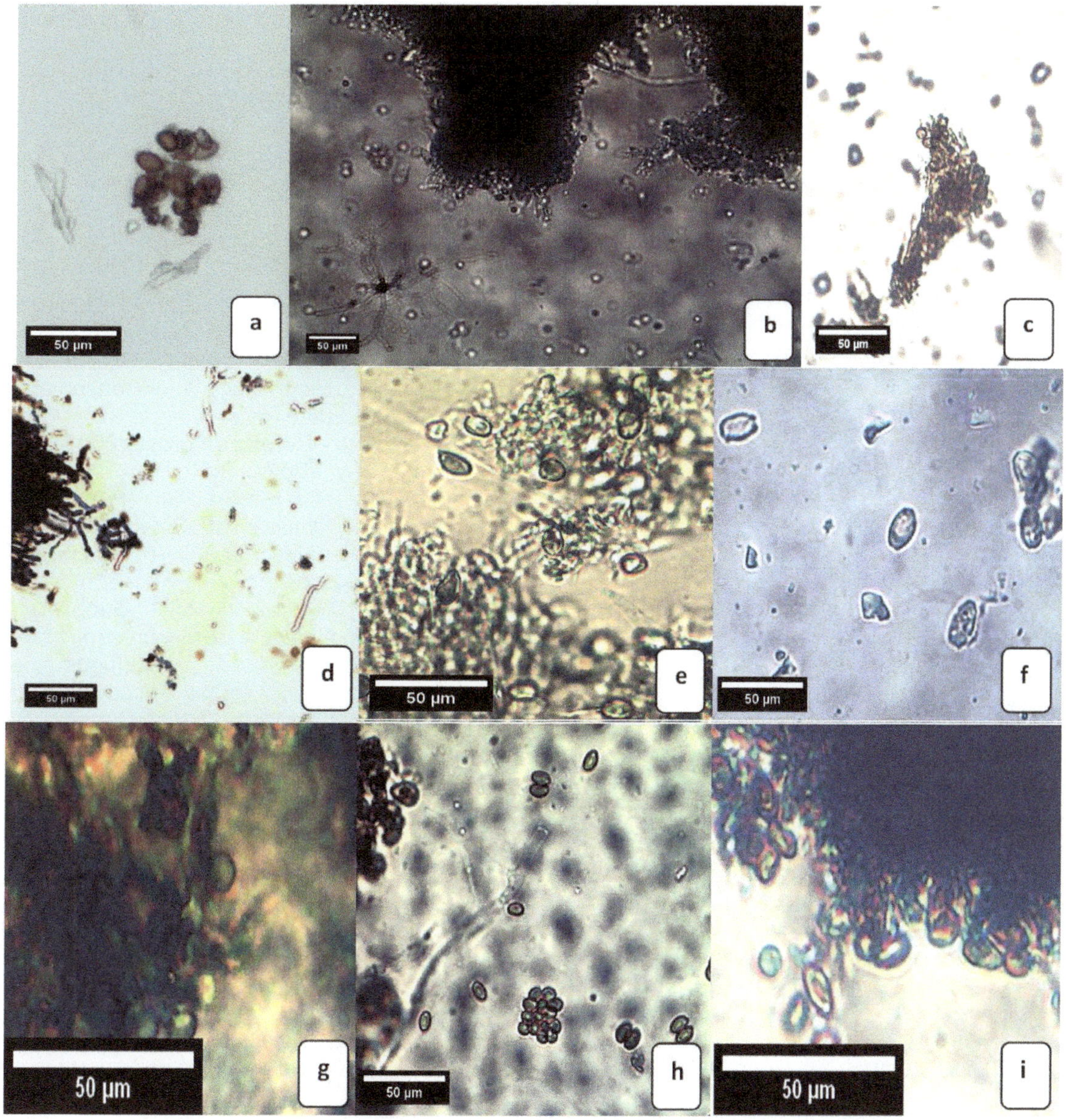

Fig. 3. Spores of the collected mushrooms: a. *Trametes versicolor;* b. *Daedaleopsis confragosa*; c. *Daedaleopsis confragosa* var. *tricolor*; d. *Pycnoporus cinnabarinus;* e. *Boletus amygdalinusa;* f. *Tuber aestivum*; g. *Megacollybia platyphylla;* h. *Panaeolus foenisecii;* i. *Armillaria mellea*

a widespread distribution and contains about 50 species [19]. The genus was also found in the Western Ghats region of India [20], whereas the species *Trametes versicolor* was also reported in and around Bangalore (Karnataka) of India and found medicinal importance [21]. The exact species of *Trametes versicolor* was also reported in Lagos State of Nigeria [22] as well as in Chungcheong Province of Korea [23]. Furthermore, the genus *Daedaleopsis* is widely distributed genus contains 6 species [19]. In this investigation, 2 species of *Daedaleopsis* were found in Dhaka district of Bangladesh. This genus was also first reported in Europe at 1791, having the antioxidant properties [24] and the same species of *Daedaleopsis confragosa* was also reported as saprophyte in Central Luzon of Philippines [25]. On the other hand, *Pycnoporus* sp. was found at Bogra district in Bangladesh. This genus contains 5 distinct species, which was saprobic on the dead wood of hardwoods [26]. It was found in some other places of Europe and North America. *Pycnoporus cinnabarinus* was also found in Northeast Ohio [27] and studied by Rasheeda et al. [28]. Furthermore, *Boletus* sp. was also found in Dhaka of Bangladesh and was first reported by Elias Magnus Fries in 1821 [18]. Almost 100 species were identified of this same genus and this genus was also found earlier in Lithuania, Poland [29]. *Boletus amygdalinus* was also described as saprophyte in nature [30,31].The phylogenic, morphological and taxonomic characteristics of *Borofutus*, a new genus of *Boletaceae* from tropical Asia were also studied in China [32]. *Tuber* sp. was collected from Dhaka district of Bangladesh. It was found in almost all European countries [33]. Furthermore, *Collybia* sp. was also found at Dhaka district of the tropical moist deciduous forest region in Bangladesh throughout the investigation. *Collybia* sp. has nearly 3 species [34]. The genus *Collybia* sp. have a widespread but rarely distributed in north temperate areas as saprophyte [35], whereas the species named *Megacollybia platyphylla* was also reported by Hughes et al. [36] as saprophyte.

Finally, *Armillaria* sp. was found in Dhaka district of the tropical moist deciduous forest region of Bangladesh throughout the investigation. It has commonly 45 species [36]. The species *Armillaria*was also reported by Ram et al. [37] and Giri et al. [38] in India, having medicinal importance. However, *Armillaria mellea* was also studied as saprophyte by Coetzee et al. in 2001 [39].

5. CONCLUSION

In the investigation, 8 genera including 9 species were identified from the 50 collected samples. The identified genera were *Trametes, Daedaleopsis, Pycnoporus, Boletus, Tuber, Collybia, Panaeolus* and *Armillaria*. In the future investigation would also needed in different seasons as well as in different forest regions to identify the new existing exotic varieties of mushroom flora, which will be also beneficial to represent the complete overview of available macro fungi of Bangladesh.

ACKNOWLEDGEMENTS

We would like to thank, the Department of Plant Pathology, Sher-e-Bangla Agricultural University, Sher-e-Bangla Nagar, Dhaka, Bangladesh for the support and help to carry-out this research work.

COMPETING INTERESTS

Authors have declared that no competing interests exist.

REFERENCES

1. Chandulal K, Gopal C, John P. Studies on biodiversity of fleshy fungi in Navsari (South Gujarat), India. International Journal of Biodiversity and Conservation. 2013;5(8):508-514.
2. FAO. State of World's Forests 2003. FAO, Rome, Italy; 2003.
3. GOB. Forestry Master Plan (Participatory Forestry). Ministry of Environment and Forests, Peoples Republic of Bangladesh, Dhaka. 1992;45.
4. Rahman MM, Motiur MR, Guogang Z, Islam KS. A review of the present threats to tropical moist deciduous sal (*Shorearobusta*) forest ecosystem of central Bangladesh. Trop. Conser. Sci. 2010;3(1):59-69.
5. Champion HG, Seth SK, Khattak GM. Forest types of Pakistan. Pakistan Forestry Institute, Peshawar, Pakistan; 1965.
6. Dhar PP, Mridha MAU. Biodiversity of arbuscular mycorrhizal fungi in different trees of madhupur forest Bangladesh. Journal of Forestry Research. 2006; 17:201-205.
7. Banglapedia. National encyclopedia of Bangladesh, Sal forest, Banglapedia; 2008. Available:http://banglapedia.search.com.bd/HT/S_0035.htm (Retrieved February 10, 2009)
8. Hailing RE. Recommendations for collecting mushroom for scientific. In Alexialdes MM, JW, (eds). Selected guide lined for ethnobotanical research. A field manual, The New York Botanical Garden Press, Bronx. 1996;135-141.
9. Molina R, O'Dell T, Luoma D, Nair MC, Devi SB. Collection and identification of agaricales, In Beneficial fungi and their cultivation (Nair MC, Balakrishanan S, eds.). Scientific Publishers SA, New Pali Road, P.O.Box, 91, Jodhpur-342001, India; 1995.
10. Arora D. Mushrooms demystified: A comprehensive guide to the fleshy fungi. Barkeley Speed Press; 1986.

11. Singer R. The Agaricales in morden taxonomy. J. Cramer, Weinheim, 4th ed. 1986;912.
12. Singer R. The Agaricales in morden taxonomy. Seven Koeltz Scientific Books, Koenigstein, Germany. 1986;624.
13. Dickinson C, John L. The colour dictionary of mushroom. Orbis Publishing, London; 1982.
14. Jorden P. The mushroom guide and identifier. Anness Publishing Limited Hermes House, London; 2000.
15. Pegler D, Spooner B. The mushroom identifier. Quintet Publishing Limited; 1997.
16. Lakhanpal TN. Mushrooms of India Boletaceae – Vol – I Studies in CryptogamicBotani (ed. Mukherjii KG), APH Publishing Corporation Delhi; 1996.
17. Rumainul MI, Aminuzzaman FM, Chowdhury MSM. Biodiversity and morphological characterization of mushrooms at the tropical moist deciduous forest region of Bangladesh. American Journal of Experimental Agriculture. 2015;8(4):235-252.
18. Fries EM. Systema Mycologicum 1. Lund, Sweden: Ex Officina Berlingiana. Retrieved 2011-08-05. 1821;19.
19. Kirk PM, Cannon PF, Minter DW, Stalpers JA. Dictionary of the Fungi (10th ed.). Wallingford, UK: CABI. 2008;394-401.
20. Thiribhuvanamala G, Prakasam V, Chandrasekar G, Sakthivel K, Veeralakshmi S, Velazhahan R, Kalaiselvi G. Biodiversity, conservation and utilization of mushroom flora from the Western Ghats Region of India. Proceedings of the 7th International Conference on Mushroom Biology and Mushroom Products (ICMBMP7). Tamil Nadu, India. 2011;155-164.
21. Pushpa H, Purushothama KB. Biodiversity of mushrooms in and around Bangalore (Karnataka), India. American-Eurasian J. Agric. & Environ. Sci. 2012;12(6):750-759.
22. Bankole PO, Adekunle AA. Studies on biodiversity of some mushrooms collected in Lagos State, Nigeria using biotechnological methods. Journal of Yeast and Fungal Research. 2012;3(4):37-48.
23. Jo JW, Kwag YN, Kim CS, Han SK. Distribution of higher fungi in Chungcheong Province, Republic of Korea. Journal of Asia-Pacific Biodiversity. 2013;6(3):347-359.
24. Bolton J. A History of fungusses. Growing about Halifax. Halifax, Canada: Husddersfield. 1791;160.
25. De Leon AM, Luangsa-ard JJD, Karunarathna SC, Hyde KD, Reyes RG,dela Cruz TEE. Species listing, distribution, and molecular identification of macrofungi in six Aeta tribal communities in Central Luzon, Philippines. Mycosphere. 2013;4(3):478-494.
26. Eggert C, LaFayette PR, Temp U, Eriksson KE, Dean JF. Molecular analysis of a laccase gene from the white rot fungus *Pycnoporus cinnabarinus*. Appl. Environ. Microbiol. 1998;64:1766-1772.
27. Bunyard BA. A survey of fungal diversity in Northeast Ohio. The Ohio Journal of Science. 2003;103(2):29-32.
28. Rasheeda K, Prasuna RG. Dairy waste 'ghee residue' influence laccases production by *Pycnoporus cinnabarinus* and its optimization. International Journal of Applied Biology and Pharmaceutical Technology. 2013;4(4):380-389.
29. Ramsbottom J. Mushrooms & Toadstools. Collins. 1953;6.
30. Thiers HD. California Mushrooms-A Field Guide to the Boletes. New York, New York: Hafner Press. 1975;261.
31. Chiu WF. The boletes of Yunnan. Mycologia. 1948;40(2):199-231.
32. Hosen MI, Feng B, Wu G, Zhu XT, Li YC, Yang ZL. *Borofutus*, a new genus of *Boletaceae* from tropical Asia: Phylogeny, morphology and taxonomy. Fungal Diversity. 2013;58:215–226.
33. Hall IR, Gordon TB, Alessandra Z. Burgundy or summer Truffle. Taming the truffle: the history, lore, and science of the ultimate mushroom. Timber Press; 2007.
34. Desjardin DE, Perry BA, Lodge DJ, Stevani CV, Nagasawa E. Luminescent *Mycena*: New and noteworthy species. Mycologia. 2010;102(2):459–77.
35. Knudsen H, Vesterhout J. Funga Nordica. Copenhagen, Denmark: Nordsvamp. 2008; 403.
36. Hughes KW, Petersen RH, Mata JL, Psurtseva NV, Kovalenko AE, Morozova OV, Lickey EB, Blanco JC, Lewis DP, Nagasawa E, Halling RE, Takehashi S, Aime MC, Bau T, Henkel T. Rep. Tottori Mycol. Inst. 2007;45:1–57.
37. Ram RC, Pandey VN, Singh HB. Morphological characterization of edible fleshy fungi from different forest regions. Indian J. Sci. Res. 2010;1(2):33-35.

38. Giri S, Biswas G, Pradhan P, Mandal SC, Acharya K. Antimicrobial activities of basidiocarps of wild edible mushrooms of West Bengal, India. International Journal of PharmTech Research. 2012;4(4):1554-1560.

39. Coetzee MPA, Wingfield BD, Harrington TC, Steimel J, Coutinho TA, Wingfield MJ. The root rots fungus *Armillaria mellea* introduced into South Africa by early Dutch settlers. Molecular Ecology. 2000;10:387-396.

18

Species Composition and Structure of *Isoberlinia* Woodland of Shika, Zaria Nigeria

H. Bello[1*] and A. O. Musa[1]

[1]*Department of Biological Sciences, Ahmadu Bello University, Zaria, Nigeria.*

Authors' contributions

This work was carried out in collaboration between both authors. Author HB designed the study, wrote the protocol and wrote the first draft of the manuscript. Author AOM managed the literature searches, analyses of the study performed the structural equation modeling and discuss the conclusion. Both authors read and approved the final manuscript.

Editor(s):
(1) Mirza Hasanuzzaman, Department of Agronomy, Faculty of Agriculture, Sher-e-Bangla Agricultural University, Bangladesh.
Reviewers:
(1) Igor Villela Marroni, Universidade Federal do Rio Grande do Sul, Brazil.
(2) Sholeh Ghollasimod, University of Birjand, Iran.
(3) Francisco Javier Peris Felipo, University Valencia, Spain.
(4) Noiha Noumi Valery, University of Ngaoundere, Cameroon.

ABSTRACT

This study was carried out to investigate the effect of farming activities on *Isoberlinia* woodland. This was achieved by assessing their composition and structural characteristics in two plots, 30 m x 30 m each, the plots were farming plot and a rested plot of minimum anthropogenic impacts (here called disturbed and undisturbed plot respectively). Data were collected from each plot, all woody species were identified and their height, diameter at breast height (dbh), basal area, volume, diversity and biomass were measured. In addition, herbaceous diversity and biomass were also determined. Dominance species were determined by the Importance Value Index (IVI). Soils were augured, collected and analyzed for physico-chemical parameters. A total of 71 species belonging to 36 families were identified. Shannon-wiener diversity index showed that the disturbed plot was 2.441 and equitability of 0.733 had higher diversity when compared to the undisturbed (2.331 and 0.685) respectively. These two plots were 66% similar in vegetal composition. Analysis of the importance value index revealed that the dominant species in the area was *Isoberlinia doka*. The

Corresponding author: Email: hbello@abu.edu.ng

total basal area and volume was very low at both plots. The woody biomass for the undisturbed was 0.015 ton/ha for the undisturbed and 0.001 ton/ha for the disturbed plot. The total herbaceous biomass for the undisturbed plot was 70.7 ton/ha and 49.7 ton/ha for the disturbed plot. The soils were characteristically alfisols and much degraded reflecting degrees of vegetal exploitation. The low woody volume and biomass though disturbing but not withstanding had high *Isoberlina* component hence, the woodland can still and better be called degraded *Isoberlinia* woodland.

Keywords: Species composition; woodland; Isoberlinia; soil type.

1. INTRODUCTION

Worldwide, mankind is facing the negative repercussions of global change [1,2]. Recent global forest resources assessment revealed that Nigeria is one of the five countries in the world with the highest annual rate of deforestation for the period 2000 – 2010 [3,4]. A major consequence of global environmental change is an increasing shortage of natural resources, especially the threat to biodiversity [5-7]. The vegetation of any area is inadvertently tampered with by human. At certain level such uses could become detrimental to a sustainable vegetation cover [8,9]. The genus *Isoberlinia* is a woody Caesalpinoideae in the family Fabaceae (legume family) of five species native to tropical Africa, two species are predominant in Nigeria's guinea savanna *I. doka* and *I. tomentosa* [10]. The trees of the Nigeria guinea are on the average, little more than 7-2m tall and also have a less compact special arrangement than the Miomba woodland with which they have similar characteristics [11]. Shika *Isoberlinia* woodlands are used for cropping and livestock grazing [12]. Subsistence farmers also collect fuel wood which may have added effect on degradation of the indigenous woodlands. FAO [12] also advocated for more detailed and standardized data on biomass and structure of vegetation units, in order to parameterize global and regional vegetation maps. Studies linking species composition to environmental parameters are sparse for West Africa savanna systems [13]. Such knowledge is however required in order to expand the understanding of ecosystem processes. This study aimed at assessing the impact of farming activities on the composition and structure of *Isoberlinia* woodlands.

2. MATERIALS AND METHODS

2.1 Study Area

The study was carried out in Shika, Zaria, Nigeria. Shika is geographically located between latitude 11°12'N and longitude 7°33'E at an altitude of 640 m above sea level [14]. Shika is located about 20 km along the Zaria -Sokoto road in Kaduna state, North Central, Nigeria. It has three distinct climatic seasons. These are the cold dry season (November –February), the hot dry season (March- May) and the wet season (June – October). The total annual rainfall ranges from 617 to 1365 mm with a 50-year average of 1041 mm. Most of the rains fall between July and September [14]. Area extent of study was approximately 15 hectares of range land under different management strategies. The management types included farmed sections, grazing lots and rested plots (i.e. un grazed and unfarmed) with minimum anthropogenic impact. The study was conducted from June to September, 2006.

2.2 Vegetation Sampling

A strategy of simple random sampling was used as the vegetation appeared on inspection to be homogeneous in structure and flourishes. The sampling unit was a plot of 30 m x 30 m (900 m^2). Two plots were sampled for each of the woodland.

2.3 Identification of Species

Species were identified on the field from recognition while critical species were determined by comparing with types and use of flora in the herbarium unit of Biological sciences Department, Ahmadu Bello University, Zaria, Nigeria.

2.4 Parameters Evaluated on Vegetation

2.4.1 Height and diameter

With each 30 m x 30 m plot, the height and dbh of all individual trees was measured in order to characterize tree structure of the plot.

2.4.2 Calculation of Tree cover and volume

For each tree individual in 30 m x 30 m plot, the basal area was calculated using πr^2 (m^2) and volume was calculated using $\pi r^2 h$ where π=22/7, r=radius, h=height of trees were measured.

The species composition of the plots was described using Importance Value Index [15].

The following parameters were measured:

1. Relative dominance = (total basal area for a species/total basal area of all species) x 100.
2. Relative density = (number of individuals of a species/total number of individuals) x 100.
3. Relative frequency = (frequency of a species/sum of all frequencies) x 100.
4. The importance value index (IVI) = relative dominance + relative density + relative frequency.

In general the relative dominance, relative frequency, and relative density vary in the range 0 – 100%, so the species and importance value indices

(IVI) vary between 0 and 300%.

2.4.3 Woody biomass

Woody biomass was calculated after Schongart (2003) as stated in Orthman [16] given the equation below:

Woody biomass=

$$\frac{height(m)}{2} \times \frac{dbh(m^2)}{2} \times \pi \times \text{wood density } (t/m^3)$$

Wood density defined as the over-drymass per unit of green volume (tons/m^3) for tropical tree species were given by Brown [17]. Data on wood density were not available for all sampled tree species. Brown [17] gives wood densities for many Africa tree species within the range of 0.5-0.79 (tm-^{3}) taking into account only those in the studied area, a mean value of 0.65 (tm-^{3}) was calculated.

2.5 Sampling of Herb Layer Data

2.5.1 Species composition

Vegetation surveys were conducted in the 5 m x 5 m (25 m^2) plot in each of woodland. The species were identified and their numbers determined in the 1 m x 1 m quadrants.

2.5.2 Herbaceous biomass

The herbs in the 1 m x 1 m quadrant were harvested with sickle for above ground biomass determination. The harvested herbs were weighed as grass and forbs separately. Harvesting was done in September at the peak period of flowering grass species. Samples were oven dried to constant weight. Dry matter yield was calculated using the formular below:

Percentage dry matter (%Dm)

$$= \frac{sub\ sample\ of\ dried\ matter\ weight \times 100}{sub\ sample\ of\ fresh\ weight}$$

$$\text{Dry matter yield} = \frac{\%Dm \times fresh\ weight}{100} (g/m^2)$$

2.6 Data Analysis

Vegetation data were analysed using the mathematical index of Shannon –wieners [18] as stated in Tiseer and Smith [19]. The Shannon Diversity Index and equitability was computed as

(H' = ∑Pi*ln Pi) where Pi=ni/N.
H' is the index of diversity,
Pi is the importance value of a species as a proportion of all species.
ni' is number of individual species.
N' is Total number of individual species.
Equitability J=H/lnN.

Microsoft Excel 2007 package was used for data analysis.

3. RESULTS

A total number of 71 species were recorded in 36 families. The families represented by the greatest number of species were Fabaceae (12 species), Combretaceae (7 species), Poaceae (7 species), 39 species were trees and shrubs while 33 herbaceous species were recorded. The predominant woody species was *Isoberlinia doka* with 307 individuals in the undisturbed plot and 224 individuals in the disturbed plot. The dominant herbaceous genus was *Setaria barbata* with 300 individuals per hectare in the undisturbed plot and *Fadogia pobeguini* with 501 individuals per hectare in the disturbed plot. A single stands of *Entada africana*, *Strychnos spinosa* has the highest height of 5.5 m and 4.5 m respectively. While *Isoberlinia doka* has the height of 2.5 m in the undisturbed plots. In the disturbed plot all height measurement was less than 2.5 m. In undisturbed plot *Lannea schimperi* contributed the highest girth of 40cm as showed in Table1b.

Importance value index (IVI) provides knowledge on important species of a plant community. Based on IVI *Isoberlinia doka, Sechium idule, Dichrostachys cineria, Ochna afzeli, Detarium microcarpum* were the dominance species in undisturbed plot whereas *Isoberlinia doka, Securidaca longepedonculata, Pilostigma thonningii, Hymenocardia acida, Dichrostachys cineria* were dominance in the disturbed plot. As showed in Table 2.

Table 1a. Plants diversity for undisturbed and disturbed *Isoberlinia* Woodland

S/no	Species	Families	Undisturbed		Disturbed	
			Woody	Herb	Woody	Herb
1.	*Isoberlinia doka*	Fabaceae (c)	307		224	
2.	*Adenodolicus paniculatus*	Fabaceaes (p)				159
3.	*Khaya senegalensis*	Meliacece	-		8	
4.	*Butryospermum paradoxum*	Sapotaceae	18		23	
5.	*Vitex doniana*	Verbenaceae	6		1	
6.	*Securidaca longepedonculata*	Polygalaceae	20		35	
7.	*Afxelia Africana*	Fabaceae (c)	-		10	
8.	*Entada Africana*	Bignoniaceae	10		26	
9.	*Stereosperrmum kunthiana*	Fabaceae (m)	-		6	
10.	*Detarium microcarpum*	Fabaceae (c)				
11.	*Cussonia barteri*	Araliaceae				
12.	*Annona senegalensis*	Annonaceae	11		16	
13.	*Piliostigma thonningii*	Fabaceae (c)	17		48	
14.	*Azadirachta indica*	Meliaceaa	-		-	
15.	*Dicrostachy scineria*	Fabaceae (m)	122		38	
16.	*Bridelia ferruginea*	Euphorbiaceae	9		1	
17.	*Lannea schimperi*	Anacadiaceae	8		10	
18.	*Terminalia avvicinoides*	Combretaceae	1		8	
19.	*Fadogia pobeguini*	Rubiaceae		16		501
20.	*Ximenia Americana*	Olaceae	-		3	
21.	*Maerua crassifolia*	Capparidaceae	-		2	
22.	*Cissus cornifolia*	Vitaceae		4		2
23.	*Vernonia kotschyana*	Asteraceae				
24.	*Cochlospermum tinctorum*	Cochlespermaceae		38		23
25.	*Phyllantus muellerianus*	Euphorbaceae	-		1	
26.	*Gardenia Aquila*	Rubiaceae	6		2	
27.	*Guiera senegalensis*	Combretaceae	2		3	
28.	*Swartzia madagascariensis*	Fabaceae (c)				
29.	*Lippia multiflora*	Verbenaceae		196		72
30.	*Strychnos spinosa*	Loganiaceae	15		12	
31.	*Astrosolen marcrozhia*	Thymelaceae				69
32.	*Costus afer*	Zingiberaceae				102
33.	*Andropogon gayana*	Poaceae		80		80
34.	*Stylochiton hypoqea*	Araceae		36		23
35.	*Setaria pallidefusca*	Poaceae		47		12
36.	*Xyris sp*	Xyridaceae		14		36
37.	*Heckhloe sp*	Poaceae		130		60
38.	*Setaria barbata*	Poaceae				
39.	*Aspilia Africana*	Asteraceae		25		189
40.	*Eragrostic atrovirens*	Poaceae		35		21
41.	*Hypharrhenia rufa*	Poaceae		52		
42.	*Gutenbergia cordifolia*	Asteraceae		3		74
43.	*Cyperus michellus*	Cyperaceae		120		4
44.	*Paspalum arbrculare*	Poaceae				25
45.	*Heeria insignis*	Anacardiaceae		6		
46.	*Aselepias gurassavica*	Asclepiadaceae				5
47.	*Waltheria indica*	Sterculiaceae		27		58

S/no	Species	Families	Undisturbed		Disturbed	
			Woody	Herb	Woody	Herb
48.	*Crotolaria lanchnosuma*	Fabaceae (p)				1
49.	*Latana camara*	Verbenaceae				6
50.	*Pericopsis laxiflora*	Fabaceae (c)				
51.	*Gmelina arborea*	Verbenaceae			1	
52.	*Terminalia mollis*	Combretaceae	22		6	
53.	*Pavetta corymbosa*	Rubiceae	2		4	
54.	*Ochna afzelii*	Ochnaceae	102			
55.	*Combretum lamprocarpum*	Combretaceae				
56.	*Ficus thonningii*	Moraceae	3			
57.	*Sechium edule*	Cucurbitaceae	145			
58.	*Scoparia dulcis*	Scrophalariaceae		4		
59.	*Tephrosia pedicellata*	Fabaceae (p)		60		
60.	*Stylosanthes auretta*	Fabaceae (p)				
61.	*Ipomea aquatic*	Convolvulaceae		32		
62.	*Ekebegia senegalensis*	Meliaceae	1			
63.	*Diospyros mespiliformis*	Ebenaceae				
64.	*Psorospermum senegalense*	Hypericaceae				
65.	*Asperagus Africana*	Liliaceae				
66.	*Hymenocardia acida*	Euphorbiaceae	2		48	
67.	*Maranthes polyandra*	Rosaceae		8		
68.	*Terminalia lanceolatum*	Combreteceae			10	
69.	*Danielia oliveri*	Fabaceae (c)	1			
70.	*Securinega virosa*	Euphorbiaceae	38			
71.	*Tacca involucrate*	Taccaceae				4
Total			941	611	1622	1546

**letters in bracket represent the sub-families of fabaceae, (c- ceasalpinioideae, p- papilionoideae, m - mimosoideae)*

Table 1b. Summary of species composition and structural characteristics for each Woodland type

Diversity measure	Undisturbed		Disturbed	
	Woody	Herb	Woody	Herb
No of family	18	17	15	15
No of species	30	25	28	25
Average dbh (m)	0.1±0.5		0.02±0.06	
Mean height (m	2.24±1.4		0.7±0.05	
Basal area (m^2)	0.43		0.05	
Volume (m^3)	9.53		0.68	
Biomass (tons)	0.54	70.7(g/m^2)	0.02	49.7(g/m^2)
Shannon index	2.33	2.63	2.44	2.68
Equitability	0.69	0.82	0.73	0.83

**dbh-diameter at breast height, N- 30 and 28 for woody average in undisturbed and disturbed plot respectively, herbaceous biomass were measured in g/m²*

In Table 3 below, the textural class of the soil in the farmed plot is loam with 45% sand, 34% silt and 21% clay. The physicochemical parameters shows average low organic matter in the farmed plot which had correspondingly high Nitrogen content of 0.53 g/kg^{-1}, other macro elements were low except for Magnesium (0.78 cm/kg). The undisturbed plot was texturally sandy loam with 49% sand, 34% silt and 21% clay. The organic carbon at 0.97 g/kg^{-1} was low. The nitrogen content of the soil at 0.53 g/kg^{-1} was high. Other element were low except again for magnesium at 0.63 cmg/kg^{-1} in all cases the sodium content 1.8 cm was low which removes any fears of solidification. The available phosphorous was low in the undisturbed and the disturbed plot.

4. DISCUSSION

There is variation in the species composition and structures of the disturbed and undisturbed plots of the woody and herbaceous layer. Mishra [20] observed greater diversity of herbs in disturbed forest than in undisturbed ones. The composition is similar to that described [21] in the central border Nigeria *Isoberlinia doka* being the most dominant and numerous giving a monospecific picture of the woodlands gives reason why it was called *Isoberlinia* woodlands [10].

Due to the less ability of *Isoberlinia sp* to regenerate after over exploitation for fuel wood or clearing of the woodland during farming, other competitive species such as *Adenodolichos penniculatus Detarium microcarpum*, *Daniella oliver, Vitex doniana, Annona senegalensis* etc comes into the woodland. The diversity of plant in all plots was high. However this was higher in the disturbed plots, this may be due to the resilient nature of most savanna plants which have a phyrophytic behavior and regenerate from buds on their below ground parts. The consequence is that cutting during farm clearing operation serve to multiply regenerating stocks. This proves a case of disturbance induced diversity [22,23].The maximum population of species of 39 in the plots is also appreciable given the area sampled. The consequence is that diversity at the level of this interference is not too adversely affected. An earlier report of sample size larger than 900 square meters, yield a total of 42 plants species [24]. The plots were also of 66% similar meaning that subsistence farming did not account for gross disappearance of species in this case. Also important value index in Figs. 1 and 2. Shows that *Isoberlinia doka* appears to be dominant in both plots. These results disclose that, the most important species in the plots have high diversity in the scale of Shannon-Weiner Index of Diversity. The IVI rank species in a way as to give an indication on which species come out as important element of the woodlands [25].

Isoberlinia doka, though dominant in numerical value had awfully low basal area plant has of late been highly exploited for fuel to the extent that yearly regrowth are even harvested. Indeed these appear to be a high skew to biomass in all plots in favor of grass and forbs. While figure for this are alarmingly high, the biomass estimate for wood are very low. The implication are that, the vegetal conditions typified by this plots can no longer supply wood either for rough construction or timber, there is progressive turning of the woodlands of Shika into an open range which favour grassing more than any other form of land use. The high herbaceous biomass cannotes high danger of annual fires which can exacerbate the degradation of the range. *Isoberlinia* species has been recorded as an endangered species [26]. The ridiculously very low volume in the plots indicated the wanton exploitation which is not pursued with conservation, indicates that no regard is given to sustainability and perpetuity. As poverty increase and population increase, the range will no longer cope with the demand for fuel wood and farm land. This may reach a catastrophy that can result to ecosystem collapse [27,28].

Table 2. The five most abundant woody species in each Woodland type according to decreasing order of the species importance value index (IVI)

Species	Undisturbed IVI(%)/200		Disturbed IVI(%)/200	
	Woody	Herb	Woody	Herb
Isoberlinia doka	39.9		42.3	
Sechium idule	15.4			
Dichrostachys cineria	13.1		6.5	
Ochna afzeli	12.9			
Detarium microcarpum	8.0			
Other woody species	110.7			
Securidaca longepedonculata			29.1	
Pilostigma thonningii			9.8	
Hymenocardia ocida			7.8	
Other woody species			104.5	
Setaria babarta		19.9		
Stylosanthes qurretta		17.4		
Lippia multiflora		12.8		
Heckloea sp		8.4		
Cyperus michellus		7.7		
Others herbs		133.8		
Fadogia pobeguini				32.4
Cochlospermun tinctorium				14.6
Aspilia Africana				12.0
Adenodolicus peniculatus				6.6
Other herbs				122.2

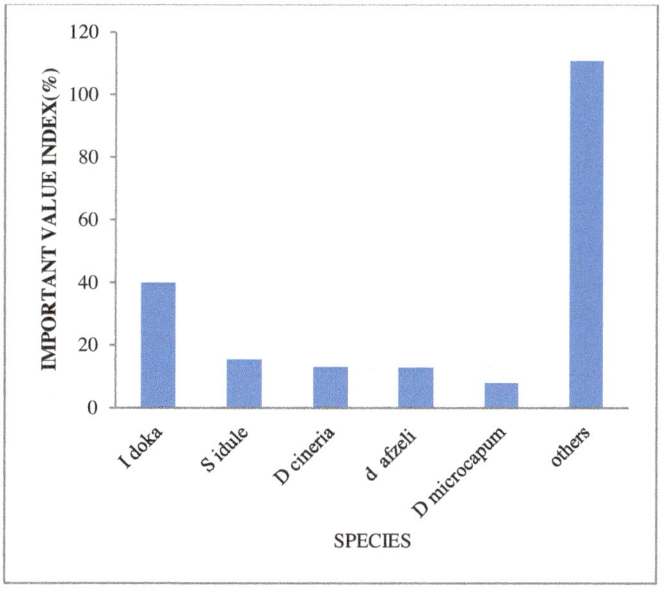

Fig. 1. Important dominants woody species in the undisturbed woodland

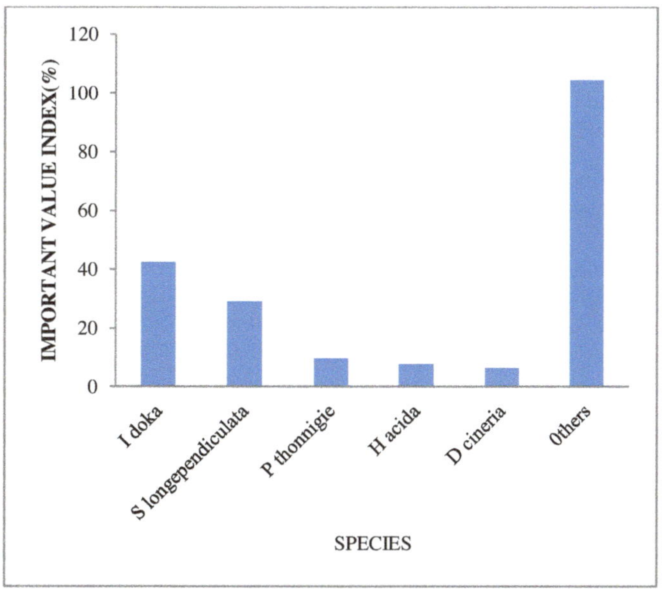

Fig. 2. Important dominants woody species in the disturbed woodland

Table 3. Physico-chemical parameters of soil in the investigated plots

Woodland type	Texture class	%o.c g/kg^{-1}	N g/kg^{-1}	P ppm	Ca cm/kg^{-1}	Mg cm/kg^{-1}	K cm/kg^{-1}	Na cm/kg-1
Undisturbed	Sandy loam	0.97	0.53	0.79	1.00	0.60	0.12	1.8
Disturbed.	Loamy	0.75	0.53	0.63	1.7	0.78	0.19	1.65

**o.c-organic carbon, N- Nitrogen, P - Phosphorus, Ca – Calcium, Mg – Magneșium, K- potassium, Na- Sodium*

The soil presented mixed parameters which indicate at best progression on the part of degradation. Some normal situation ware noted such as P^{H} at 5.8-6.0 range for the soil. The low sodium entails that no immediate solidification problems are envisage with soils. The low phosphorus level is characteristics of alfisol of the guinea savanna [29]. The low organic matter and relatively high Nitrogen could suggest that rapid mineralization occurs in the soils. Perhaps the soil nitrogen level is not commensurate the rapid organic carbon loss. Reduction in nitrogen and available phosphorus showed that the major soil nutrient has been noted to be limited in disturbed soil compared to undisturbed [30]. The low organic matter carbon status could also be due to lack of replenishment of soils with organic matter as crop residues are used as fuels. The rapid removal of trees can also be another factor in the low soil organic Carbon. This loss of fertility in either the undisturbed or the farmed plots can note a situation where plants regeneration and production will be painfully slow and low. With unabated use of such lands perhaps with fertilizer practices, the consequences would be degradation. This may aggravate the already poor vegetal state and escalate/set stage for desertification [31].

5. CONCLUSIONS

This study revealed that deforestation problem in Shika, zaria has attained a critical level which can no longer be ignored. Any forms of anthropogenic activity detrimental to biodiversity need to be checked against background of sustainability. The prevailing trends of exploitation certainly connote danger in the not too distance future given emerging trends of desertification, drought and global warming. Therefore proper vegetation management should be adopted in order to maintain a balance ecosystem and a sustainable resources base for the benefit of the present and future generations.

COMPETING INTERESTS

Authors have declared that no competing interests exist.

REFERENCES

1. Walker S, Steffin W, Caradelle J. The terrestrial biosphere and global change. Cambridge University Press, Cambridge. 1991;304.
2. IPCC. Land use, land-use change, and forestry: Summary for policymakers. WMO, UNEP, Intergovernmental Panel on Climate Change. 2000;375.
3. FRA. Global Forest Resources Assessment 2010, Main report. FAO Forestry Paper, 163, FAO Rome. 2010; 340.
4. Aigbe HI, Omokhua GE. Tree species composition and diversity in oban forest reserve, Nigeria. Journal of Agricultural Studies. 2015;3:1.
5. Thomas DC, Cameron A, Green RE, Bakkenes M, Beaumont LB, Collingham YC, et al. Extinction risk from climate change. Nature. 2004;427:145-148.
6. IUCN. Plants under pressure –a global assessment. The first report of the IUCN Sampled Red List Index for Plants. Royal Botanic Gardens, Kew, UK, Natural History Museum, London, and International Union for Conservation of Nature. 2010;15.
7. Udofia SI, Owoh PW, Attah VI, Thomas, AD. Assessment of plant species composition in Ayan nsit sacred forest of Akwa Ibom state, Nigeria. Nigerian Journal of Agriculture, Food and Environment. 2014;10(2):34-37.
8. Jones G. Vegetation productivity. longman Ltd. New York. 1979;1-4:8-34.
9. Salami AT, Balogun BO, Oloyede- Kosoko SOA. Geospatial data application in the assessment of population impact on a tropical lowland rainforest of Southwest Nigeria. Geo Observateur. 2007;16:31-37.
10. Jackson G. Vegetation around the city and nearby villages of Zaria. Mortimore MJ(ed). Zaria and its region occasional paper no 4.14th Annual conference of the Nigeria Geographical Association, Zaria. 1970;61-71.
11. White F. The vegetation of Africa. A descriptive memory to accompany the UNESCO/AETFAT, UNSO vegetation map of Africa, UNESCO. Paris. 1983;356.
12. FAO .Global forest resources assessment. Main report. FAO, Forestry Paper, 140. FAO Rome. 2010;479.
13. Hahn K. Die planzengesellschaften der savannah in sudosten Burkina fasos (Westa-africa) ihre Beeinflussunadurch den menschen und die naturran lichen gegebenheiten. Dissertation, universitat frank Furt am main. Germany. 1996;3-79.
14. Akpa GN, Ifut JO, Mohammed F. Indegenous management of dystocia in ruminant livestock of Northern guinea savannah of Nigeria. J. Anim. Prod. 2002; 29:264-270.
15. Van Andel LT. Floristic composition and diversity of mixed primary and secondary forests in northwest Guyana. Biodiversity and Conservation. 2001;10:1645-1682.
16. Orthman B. Vegetation ecology of a woodland savanna mosaic in central Benin

(W. Africa) Ecosystem analysis with a focus on the impact of selective logging. Ph. D. Thesis, University of Rostock. 2005; 22

17. Brown S. Estimating biomass and biomass change of tropical forests: A primer. Forestry paper 134, FAO, Rome. 1997;47-72.
18. Shannon CE, Weiner W. The mathematical theory of communication. University of Illinos Press Urban. 1986;125.
19. Tisser FA, Smith RO. Beta diversity of Angiosperm flora in relation to exploitation trend in the savannah region. Nigeria Journal of Tropical Bioscience. 2004;9:46-50.
20. Mishra RK, Upadhyay VP, Mohanty RC. Vegetation ecology of the Similipal Biosphere Reserve, Orissa, India. Appl Ecol Env Res. 2008;6:89-99.
21. Kershaw KA. A survey of the vegetation Zaria province, Northen Nigeria Vegetation. 1968;30:244-268.
22. Hannson L. Technical report no 1 indicators of biodiversity Recent approaches and some general suggestions. Europe. 2000;1.
23. Laidlaw M, Kitching R, Goodall K, Small A, Stork N. Temporal and spatial variation in an Australian tropical rainforest. Austral Ecol. 2007;32:10–20.
24. Obom RM, Isichei AO. Composition, structure and relations of isoberlinia wood lands in Nigeria. Nigeria Journal of Botany. 1998;11:67-78.
25. Munishi PKT, Philipina F, Temu RPC. Pima NE. Tree species composition and local use in agricultural landscapes of west Usambaras Tanzania. African Journal of Ecology. 2008;46:66–73.
26. Famuyide OO, Popoola L, Owonaubi JJ. An evaluation of preferred Forest Resources in selected areas of Nigeria. In: Forestry and Challenge of sustainable Livelihood. A proceeding, of the 28th Annual conference of FAN held in Akure Ondo State. 2004;327-331.
27. FAO. Food Agricultural Organisation: Land Degradation and Food Supply issues: United Nation, Rome. 1999;3.
28. FAO. State of the World's Forests, 2011, FAO Rome. 2011;164.
29. Adeayel KB. Nutrient status of Nigeria soil to forage production. A paper presented at the pasture production-training course held at NAPRI/ABU Zaria. 31st OCT-2nd Nov.1989.
30. Grossman RB, Reinsch TG. Bulk density and linear extensibility. In: Dane JH, Topp GC. (eds). Methods of Soil Analysis part 4. Physical methods. Soil Sci. Soc. Am. Book Series No 5 ASA and SSSA, Madison, WI 2002;201–228.
31. Sivakumar MVR, Manu A, Virma SA, Konemasu ET. Relation between climate and science of soil of the tropics SSSA Special Publication Myths. 1992;29:9-113.

19

Investigation of Morpho-chemical Characteristics and Yield Potential of Five Local Rice Cultivars

Md. Sariful Islam[1*], Bikash Chandra Sarker[1], K. M. Manirul Islam[2], Shah Md. Shakil Ahmed[2] and Najim Uddin[1]

[1]Department of Agricultural Chemistry, Hajee Mohammad Danesh Science and Technology University, Dinajpur-5200, Bangladesh.
[2]Faculty of Agriculture, Hajee Mohammad Danesh Science and Technology University, Dinajpur-5200, Bangladesh.

Authors' contributions

This work was carried out in collaboration between all authors. Author MSI wrote the protocol, conducted the field experiments and wrote the first draft of the manuscript. Author BCS designed the study and supervised the research work. Authors KMMI and SMSA managed the literature searches, collected data from field, analyzed the data of the research and help in manuscript preparation. Author NU overviewed the research and co-supervised the work. All authors read and approved the final manuscript.

Editor(s):
(1) Daniele De Wrachien, Department of Agricultural and Environmental Sciences of the State University of Milan, Italy.
Reviewers:
(1) Anonymous, University of Tsukuba, Japan.
(2) Bahaa El Din Mekki, National Research Centre, Egypt.
(3) Shelley Gupta, Pune University, Pune, India.
(4) Renata Kieloch, Institute of Soil Science and Plant Cultivation, Poland.
(5) Roberto b. Barba jr, Catanduanes State University, Philippines.
(6) Swarnalata Das, Orissa University of Agriculture and Technology, India.

ABSTRACT

Aims: To study morpho-chemical, yield and yield contributing characteristics of five local rice cultivars viz. Red Swarna, Jota pari, Pariza, Sugandhi and ShaitaBoro.
Study Design: The experiment was laid out in Randomized Complete Block Design with three replications.
Place and Duration of Study: A field experiment was conducted at the farm laboratory of Agricultural Chemistry Department, Hajee Mohammad Danesh Science and Technology

**Corresponding author: Email: sarif.islam63@gmail.com*

University, Dinajpur, Bangladesh during the period from February to May, 2014.
Methodology: The total experimental area was divided into 3 blocks. Each block was then subdivided into 5 plots. The plot unit area was 2 m x 2 m. The replications were separated from one another by one meter border. The border between the individual plot was 0.5 m. All the intercultural operations were done as when necessary. Data on different characteristics were collected at different growth stages.
Results: Results revealed that morphological, physiological; yield and yield attributes were significantly different among the cultivars studied. The Sugandhi produced higher number of grains per panicle and bolder grains which resulted higher grain yield. Further, Sugandhi showed superiority in yield contributing characteristics: total tillers number, filled grain per panicle over other local cultivars and it also produced higher grain yield (4.86 t per ha).
Conclusion: It is revealed that Sugandhi had the higher yielding ability than those of other local cultivars.

Keywords: Investigation; morpho-chemical characteristics; yield potential; local rice cultivars.

1. INTRODUCTION

Bangladesh is an agricultural country and still hard working for rapid development of its economy. Approximately, 22 percent of Gross Domestic Product (GDP) has been derived from agriculture sector while rice contributes 9.5 percent to the agricultural GDP [1]. Rice is the staple food of about 135 million people of Bangladesh which provides nearly 48% of rural employment, about two-thirds of total calorie supply and about one-half of the total protein intakes of an average person in the country. Rice production contributes one-half of the agricultural GDP and one-sixth of the national income in Bangladesh [2]. Rice provides 20% of the world's dietary energy supply which is also a good source of thiamine, riboflavin, niacin and dietary fibre [3]. Unlike the other major cereals, more than 90% of rice is consumed by humans [4]. More than 95% of population consumes rice and it alone provides 76% of calorie and 66% of total protein requirement of daily food intake [5]. It is necessary to enhance the growth of rice production through increasing land productivity to meet the increasing food demand for the vast population of the country as the country has serious land constraints.

Rice in Bangladesh is grown in three distinct seasons: Boro (January to June), Aus (April to August) and Aman (August to December) [6]. It provides nearly 95 percent of the total food requirements, but there is still a need to increase production to feed the growing population which increases at the rate of 1.32 percent per annum [7].

Two genotypes of rice cultivars namely traditional (local) and modern (high yielding) are cultivated in Bangladesh. Local rice variety has a large contribution to meet the demand of people in Bangladesh. Recently the farmers of Bangladesh produce a huge amount of rice from local cultivars. Though the soil and climate of our country is quite suitable for the production of rice, still it is facing many problems of which the poor yielding inherent capability of our local varieties is the most important one. Poor plant type such as tall plants, long and droopy leaves, weak culms, susceptible to lodging are the main causes of the low yield of the local varieties. Several of local rice landraces are cultivated in different AEZ (Agro Ecological Zone) in different seasons in the country. Among those, there are various types depending on their photosensitivity, crop duration, grain size, aroma, chlorophyll, minerals and vitamins content etc. For the additional food requirement, in these days of alarming population increase in Bangladesh, we cannot afford to grow those local cultivars unless their yield per unit area is improved.

Rice yield is mainly dependent on producing ability of dry matter before heading. [8] Showed that the degree of heterosis in dry matter accumulation was relatively lower at the later growth stage than at the early growth stage. However, photosynthetic rate of the same genotypes showed the marked difference during growth stage [9]. It was also reported that the physiological traits, such as photosynthetic rate [10], nitrogen uptake [11], floret number per unit area [12], which are closely related to the yield, showed wide variation over genotypes and environments. Local rice is superior compared to modern rice. It has a better flavour, is more nutritious, can be grown continuously throughout the year, is easier to plant and more economical, especially if grown organically. Some local cultivers also have medicinal properties for common diseases such as stomachache, cough, metabolic acceleration, and others. The objective of the research was comparison of local rice cultivars with respect to selected morpho-chemical features and yield potential as well as determination of cultivar which is the most useful for cultivation in Northwest Bangladesh.

2. MATERIALS AND METHODS

The experiment was conducted in the farm laboratory of Agricultural Chemistry Department, Hajee Mohammad Danesh Science and Technology University (HSTU), Dinajpur during Boro season, from February to May, 2014. The treatments of this experiment are five rice cultivars *viz.*: *Red Swarna, Jota pari, Pariza, Sugandhi and ShaitaBoro*. These cultivars are well adapted in the environmental condition of the research area and highly practiced by the local farmers. The experiment was laid out in a Randomized Complete Block Design (RCBD) with three replications. A total of 15 plots were prepared. The total experimental area was divided into 3 blocks. Each block was then subdivided into 5 plots. The size of each plot was 2 m x 2 m. The replications were separated from one another by one meter border. The border between the individual plots was 0.5 m. All the intercultural operations were done as when necessary. Observations were recorded on different characteristics such as: 1. Morphological characteristics: plant height (cm), number of leaves per plant, and root weight (gm) 2. yield and yield components: effective tillers per hill, non-effective tillers per hill, total number of tillers per hill, filled grain per panicle, non-filled grain per panicle, 1000-grain weight (gm), grain yield (t per ha), straw yield (t per ha) and harvest index. 3. Leaf chemical constituents: Chlorophyll i. chlorophyll-a ii. chlorophyll-b iii. carotenoids, phosphorus, potassium, calcium and magnesium.

Harvest index is the relationship between economic yield and biological yield [13]. It was calculated by using the following formula:

$$\text{Harvest index (\%)} = \frac{\text{Economic yield}}{\text{Biological yield}} \times 100$$

2.1 Determination of Chlorophyll

For measuring chlorophyll exactly 0.1 g fresh leaf tissues of rice flag leaves were taken in a test tube containing 10 mL of 80% acetone. It was then shaken overnight using an electric horizontal shaker. The optical density or absorbance of the supernatant was measured by using UV-visible spectrophotometer at different wavelengths. As 80% acetone was used in this study, the absorbance was measured at 663 nm, 646 nm and 470 nm wavelength for chlorophyll-a, chlorophyll-b, total chlorophyll and carotenoids content. The concentration of chlorophyll-a (Chl-a), chlorophyll-b (Chl-b), total chlorophyll and total carotenoids were measured by using following formula.

Chl-a= 12.21 A_{663}-2.81A_{646} (μg ml of plant extract[1] or mg per g fresh weight)

Chl-b=20.13 A_{646} – 5.03 A_{663}

Total chlorophyll = 17.76 (A_{646}) + 7.34 (A_{663})

Total carotenoids,

$$C_{x+c} = \frac{1000\ A_{470} - 2.05\ C_a - 114.8\ C_b}{245}$$

2.2 Determination of Phosphorus, Potassium, Calcium and Magnesium

For leaf chemical composition the selected leaves during pre-flowering stage different replicated cultivars were collected. The leaves were carefully dried in an oven at 60°C for 72 hours. The leaves samples were ground and the extract was prepared by wet oxidation method using Di-acid mixture, the extraction was used for chemical nutrient estimation. Calcium and Magnesium were measured by complexometric method. Leaf Phosphorus and Potassium were measured by spectrophotometer using $SnCl_2$ as a reducing agent.

2.3 Statistical Analysis

The data collected on different parameters under the experiment were statistically analyzed to obtain the level of significance using the MSTATC computer software. If the characteristics under study were significant the differences between pairs of means were compared by Duncan's Multiple Range Test (DMRT) [14].

3. RESULTS AND DISCUSSION

3.1 Morphological Characteristics

Plant height of different rice cultivars varied markedly amongst themselves. Among the five cultivars studied, the plant height 86.44 cm was recorded in Sugandhi at seedling stage and 95.33 cm at the harvesting stage. Lowest plant height was recorded in Red Swarna (42.66 cm) at seedling stage and the highest plant height ShaitaBoro (107.3 cm) at harvesting stage which is similar to Red swarna (104.0 cm). Among these five cultivars, the highest leaf number (7.1) at vegetative stage and (8.4) at harvesting stage was recorded in Sugandhi which is similar to Jotapari (6.5) and (6.7).The lowest leaves number was recorded in ShaitaBoro (3.7) at vegetative stage and (4.3) at harvesting stage. The highest dry root weight was observed in Red Swarna (5.20 g) followed by Jotapari (4.91 g), while the lowest root weight was found in ShaitaBoro (1.20 g) after harvest. Considering root weight of examined cultivars, their variability could be attributed to its genetic potential (Table 1).

3.2 Yield and Yield Attributes

Results revealed that in general high yielding cultivars produced higher number of effective tillers per hill than low yielding ones. The highest effective tillers per hill were observed in Sugandhi (10.00), while the differences among the other cultivars Red Swarna, Jota pari, Pariza and ShaitaBoro were not significant. The highest non-effective tillers per hill (2.00) were produced in Jota pari and Pariza cultivars, while the lowest (0.33) was produced with Sugandhi cultivar. The highest number of tillers was produced by Sugandhi (11.33) and Jota pari and Pariza showed statistically same (10.67, 10.00). In contrast, both Red Swarna and ShaitaBoro produced the lowest number (9.33) of tillers per hill at harvesting stage. The highest filled grain number per panicle was recorded in Sugandhi (137.0) while the lowest filled grain number per panicle was in Pariza (76.67). The highest non-filled grain number per panicle was recorded in ShaitaBoro (40.67) while the lowest non- filled grain number per panicle was in Pariza and Jota pari (25.67) (Table 2).

The research also revealed that the highest 1000-grain weight was recorded in Pariza (20.57 g) followed by Sugandhi (20.13 g), and Jota pari (19.67 g), but the differences between them were not significant. The highest grain weight was recorded in Sugandhi (4.86 t per ha) followed by Jota pari (4.73 t per ha) and Red Swarna (4.63 t per ha), but there were insignificant differences between the three cultivars. The lowest grain weight was recorded in ShaitaBoro (2.31 t per ha). The higher yield in Sugandhi might be due to the production of higher number of effective tillers per hill and higher number of filled grain per panicle. Similar results were also reported by [15].

The highest straw yield was recorded in Jota pari cultivar (15.42 t per ha), followed by Sugandhi cultivar (15.13). The lowest straw yield was recorded in ShaitaBoro (7.5 t per ha). [16] reported that grain yield was positively correlated with biological yield in rice. Similar result was also reported by [17] in rice. The highest harvest index was recorded for Red Swarna (27.49%) and the lowest harvest index was recorded in ShaitaBoro (8.93%). The other variety showed intermediate values. HI is the measure of the efficiency of conversion of photosynthate into economic yield of a crop plant (Table 2).

3.3 Leaf Chemical Constituents

3.3.1 Chlorophyll

Leaf chlorophyll content is one of the important physiological traits closely related to photosynthetic ability in rice. Undoubtedly, understanding the genetic mechanisms underlying the leaf chlorophyll content across different developmental stages of rice has significant implications for improving photosynthetic ability in rice.

3.3.1.1 Chlorophyll-a content

Chlorophyll-a is the primary photosynthetic pigment. It absorbs most energy from wavelengths of violet-blue and orange-red light. Chlorophyll-a is essential for photosynthesis in eukaryotes, cyanobacteria and pro-chlorophytes because of its role as primary electron donor in the electron transport chain. In this study, a significant difference in chlorophyll-a content was found amongst the booting stage of selected five local rice cultivars as shown in Table 3. The highest chlorophyll-a content was found in Pariza (24.69 mg per g), while the lowest chlorophyll-a content was observed in Jota pari (18.32 mg per g).

3.3.1.2 Chlorophyll-b content

In this study, a significant variation in chlorophyll-b synthesis was evident in the booting stage of selected local rice cultivars during the study period. The cultivar Pariza was found to have more chlorophyll-b content during the experimental period. The highest chlorophyll-b content was found in Pariza (7.90 mg per g) and the lowest chlorophyll-b content was found in Jota pari (4.08). Chlorophyll-b contents in flag leaf of five local rice cultivars (Table 3). There is a strong relationship between temperature and plant chlorophyll synthesis.

3.3.1.3 Total chlorophyll content

In this study, the total chlorophyll content of selected rice cultivars did not varied significantly during the experimental period as shown in Table 3. At the booting stage, the highest total chlorophyll content was found in Pariza (32.59 mg per g) which is statistically similar with Sugandhi (28.05 mg per g), ShaitaBoro (27.82 mg per g). The lowest total chlorophyll content was observed in Jota pari (22.40 mg per g) which is statistically similar with Red Swarna (24.52 mg per g).

3.3.2 Total carotenoids content

The highest carotenoids content was found in Pariza (10.07 mg per g) which is statistically similar with other varieties. Carotenoids are multifunctional compounds serving as structural components of light-harvesting complexes, accessory pigments for light harvesting, substrates for abscisic acid synthesis, and components of photo protection and scavengers of oxygen. Carotenoids contents in flag leaf of five local rice cultivars were not significant (Table 3).

3.3.3 Nutrient contents in leaves of different rice cultivars

Nutrient elements present in leaf samples of five local rice cultivars were presented in the following Figs.1-4.

3.3.3.1 Phosphorus

The highest Phosphorus (P) content was observed in Sugandhi (0.73 ppm) followed by Jota Pari (0.52 ppm) and Red Swarna (0.51 ppm). On the other hand, the lowest P content was recorded in ShaitaBoro (0.37 ppm) followed by Pariza (0.39 ppm) (Fig. 1).

Table 1. Morphological parameters of five local rice cultivars

Cultivars	Plant height		Leaf number per plant		Root weight per plant
	Vegetative growth stage	Harvesting stage	Vegetative growth stage	Harvesting stage	
Red swarna	42.66c	104.0a	4.0b	5.3b	5.20a
Jota pari	60.22b	98.03ab	6.5a	6.7a	4.91a
Pariza	44.37c	75.33c	5.7b	5.2b	2.34b
Sugandhi	86.44a	95.33ab	7.1a	8.4a	1.81b
ShaitaBoro	65.4b	107.3a	3.7c	4.3c	1.20b
LSD	7.997	12.11	5.767	4.72	2.04
CV	7.10	6.73	8.19	6.63	34.98

** Values marked by the same letter do not differ significantly*

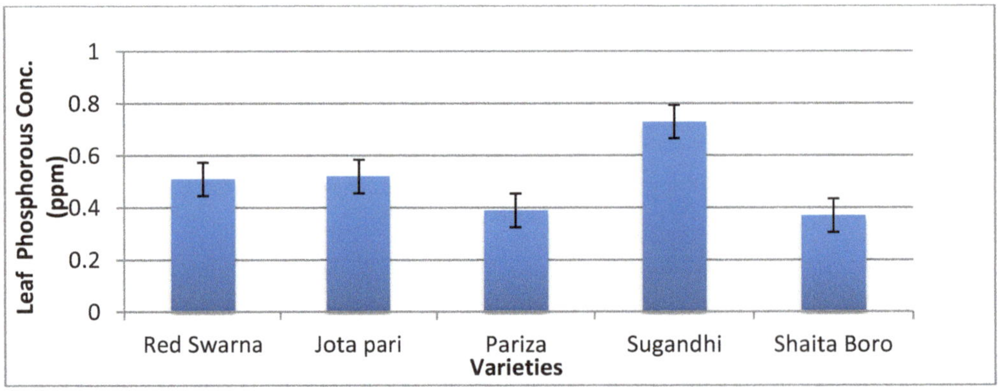

Fig. 1. Leaf phosphorus concentration of different rice varieties

Table 2. Yield and yield attributes of five rice cultivars

Cultivar	Number of effective tillers per hill	Number of non-effective tillers per hill	Total number of tillers per hill	Filled grain Per panicle	Non-filled grain per panicle	1000-grain weight (gm)	Grain yield (t per ha)	Straw yield (t per ha)	Harvest Index
Red Swarna	8.00b	1.33ab	9.33bc	130.0a	39.33a	16.73b	4.63a	12.29ab	27.36a
Jota pari	8.66b	2.00a	10.67a	123.3ab	25.67c	19.67a	4.73a	15.42a	23.47c
Pariza	8.00b	2.00a	10.00ab	76.67d	25.67c	20.57a	4.30ab	11.88ab	26.58a
Sugandhi	10.00a	0.33c	11.33a	137.0a	29.67b	20.13a	4.86a	15.13a	24.31b
ShaitaBoro	8.33b	1.00ab	9.33bc	112.3b	40.67a	16.60b	2.31c	7.50b	23.55c
LSD	2.20	1.57	0.47	56.71	16.18	9.17	1.57	4.80	5.61
CV	13.28	62.7	7.75	26.00	26.68	9.60	18.03	20.51	16.98

** Values marked by the same letter do not differ significantly*

Table 3. Flag leaf chlorophyll content of five local rice cultivars

Cultivar	Chlorophyll-a	Chlorophyll-b	Total chlorophyll	Carotenoids
Red swarna	19.21a	5.31ab	24.52b	7.207a
Jota pari	18.32a	4.08b	22.40b	7.213a
Pariza	24.69a	7.90a	32.59a	10.07a
Sugandhi	21.72a	6.33ab	28.05a	8.727a
ShaitaBoro	22.56a	5.26ab	27.82a	10.01a
LSD	10.01	4.14	12.86	3.509
CV	24.95	36.68	25.01	21.51

** Values marked by the same letter do not differ significantly*

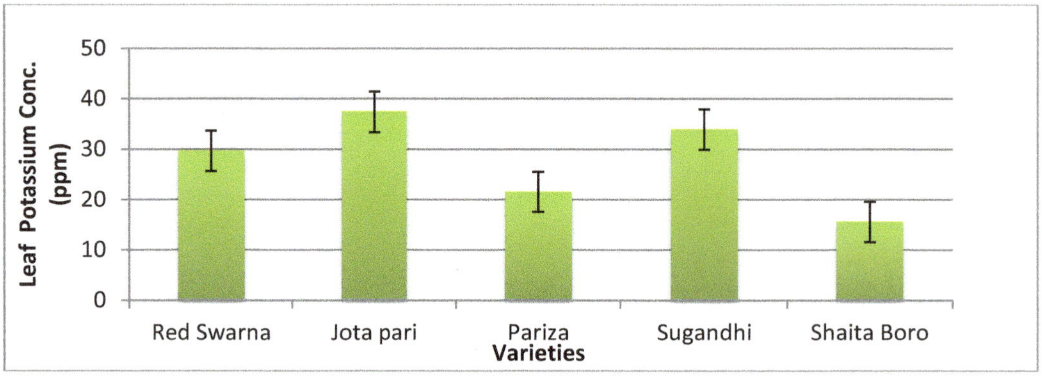

Fig. 2. Leaf potassium concentration of different rice varieties

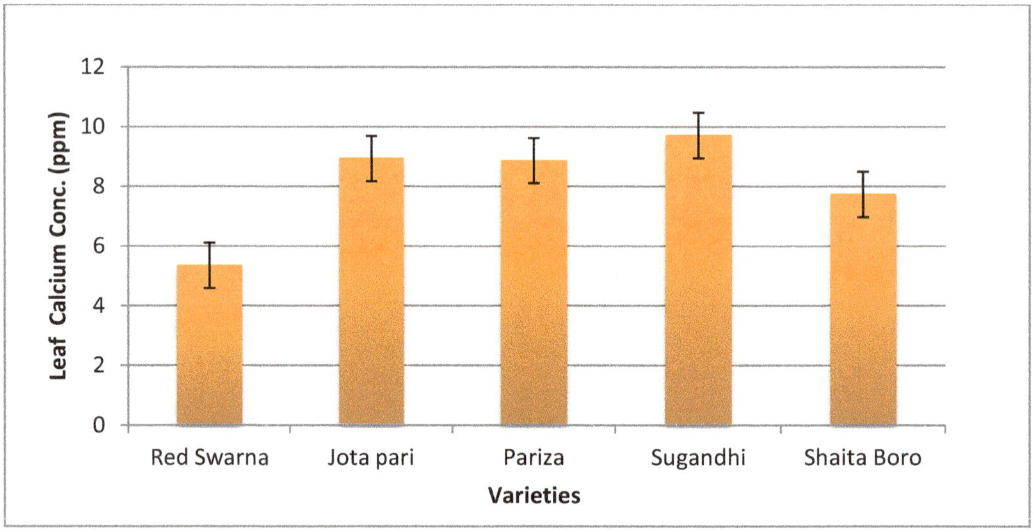

Fig. 3. Leaf calcium concentration of different rice varieties

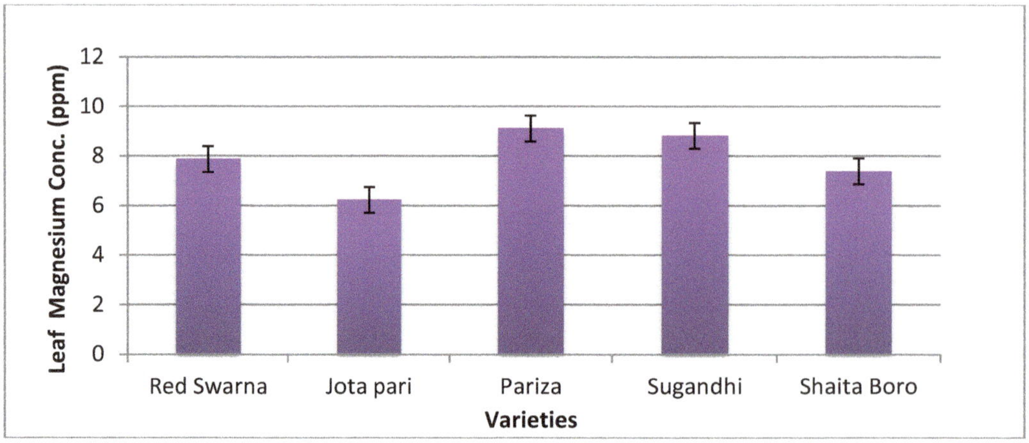

Fig. 4. Leaf magnesium concentration of different rice varieties

3.3.3.2 Potassium

The variation in potassium (K) content in leaves among the cultivars was recorded in Fig. 2. The highest K content was found in Jota pari (37.38 ppm) followed by Sugandhi (33.92 ppm). On the other hand, the lowest K content was recorded in ShaitaBoro (15.54 ppm). The status of K improved the rice straw quality.

3.3.3.3 Calcium

The variation in calcium (Ca) content in leaves among the cultivars was recorded in Fig. 3. The highest Ca content was recorded in Sugandhi (9.72 mg%) which was statistically similar with Jota pari (8.94 mg%) and Pariza (8.87 mg%). On the other hand, the lowest Ca content was recorded in Red Swarna (5.36 mg %). Secondary nutrient plays an important role in the plant life. Calcium is an essential part of cell structure and plays its role in cell division. It is absorbed by the plants in the form of Ca ion i.e. Ca^{2+}.

3.3.3.4 Magnesium

The test of Magnesium (Mg) content in leaves among the rice cultivars was presented in Fig. 4. The highest Mg content was recorded in Pariza (9.11 mg%) which is statistically similar with Sugandhi (8.82 mg%). On the other hand, the lowest Mg content was recorded in Jota pari (6.23 mg%). Magnesium is a integral part of chlorophyll and thus, linked with photosynthesis.

4. CONCLUSION

From the results it can be concluded that Sugandhi showed the highest yield potential than those of other four local cultivars. Among the local rice cultivars, Sugandhi had the best yielding ability than those of others. Therefore, it can be suggested that, Sugandhi may be selected for cultivation in Bangladesh using modern agronomic practices due to its high genetic yield potential among the other local cultivars studied.

ACKNOWLEDGEMENT

Authors are thankful to all teachers and staffs of Department of Agricultural Chemistry, Hajee Mohammad Danesh Science and Technology University, Dinajpur, Bangladesh, for their co-ordination and inspirations during conducting the research and writing manuscript.

COMPETING INTERESTS

Authors have declared that no competing interests exist.

REFERENCES

1. BBS (Bangladesh Bureau of Statistics). Yearbook of agricultural statistics of Bangladesh. Bangladesh Bureau of Statistics, Statistics Division, Ministry of Planning, Government of the People's Republic of Bangladesh; 2008.
2. BRRI (Bangladesh Rice Research Institute). Bangladesh rice knowledge bank. Bangladesh Rice Research Institute, Joydebpur, Gazipur, Bangladesh; 2011.
3. FAO. Paper presented by the Director in 'International year of rice 2004'. Food and Nutrition Division, Food and Agriculture Organization of the United Nations, Rome, Italy; 2004.
4. Tuteja N, Gill SS, Tiburcio AF, Tuteja R. Improving crop resistance to abiotic stress. Wiley Online Library. 2012;1 and 2.
5. Bhuiyan NI, Paul DNR, Jabber MA. Feeding the extra millions by 2025. Challenges for rice research and extension in Bangladesh, National Workshop on Rice Research and Extension in Bangladesh, Bangladesh Rice Research Institute, Gazipur; 2002.
6. David C, Otsuka K. Modern rice technology and income distribution in Asia. Lynne Rienner Publishers; 1994.
7. MoF (Ministry of Finance). Bangladesh economic survey 2010. Economic Division, Ministry of Finance, Government of the People's Republic of Bangladesh, Dhaka; 2010.
8. Sarker MAZ, Murayama S, Ishimine Y, Tsuzuki E. Physio-morphological characters of F1 hybrids of rice (*Oryza sativa L.*) in Japonica-Indica crosses. Plant Production Science. 2001;4(3):196-201.
9. Tang JJ, Chen X, Katsuyoshi S. Varietal differences in photosynthetic characters and chlorophyll fluorescence induction kinetics parameters among intergeneric progeny derived from *Oryza*×*Sorghum*, its parents and hybrid rice. Journal of Zhejiang University Science. 2002;3(1): 113–117.
10. Sheng T, Qian Q, Dali Z, Yasufumi K, Kan F, Danian H, Lihuang Z. QTL analysis of leaf photosynthetic rate and related physiological traits in rice (*Oryza sativa* L.). Acta Agronomica Sinica. 2004;135(1):1-7.

11. Shan YH, Wang YL, Pan XB. Mapping of QTLs for nitrogen use efficiency and related traits in rice (*Oryza sativa* L.). Agricultural Sciences in China. 2005; 4(10):721–727.
12. Yoshida H, Horie T, Shiraiwa T. A model explaining genotypic and environmental variation of rice spikelet number per unit area measured by cross-locational experiments in Asia. Field Crops Research. 2006;97(2-3):337–343.
13. Gardner FP, Pearce RB, Mistecell RI. Physiology of crop plants. Iowa State University. Press, Iowa. 1985;66.
14. Gomez KA, Gomez AA. Statistical procedures for agricultural research. 2nd edition. John Wiley and Sons, New York. 1984;680.
15. Dutta RK, Mia MAB, Khanam S. Plant architecture and growth characteristics of fine grain and aromatic rice and their relations with yield. Crop Physiology Division, Bangladesh Institute of Nuclear Agriculture, Mymensingh, Bangladesh; 2002.
16. Chowdhury SA, Majid MA, Haque KS, Islam M, Rahman MM. Effect of variety on yield and nutritive value of rice straw. Asian-Australasian Journal of Animal Sciences. 1995;8(4):329-335.
17. Munshi RU. A comparative morpho-physiological study between two local and two modern rice cultivars. M.S. Thesis, Department of Crop Botany, Bangladesh Agricultural University, Mymensingh; 2005.

20

The Impact of Gas Flaring on Plant Diversity in Ibeno Local Government Area

M. A. Abua[1] and S. W. Ashua[2*]

[1]Department of Geography and Environmental Science, University of Calabar, Calabar, Nigeria.
[2]Department of Geography, Cross River State College of Education, Akamkpa, P.M.B. 1171, Calabar, Nigeria.

Authors' contributions

This work was carried out in collaboration between all authors. Both authors went to the field. Author AMA designed the study, wrote the protocol, and the first draft of the manuscript. Author ASW performed the analysis, discuss the conclusion. Both authors read and approved the final manuscript.

Editor(s):
(1) Bin Gao, Department of Agricultural and Biological Engineering, University of Florida, USA.
(2) Alejandro E. Ferrari, Department of Science and Technology, UNQ, Universidad Nacional de Quilmes, Argentina.
Reviewers:
(1) Chis Timur Vasile, Ovidius University Constanta, Romania.
(2) José Goldemberg , University of São Paulo's, Brazil.
(3) Anonymous, Western Delta University, Nigeria.
(4) Anonymous, Delta State University, Nigeria.

ABSTRACT

The study examined the impact of gas flaring on plant diversity in Qua-Iboe Terminal (Q.I.T). A Purposive sampling technique was adopted in choosing the direction of transects for data on plant species. A quadrant of 50 mx50 m was gridded into 10 m^2. On each transect, five quadrants were laid at an interval of 50 m. From the flare site, the first quadrant at an interval of 50 m, the second at 100 m, third at 150 m, and so on to the fifth at 250 m on both sides of the transect respectively. Plant species diversity were obtained by counting the different species found within the quadrant. At flare site; 21, 27, 34, 38 and 42, while at control site; 101, 102, 106, 107 and 109 respectively. The means of the plant species at flare site is 32.40, at the control site is 105.00. The standard deviation for the flare site is 8.44 and for the control site is 3.39. Paired Sample Statistical technique was employed to analyzed the data. The results showed strongly positive correlation co-efficient of 0.97 between plant species diversity at gas flare site and those at the control site. It was

**Corresponding author: Email: ashuasonny@yahoo.com*

observed that at Qua-Iboe flare site, great lost of plant species had taken place than the control site. Field Observation revealed that some plant species proved more resistant to gas flaring than others. Furthermore, from the analysis, we can conclude that there is statistically significant difference between plant species diversity in the flare site and the control site due to gas flaring. The results of the study is in line with theoretical postulations that increase in distance leads to a corresponding increase in plant species from the flare site. The study recommended that; they should be massive afforestation programme, gases should be harnessed for use as liquefied natural gas or re-injected into the (source) earth, and government should enforce laws against gas flaring.

Keywords: Impact; gas flaring; Niger delta; Ibeno; plants diversity loss.

1. INTRODUCTION

The Niger Delta mangrove and rainforest belt is one of the most fragile ecosystems in the world. It is the most oil- impacted environment and polluted area in the world [1,2]. Everywhere in the world, plant species are used for medicinal purpose, food, ornamentals, carbon sinkers and as building materials. In this vein, the essence of plants to man cannot be underscored. Plant species diversity is the measure of the number of different plant species present in an area with reference to their physical characteristics and genetics composition [3]. The plant species in the Niger Delta region are undergoing serious threat for more than forty years owing to the exploration and exploitation of oil and gas by multinational companies domiciled in the area [2].

Though, natural gas is an important source of energy to man, its extraction has impacted negatively on the environment. Flaring is adopted during the process of petroleum refining and chemical processing for safe disposal of waste gases, process upsets, plant start-up or shutdown and process emergencies. Flaring is a high-temperature oxidation process used to burn combustible components, mostly hydrocarbons, of waste gases from industrial operations. Gas flaring is the combustion of associated gas produced with crude oil or from gas fields [4,5]. The process or act of burning away associated gas into the atmosphere during the drilling of crude oil is called gas flaring. Natural gas can be found separately or in association with oil. Nigeria has immense reserves of both association and un-association natural gas [6]. Nigeria is known to be one of the highest countries on the earth for flaring gases. Recent report put the rate of gas flaring in Nigeria at 32.6 percent, against 0.6 percent and zero percent by U.S.A and Netherlands respectively. This menace is evident during exploration and exploitation activities of multinational oil and gas companies like Shell, Chevron, ExxonMobil, Total and Elf (Friends of the [7]. Nigeria and Russia are the highest gas flaring countries in all the world. Gas flaring is a major source of greenhouse gases (GHG) contributing to global warming which could accelerate the problem of climatic change and harsh living conditions on earth, if not checked. Flaring releases carbon dioxide and methane, the two major greenhouse gases. However, of these two, methane is actually more harmful than carbon dioxide. Gas flaring for instance, has negative effects on the immediate environment, particularly on plant growth and wildlife.

Scholars like [8-19] have attempted to address the perennial problems associated with the causes of gas flaring and oil spills. Some international institutions like the [20-22], and the [23], have equally reported facts on what constitute the causes and effects of oil and gas on a particular environment like the Niger Delta region. [24] have shown that gas flaring significantly affects not only the microclimate but also the soil physico-chemical properties and maize yield size. For optimum yield of maize within the Niger Delta, cultivation must be within 2kilometres of the bund wall of the flared site. They further observed that the sand content of the soil, pH, bulk density, air and soil temperatures increased toward the flare site. Research findings show that there is indeed a correlation between gas flaring and change in maize yield [24]. This study explores the impact of gas flaring on plant species diversity in Ibeno, Akwa-Ibom State, South-South, Nigeria.

2. MATERIALS AND METHODS

2.1 Sampling Technique and Analytical Tool

A reconnaissance survey was carried out to identify the flare site for easy transect and quadrant laying. A cutlass was used to open up the transect and the tape was used to determine the distance of the quadrant. The quadrants were established at 50m interval along the transect. A Purposive sampling technique was adopted in choosing the direction of transect. Four transects were established in each of the sampling site (that is, the flare site, the control site, against the prevailing wind and the wind direction. While, a quadrant of 50 mx50 m was measured. The quadrant size was further gridded into 10 m^2 by measuring dimensions of 10 mx10 m with a measuring tape and pegs to subdivide the sampled area from which the gridded cells were randomly sampled. The procedure was repeated for all the sampled quadrants based on the interval of 50 m. The average of each variable measured was calculated from the 20 sampling units. Also, a plant taxonomist assisted in identifying and naming plant species found within the quadrant. A Data on plant species diversity were collected. While a quadrant of 50 mx50 m was gridded into 10 m^2. On each transects, five quadrants were laid at the interval of 50 m. From the flare site, the first quadrant at the interval of 50 m , the second at 100 m, third at 150 m, and so on to the fifth at 250 m on both side of the transect respectively. Data on plant species diversity were counted on each quadrant and recorded. The means of plant species counted in the each quadrant were obtained. Names of plant species identified are found in Table 3

The paired t' test statistics was employed to analyze the means difference in plant species between the flare site and the control site. The Simpson's index of diversity was used to measure the extent of homogeneity or otherwise of plant community with respect to its floristic composition. This index was used to compare two or more sampled areas of vegetation for similarity or diversity and is given in the form:

$$Si = 1\text{-}D$$

Where:

Si = Simpson's index of Diversity

$$D = \sum(n_1/N)^2$$

Where:

n_1 = some specified importance value for a species
N = Importance value for all species
D = is an index of dominance or homogenity.

3. THE STUDY AREA

The study area is Ibeno local Government area in Akwa-Ibom State, South-South region of Nigeria. It lies within the Niger-Delta oil-rich region. The area is located between Longitude 7°.35[1]E and 7°.45[1]E of the Greenwich meridian; Latitude 6° .10[1]N and 8°.11[1]N of the equator. The climate falls within tropical equatorial (Af) type, characterized by longer wet season and shorter dry season. Temperatures and relative humidity are high throughout the year with abundance annual rainfall [25,26]. The mangrove and rainforest vegetation stretched outward along the Atlantic coast to the land (Fig. 1).

4. RESULTS AND DISCUSSION

4.1 Impact of Gas Flaring on Plant Density

Table 1 shows measurement of plant species densities at the flare site and control site. Results revealed that the values of species density at Qua-Iboe terminal (Q.I.T.) flared site were 21 plant species at 50 m, 27 plant species at 100 m, 34 plant species at 150 m, 38 plant species at 200 m and 42 plant species at 250 m. while, at the control site measurement were 101 plant species, 102 plant species, 106 plant species, 107 plant species and 109 plant species at 50 m, 100 m, 150 m, 200 m and 250 m respectively about 2 km away fromthe flare site. Plant species density at the direction of the prevailing wind ranges from 20 plant species, 25 plant species, 33 plant species, 36 plant species and 40 plant species away from the flare site (Table 2). This was observed at the distances of 50 m, 100 m, 150 m, 200 m and 250 m. At the opposite direction of the prevailing wind, plant density were 23, 29, 35, 40 and 45 higher than the direction of the prevailing wind, (Table 2). The results in Table 1 and 2 reveals that apart from the control site, plant species density were higher at 250 m and lower at 50 m quadrant zone. At 50m distance from the flare site, the impact of gas flared was more on plants species than quadrant zones farther away from the flare site. In Table 4, the Simpson's index employed to measure the extent of homogeneity of plant community with respect to its floristic composition showed an index less than one across the transects. This implies that the plant species in the study area are heterogeneously distributed. The Simpson's index of plant community against prevailing wind (APW) is 0.997, at wind direction

(WD) is 0.996, at flare site (FS) is 0.997 and at the control site (CS) is 0.911 respectively. In other words, there is no consociation in the plant community. Consociation means all individual plants belong to the same species.

Table 1. Plant species density on flare site and control site

Distance from flare site	Flare site	Control site
50 m	21	101
100 m	27	102
150 m	34	106
200 m	38	107
250 m	42	109

Source: Author's field work, 2011

Table 2. The result of plant species count At Qua Iboe flare (Ibeno)

Distance	Wind direction	Against prevailing wind	Average count
50 m	20	23	21.5
100 m	25	29	32
150 m	33	35	34
200 m	36	40	38
250 m	40	45	42

Source: Author's field work, 2011

Table 5a revealed the mean of 32.40 and 105.00 at the flare site and control site, with the standard deviation of 8.44 and 3.39 respectively. It was also established in Table 5b that a strong positive correlation existed between plant species density and gas flaring at both the flare site and the control site. Plant species density at the flare site was relatively fewer than those found at the control site due to gaseous pollutant released into the air. This finding alludes to the scientific survey of the vegetation, particularly palm trees, which revealed that no plant could survive within 100 m of the flare in Ijaw oil-producing communities. Some plants were found to flower at 120 m, while others survived at 150m from the flare site [2]. In this study, the analysis revealed that plants like; *Acrostichum aureum* L., *Anthocleista vogelii, Pranch Avicennia specie, Rhizophora species and Eichornia crassipes* , *Lophira alata* were found to survive around the flare site at distance of 50 m.

The study also revealed that as distance increases away from the gas flare sites, a dramatic increase in plant species were observed [Fig. 2]. Conversely, plant species decrease with proximity to the flare site. Different species responded differently to gas flaring. This was seen in their height, structure, composition and adaptability. Plants like *Nypa fruticans, Lophira alata* and *Rhizophora species proved more resistant and adaptable to gas flaring.* Species density were higher along the control transect than the two sites adjacent to the gas flaring site. Result of analysis with paired sample t-test statistics in Table 4c showed that there is statistically significant difference between plant species in the flare site and the control site. This is consequent upon the nefarious activities of gas flaring activities. Implying that there is significant difference between plant species diversity at the flare site than at the control site of the study area. This is consequent upon the nefarious activities of gas flaring activities going on in the region. It was further observed that plant species density did not vary significantly at the control site. The impact of gas flared were less significant at the control site.

Table 3. Plant species in Ibeno L.G.A. Akwa-Ibom State South-Eastern Nigeria

S/no	Plant species
1	*Dictamoius albus Albiflorus*
2	*Raphia hookeri, Raphia Africana*
3	*Puereria phaseoloides*
4	*Ocimum basilicum*
5	*Nypa fruticans*
6	*Napoleonaea imperialis*
7	*Myrianthus arboreus Beavois*
8	*Musanga R. Brown*
9	*Panicum maximum*
10	*Phoenix reclinata Jacq.Fragm*
11	*Musa paradisiacal*
12	*Aspilia Africana*
13	*Andropogon gayanus*
14	*Allamanda L.*
15	*Acalyphia L.*
16	*Aframomum melegueta*
17	*Encephalartos Lehm*
18	*Elaeis guineenis*
19	*Colocasia esculenta*
20	*Cocos nucifera*
21	*Acrostichum aureum L.*
22	*Anthocleista vogelii Pranch*
23	*Avicennia species*
24	*Rhizophora species.*
25	*Eichornia crassipes*
26	*Lophira alata*

Source: [27]

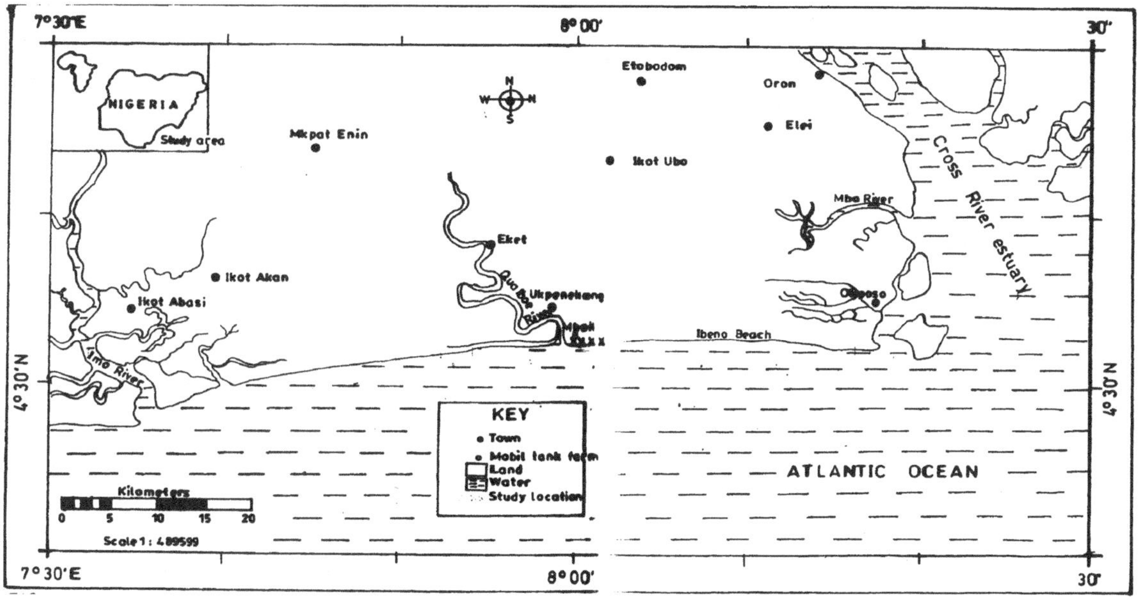

Fig. 1. Map of the study area (Ibeno)

Table 4. Simpson's Index

S/no	Against Prevailing wind (APW)			Wind direction (WD)			Flare site (FS)			Control site (CS)		
	n.	D	1-D	n	D	1-D	n	D	1-D	n	D	1-D
1	5	.001	.999	4	.001	.999	3	.000	1.000	20	.001	.962
2	25	.021	.979	15	.009	.991	16	.010	.990	19	.001	.964
3	5	.029	.999	3	.000	1.000	2	.000	1.000	18	.001	.966
4	6	.030	.999	4	.001	.999	3	.000	1.000	16	.001	.970
5	30	.170	.971	40	.067	.933	22	.018	.982	50	.009	.905
6	2	.090	1.000	1	.000	1.000	1	.000	1.000	21	.002	.960
7	2	.010	1.000	0	.000	1.000	2	.000	1.000	19	.001	.964
8	4	.023	.999	2	.000	1.000	3	.000	1.000	20	.010	.286
9	2	.010	1.000	1	.000	1.000	2	.000	1.000	21	.002	.960
10	5	.029	.999	2	.000	1.000	3	.000	1.000	14	.001	.973
11	3	.017	1.000	0	.000	1.000	2	.000	1.000	12	.001	.977
12	2	.010	1.000	1	.000	1.000	2	.000	1.000	19	.460	.321
13	4	.023	.999	2	.000	1.000	10	.004	.996	22	.002	.958
14	5	.029	.999	4	.001	.999	6	.001	.999	20	.001	.962
15	15	.087	.992	12	.006	.994	17	.011	.989	25	.002	.952
16	9	.052	.997	7	.002	.998	12	.005	.959	17	.001	.968
17	3	.017	1.000	1	.000	1.000	1	.000	1.000	16	.001	.970
18	11	.063	.996	12	.006	.994	12	.005	.995	18	.001	.966
19	7	.040	.998	8	.003	.997	6	.001	.999	19	.001	.964
20	8	.046	.998	12	.006	.994	10	.004	.996	21	.002	.960
21	2	.010	1.000	1	.000	1.000	3	.000	1.000	15	.001	.971
22	4	.023	.999	2	.000	1.000	4	.001	.999	15	.001	.971
23	2	.010	1.000	6	.002	.998	4	.001	.999	16	.001	.970
24	3	.017	1.000	5	.001	.999	6	.001	.999	24	.002	.954
25	2	.010	1.000	4	.001	.999	5	.001	.999	23	.002	.956
26	6	.030	.999	5	.001	.999	5	.001	.999	25	.002	.952
TOTAL	172	.037	.997	154	.004	.996	162	.038	.997	525	.039	.911

Table 5a. Paired samples statistics

		Mean	N	Std. deviation	Std. error mean
Pair 1	Flaresite	32.4000	5	8.44393	3.77624
	controlsite	105.0000	5	3.39116	1.51658

Table 5b. Paired samples correlations

		N	Correlation	Sig.
Pair 1	Flaresite & controlsite	5	.987	.002

Table 5c. Paired samples test

		Paired differences			t	Df	Sig. (2 tailed)
		Mean	Std. deviation	Std. error mean			
Pair 1	Flaresite - controlsite	-72.60000	5.12835	2.29347	-31.655	4	.000

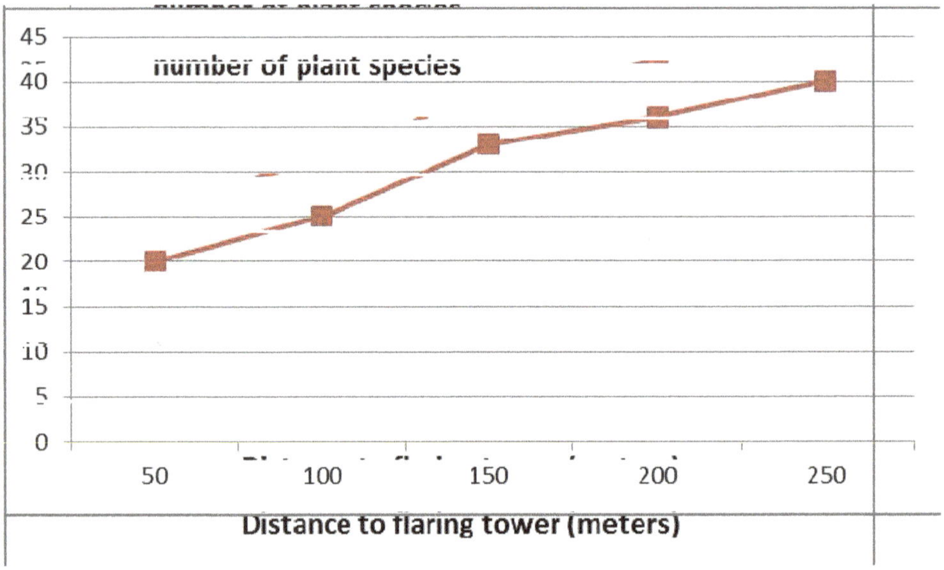

Fig. 2. Distance to flaring tower (meters)

5. CONCLUSION

Analysis of the impact of gas flaring on plant diversity in Ibeno revealed that gas flare has greater impact on plant species closer to the flare site at 50m than at other zones where the quadrant was laid. As distance increases from the flare site, more plants species were found. The study further showed that certain plant species were found to be more resistant to gas flaring at 50 m and 100 m such as; *Nypa fruticans Acrosticum aureum and Eichornia crassipes.* Plant species are heterogeneously distributed in the region as revealed by the Simpson's index analysis of specie diversity.

The Niger Delta environment ecosystem is the most fragile in the world. It has being under serious threat by the nefarious activities of gas flaring carried out there. More awareness campaign should be embarked upon to assuage the long trend of gas flaring. The study further suggests that plant species that prove resistant to gas flaring should be planted, laws against gas flaring must be enforced and re-injection techniques should be implemented.

COMPETING INTERESTS

Authors have declared that no competing interests exist.

REFERENCES

1. Kia M. The Struggles for the Survival of 12 million People in the Niger Delta. Environmental experts from UK, US and Nigeria affirmed that the Delta is the most polluted area in the World; 2009. Available:http://nigerdeltasolidarity.wordpress.com (Accessed June, 2014).
2. Raji A, Abejide Ts. An assessment of environmental problems associated with pollution and gas flaring in the Niger Delta Region, Nigeria 1960-2000. Arabian Journal of Business and Management Review. 2013;3(3):48-62.
3. Enger ED, Smith BF. Environmental Science: A study of Interrelationships. New York: MacGrawHill; 2008.
4. Adole T. A Geographic Information System (GIS) based assessment of the impacts of gas flaring on vegetation cover in Delta State, Nigeria. Master's Thesis, Environmental Sciences University of East Anglia, Norwich; 2011.
5. Ubani EC, Onyejekwe IM. Environmental Impact analysis of gas flaring in the Niger Delta region of Nigeria. America Journal of Scientific and Industrial Research. 2013; 4(2):246.252.
6. Ade A. Introductory environmental resource management. Ibadan: Daybis limited; 2002.
7. Environmental Right Action, Friends of the Earth. The Shell Report: 'Continue Abuses in Nigeria 10 years after Ken-Saro-Wiwa. Benin City, Nigeria; 2008.
8. Abdulkareem AS, Odigure JO, Abenege S. Predictive model for pollutant dispersion from gas flaring: A case study of oil producing area of Nigeria. Energy Sources. 2009;31(12):1004-1015. DOI:10.1080/15567030801909318
9. Aghalino SO. 'Petroleum exploitation and the agitation for compensation by oil producing communities in Nigeria', Geo-Studies Forum. 2002;11-20.
10. Aghalino SO. Gas flaring, environmental pollution and abatement measures in Nigeria, 1969-2001. Journal of Sustainable Development in Africa. 2009;11(4):219-238.
11. Akujuru V. The professional duty of valuer in oil pollution compensation valuation. Paper presented at the NIESU 22nd Annual Conference (Port Harcourt). 1992; 394.
12. Augustine OI, Sanford WW. The effects of waste Gas Flare on the surface vegetations in South- Eastern Nigeria. Journal of Applied Ecology. 1976;13:173.
13. Awunudiogba G. Adaptive strategies of small fishermen to environmental degradation induced by oil pollution in the Niger Delta,' Michael & Apollo eds. Human Impact on Environment and Sustainable Development. England: Ashgate Publishing; 2003.
14. Brouwer R. Nutritive influences on the distribution of dry matter in the plant. Netherland Journal of Agric Science. 1971; 10:361-76.
15. Emoyan OO, Akpogorie IA, Akporhonor EE. The oil and gas industry and the Niger Delta: Implication for the Environment. Journal of Applied Science and Environmental Management. 2008; 12(3):30- 31.
16. Fagbami AA, Udo EJ, Odu CTI. Vegetation and damage in an oil field in the Niger Delta of Nigeria. Journal of Tropical Ecology. 1988;4(1):61-64.
17. Frynas JG. Oil in Nigeria: Conflict and litigation between oil companies and village communities, Harmburg: Lit Verlag. 2000;27.
18. Frynas JG. Corporate and State Responses to Anti-Oil Protest in the Niger Delta. African Affairs. 2001;100(27-54):24-40.
19. Orimoogunje OOI, Ayanlade A, Akinkuolie TA, Odiong AU. Perception on effect of gas flaring on the environment. Research Journal of Environmental and Earth Sciences. 2010;2(4):188-193.
20. World Bank Document. Republic of Nigeria, Niger Delta Social and Economic Conflict Analysis. Sustainable Development Department African Region, 2008;12-14.
21. World Bank. Gas Flaring Reductions Avoid 30 Million Tons of Carbon Dioxide Emissions in 2010. World Bank Press Release, Washington; 2011.
22. World Bank. World Bank GGFR Partners Unlock Values of Wasted Gas, World Bank. 2009;14.
23. World Bank. *Memorandum of the* President of the IDA and the IFC to the executive directors on an interim strategy update for the Federal Republic of Nigeria, New York; 2002.
24. Odjugo PAO, Osemwenkhae EJ. Natural gas flaring affects microclimate and

reduces maize (*Zea mays*) yield. International Journal of Agriculture & Biology. 2009;11(4):408-412.

25. Inyang PEB. Pollution: A factor in the climate of Calabar and environs: Geographic studies published by department of geography, University of Calabar for the 23rd Nigerian Geographic Association Conference, Calabar. 16th-21st March; 1980.
26. Oguntoyinbo JS. *Climatology of West Africa*. New Jersey: Library of Congress Cataloguing in publication; 1978.
27. Nyananyo BL. Plants from the Niger Delta. Rivers State, Nigeria: Onyoma Research Publications; 2006.

21

Bioindicators: Study on Uptake and Accumulation of Heavy Metals in Plant Leaves of State Highway Road, Bagalkot, India

N. M. Rolli[1*], S. B. Gadi[2] and T. P. Giraddi[1]

[1]BLDEA's Degree College, Jamkhandi, 587301, Karnataka, India.
[2]Department of Botany, JSS College, Dharwad, 580004, Karnataka, India.

Authors' contributions

This work was carried out in collaboration between all authors. All authors read and approved the final manuscript.

Editor(s):
(1) Alejandro E. Ferrari, Department of Science and Technology, UNQ, Universidad Nacional de Quilmes, Argentina.
(2) Krzysztof Skowron, Department of Microbiology, Nicolaus Copernicus University in Toruń, Collegium Medicum of L. Rydygier in Bydgoszcz, Poland.
(3) Daniele de Wrachien, Department of Agricultural and Environmental Sciences of the State University of Milan, Italy.
Reviewers:
(1) Isiaka A. Ogunwande, Lagos State University, Lagos, Nigeria.
(2) Shruti Murthy, Bangalore University, India.
(3) Rajaram Pandurang Dhok, Savitribai Phule Pune University, Pune, India.

ABSTRACT

In this study, Caesalpinia (*Caesalpinia pulcherrima*) and grass (*Cyndon dactylon*) was evaluated as the bioindicators of heavy metals such as the Lead (Pb), Copper (Cu), Cadmium (Cd), Manganese (Mn), Zinc (Zn), Chromium (Cr) and Nickel (Ni) contaminated in Bagalkot and along the state high way upto Mudhol. The soil samples at depth (0-20 cm) and caesalpinia and grass leaves were taken from different sampling stations namely Navnagar bypass road (S_1), Gaddanakeri cross (S_2), Tulasigeri (S_3), Kaladagi (S_4), Lokapur (S_5), Chichakhandi (S_6) and Mudhol (S_7). The concentrations of Pb, Cu, Cd, Mn, Zn, Cr and Ni were measured using GBC- 932 plus Atomic Absorption Spectrophotometer (Austrelia). The results of the study shows that the concentrations of heavy metals in caesalpinia ranged from Pb 20.36 to 29.39 μg/gmm, Cu 3.92 – 5.94 μg/gm, Zn 24.40 to 35.7 μg/gm, Cd 1.01 to 1.78 μg/gm, Mn 27.01 to 69.10 μg/gm, Cr 1.20 to 7.8 μg/gm and Ni 7.9 to 13.1 μg/gm. In grass heavy metal ranges between for Pb 20.16 to 28.01, Cu 3.95 to 5.76

**Corresponding author: E-mail: drnmrolli@rediffmail.com*

μg/gm, Zn 24.50 to 35.8 μg/gm, Cd 1.15 to 1.52 μg/gm, Mn 28.91 to 72.51 μg/gm, Cr 1.28 to 8.0 μg/gm, Ni 8.1 to 15.1 μg/gm. Similarly in roadside soil heavy metal ranges between Pb 81.91 to 139.8 μg/gm, Cu 39.54 to 58.58 μg/gm, Zn 32.29 to 381.54 μg/gm, Cd 1.51 to 2.08 μg/gm, Mn 1257.9 to 2051 μg/gm, Cr 131.9 to 951.2 μg/gm and Ni 69.53 to 108.6 μg/gm. According to these results the concentration of heavy metals in grass was found high as compared to caesalpinia. Thus, compared to the metal accumulation potential grass is said to be heavy metal accumulator. The variation in heavy metal concentrations is due to changes in traffic density and anthropogenic activities. Thus, generally it is considered that grass and caesalpinia are good bioindicators and can be used in air pollution monitoring studies.

Keywords: Heavy metals; bioindicators; accumulation; anthropogenic activities.

1. INTRODUCTION

Environment pollution with toxic metals has increased since increasing industrialization and human activities [1]. Emissions of various pollutants into the atmosphere have a lot of harmful effects on plants growth [2]. Rapid urbanization, unregulated industrialization growing transport intensity and agricultural practices have created heavy metals concentrations in serious stage [3]. Plants have a major contributions in purifying land water and environmental air. Entrance of heavy metals may occur in human and animal food chain as a result of their uptake by edible plant grown in contaminated soil [4]. The toxic and hazardous effects of some heavy metals on human health are very significant and may cause many fatal diseases.

Lead (Pb) is one of the heavy metal that is responsible for anemia, neurological disorder, hyperactivity and changes in blood enzymes in human body [5]. Cadmium (Cd) and Zn are important toxic metals and longtime exposure of which may cause renal, pulmonary, hepatic, skeletal, reproductive and many other carcinogenic effects [6,7]. It is widely recognized that the principal reasons of heavy metals (Pb, Cu and Cd) derived from traffic congestion, long-range transport and household heating [8]. The spreading of contaminants is influenced by meteorological parameters such as rainfall, wind and traffic intensity [4]. The same meteorological conditions affect the concentration of same contaminants in the roadside soil [8]. The traffic density determines the lead level in soil and vegetation [9-11].

Soil samples and vegetation are the most economic and reasonable ways for assessing heavy metal status in the atmosphere [12]. Acacia [13], grass [14], plants [15], and other organisms such as fish [16] have also have been used for monitoring. In order to assess contamination by metals in the vicinity of a highway, several studies have been carried out with the different compartments: Study of global deposits, roadside soil and vegetation [8]. Information on accumulation of heavy metal on roadside soil of this city due to highway traffic and vehicles is very limited [17], but this could be the new threat for agriculture. Determination of heavy metal accumulation in roadside soil may be an index of the environmental pollution of Bagalkot city. Keeping this view in mind, the research was conducted to know the heavy metal accumulation of roadside soil, grass and caesalpinia of Bagalkot city and along the state highway from Bagalkot to Mudhol.

2. MATERIALS AND METHODS

Bagalkot is the city of Northern region of Karnataka at latitude 16° 04 ¹ N to 16° 21 ¹ North and longitude 75° 26 ¹ E to 76° 02 ¹ Eastern. The city is suffered from high traffic density caused by vehicles. Grass, caesalpinia and soil were collected during 2013, which were three meters away from the State High way (Fig. 1 and Table 1) passing through Navanagar. Grass and caesalpinia samples were collected from each site at three random spots those were spaced approximately at one meter interval. The leaves were clipped with stainless steel scissors. All the samples of each site were combined to give composite samples of about 300 to 500 gm. The leaves of Grass (*Cyndon dactylon*) and Caesalpinia (*Caesalpinia pulcherrima*) samples were dried at 80℃ for 48 hrs fine by powdered and sieved through 0.2 mm sieve. One gram sample was digested using Gerhardt digestion unit using mixed acid digestion method [18]. The digested material was diluted with double distilled water and filtered through Whattman paper 41 and made upto 100 ml.

Table 1. Sampling stations

Station no.	Sampling station	Nature of stations
Control	**Navanagar (Bagalkot)**	**Unpolluted urban area- vehicular movement is negligible, unpolluted area with less disturbance**
1	Navanagar Bypass road	Vehicular movement is high
2	Gaddankeri cross	Vehicular movement is high. Agricultural fields on either side of the road
3	Tulasigeri	Bricks factories around the Gaddankeri cross, vehicular traffic is high
4	Kaladagi	Vehicular movement is high. Agricultural fields on either side of the road
5	Lokapur	Vehicular movement is high. Agricultural fields on either side of the road
6	Chichakhandi	Vehicular movement is high. Agricultural fields on either side of the road
7	Mudhol	Vehicular movement is high. Agricultural fields on either side of the road

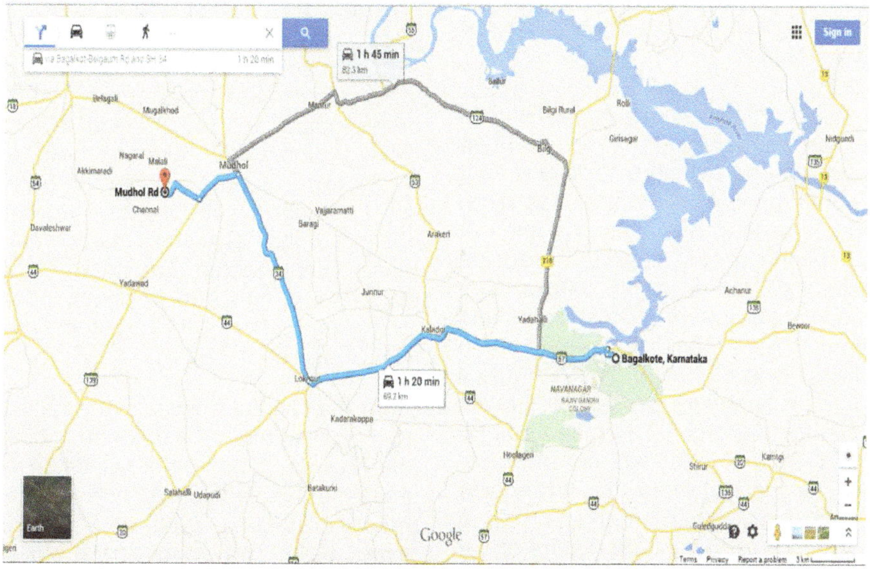

Fig. 1. Map showing the state highway of Bagalkot

Similarly, soil samples were dried, powdered and sieved through 0.2 mm sieve. One gram of sample was digested Gerhardt digestion unit according to Allen et al. [18] method (mixed acid digestion method. The results were in triplicates. The resulting extracts were diluted and filtered through Whattman No 41 paper and made upto 100 ml using double distilled water and analyzed for heavy metals viz Pb (Lead), Cd (Cadmium), Cu (Copper), Zn (Zinc), Mn (Manganese), Ni (Nickel) and Cr (Chromium) with GBC-932 plus Atomic Absorption Spectrophotometer (AAS) (Austrelia)with an air / acetylene flame and metal hollow cathode lamps. Respective wavelengths were used for the estimation of solutions for heavy metals were purchased from Siscochemical Laboratory Bombay (1000 mg/l). For Cd 228.8 nm, Pb 217.0 nm, Ni 232.0 nm, Cu 324.7 nm, Zn 213.9 nm, Mn 279.5 nm and Cr 221.8 nm filters have been used for the estimation using AAS. The working standards were prepared by serial dilution of standard stock solutions and were used for the calibration of the instrument [18].

3. RESULTS AND DISCUSSION

Environmental pollution with toxic metals has increased dramatically since the increasing of vehicular exhausts [19]. The pollution by heavy metals such as Lead, Cadmium, Nickel, Chromium etc is a problem of concern [20]. So it is now necessary to conduct this study and to determine the kinds of environmental pollution and how for these exhibit and act as an efficient bioindicator in reducing the degree of pollution in the environment. The samples were collected along the state highway between Bagalkot to Mudhol for the estimation of heavy metals (Fig. 1). Correlation coefficient of heavy metals in roadside soil, grass and caesalpinia plant samples are given in the Tables 3 and 4. The analysis showed high levels of heavy metals in the leaves of caesalpinia and grass collected from the polluted as compared to less polluted sites of the city (Control).

The ranges and arithmetic mean of heavy metal concentration of soil, grass and caesalpinia samples of state high-way and control sites are presented in the Table 2.

Lead is one of the major heavy metal and is considered as an environmental toxic pollutant [21]. Lead is considered as more toxic pollutant which brings about many physiological disorders. The main sources of lead are exhaust fumes of automobiles, chimneys of factory, roadside vehicles pollution [22]. The results shows that soils tend to accumulate more Pb then grass and caesalpinia leaves. The highest Pb level found in the soil (roadside) was 139.8 μg/gm, in grass it was 28.01 μg/gm, similarly in caesalpinia leaves it was found that 29.39 μg/gm. The mean soil Pb level was 95.71 μg/gm indicated considerable contamination of metal in the roadside environment; however, the control soil has a baseline level of 70.50 μg/gm. The results of our analysis represents, there should be significant differences for lead between sampling sites in control and polluted areas. The lead is rapidly washed onto the soil by rain water from the surface also by the death and decomposition of the plant [23,24]. Many literatures have notified the contents of heavy metals were higher in plant samples collected along the roadside of highways [8].

Copper is also one of the major heavy metal considered as an environmental pollutant. Copper is a micronutrient but toxic when it is in excess concentration. The source of copper being due to corrosion of metallic parts of cars derived from engine wear, brushing and bearing metals [25]. The mean Cu level in roadside soil (49.71 μg/gm) found to be much higher than the grass (4.87 μg/gm) and caesalpinia (4.70 μg/gm).

The main source of Zn on the roadside is because of automobile traffic i.e break lining looses of oil and cooling liquid [26]. Arithmetic mean of Zn of the roadside soil from Bagalkot to Mudhol state highway owes relatively high level of 188.31 μg/gm with a range of 32.29 – 381.54 μg/gm. the range of Zn 24.40 – 35.70 μg/gm in caesalpinia and grass 24.50 – 35.8 μg/gm. however, the Zn content in the test plant samples of polluted area are below the allowed limit (15 – 100 μg/gm) [18].

The main sources of environmental cadmium pollution are vehicle wheels, mineral oils increases Cd levels of environment and plants [27]. Cd dispersed in the natural environment through anthropogenic activities as well as natural rock mineralization process and thus, plants can easily absorbs Cd from the soil and also transport to the shoot levels. Cadmium induces complex changes in plants genetical, physiological and biochemical levels. The values were found little higher than allowed limit (1 μg/gm) in plants, i.e the Cd level in roadside soil averaged about 2.0 μg/gm and was observed that Cd was the lowest among the seven metals tested. The mean Cd in caesalpinia was 1.41 μg/gm and in grass, it is found to be 1.57 μg/gm. These findings were in confirmation with Shafia et al. [28].

From the results it was found that the soil, grass and caesalpinia contained high level of Manganese than other limits. Roadside soil, grass and caesalpinia had average 1528.34 μg/gm, 56.606 μg/gm and 56.10 μg/gm respectively. The attribution of Mn content is the lithogenic factor apart from vehicular pollution [29].

Chromium is considered as a serious concern with respect to health of biotics particularly in detrimental to plant growth and development. Cement, limestone industries are the major causes for the high influx of chromium into the atmosphere [29]. Significant differences in our analysis for chromium in soil and in plant samples are found. Chromium level was too high in roadside soil (315.54 μg/gm against control 110.41 μg/gm. In caesalpinia it was found to be 4.61 μg/gm, in grass it is found that 4.90 μg/gm against control (zero). Thus, the Cr content in soil and test plant samples of polluted area are above the maximum allowed limit (soil- 10-200 μg/gm, plant- 0.05-0.5 μg/gm).

Table 2. Metal accumulation profile in caesalpinia, grass and soil

S. no	Heavy metals	Control (µg g-1 dry wt.)	Roadside Caesalpinia (µg g-1 dry wt.)		Roadside grass (µg g-1 dry wt.)		Control (µg g-1 dry wt.)	Roadside soil (µg g-1 dry wt.)	
			Range	Mean ± SE	Range	Mean±SE		Range	Mean± SE
1	Lead	18.46	20.36-29.39	23.73±1.84	20.16-28.01	23.736±1.84	70.50	81.91-139.8	95.71±8.71
2	Copper	2.15	3.92-5.94	4.70±0.35	3.95-5.76	4.87±0.35	34.91	39.54-58.58	49.71±3.51
3	Zinc	16.19	24.40-35.7	32.84±3.10	24.50-35.8	32.87±3.11	29.84	32.29-381.54	188.3±54.28
4	Cadmium	0.82	1.01-1.78	1.41±0.08	1.15-1.52	1.571±0.082	2.16	1.51-2.08	2.0±0.19
5	Manganese	15.72	27.01-69.10	56.10±7.30	28.91-72.51	56.606±7.36	1254.1	1257.9-2051.5	1528.3±26.5
6	Chromium	N.D	1.20-7.8	4.61±2.20	1.28-8.00	4.90±2.23	110.41	131.9-951.2	315.54±2.46
7	Nickel	6.70	7.9-13.1	9.10±1.40	8.1-15.1	10.51±1.411	69.38	69.53-108.6	85.91±5.91

ND- Not detectable; Mean values ± standard error

Table 3. Correlation coefficient of heavy metals in roadside soil and grass

S. no.	Metal	r. value
1	Lead	0.628*
2	Copper	0.481*
3	Zinc	0.829*
4	Cadmium	0.748*
5	Manganese	0.529*
6	Nickel	0.331*
7	Chromium	0.215*

r= 0.60 and above have significant correlation

Table 4. Correlation coefficient of heavy metals in roadside soil and caesalpinia

S. no.	Metal	r. value
1	Lead	0.627*
2	Copper	0.480*
3	Zinc	0.824*
4	Cadmium	0.740*
5	Manganese	0.520*
6	Nickel	0.321*
7	Chromium	0.210*

r= 0.60 and above have significant correlation

Nickel is a micronutrient required at very low concentration by the plants normally in Ni^{+2} state, at high concentration Ni inhibits plant growth. The maximum Ni content was found 10.5 μg/gm in grass than caesalpinia with 9.10 μg/gm. Generally it is concluded that Ni content in test plant samples of polluted area are higher than the normal values (0.1 – 5.0 μg/gm) in plants [18].

Simple correlations ($r > 0.60$) between the metal Pb, Cu, Ni, Mn, Fe, Zn and Cd were calculated and are given in the Tables 3 and 4. Significant correlations between the metal levels in roadside soil, caesalpinia and grass were found in Zn, Cd and Ni ($r > 0.06$). It may be indicating the bioconcentration of these metals in soil, grass and caesalpinia, in addition to aerial deposition. This may attributed to the favorable root environment [30] i.e soil conditions might have favored their absorption. Correlations in case of chromium, nickel, manganese and copper contents between soil, grass and caesalpinia were low due to low bioavailability of these metals owing to unfavorable root environment. Whatever excess content of these metals found in grass and caesalpinia was presumed to be due to the aerial decomposition contributed by motor vehicles.

Thus, the order of increment of heavy metals in roadside soil is as follows: Mn > Cr > Zn > Pb > Ni > Cu > Cd, whereas in grass: Mn> Zn> Pb> Ni> Cr> Cu> Cd and in caesalpinia: Mn>Zn > Pb> Ni> Cu> Cr > Cd. The elevated levels of heavy metals in the roadside soil, grass and caesalpinia is an evidence of airborne pollutants of roadside environment along the state high way of Bagalkot to Mudhol. Soils due to their cation exchange capacity (CEC), complexing organic substances, oxides and carbonates, have high retention capacity for the heavy metals [31]. Thus, contamination levels increase continuously as long as the nearby sources remain active. During the last decade, the Bagalakot to Mudhol state high way has witnessed sharp increase in vehicle number due to urbanization. Similar observation in Caesalpinia in Madhurai city of southern region of Tamil Nadu [32].

In soil, the lesser mobility of metals and its accumulation on a long-term basis, leads to overall higher contamination level of metals; whereas, in roadside grass, caesalpinia it represents more accumulation due to turnover of plant materials (like new growths, the senescence followed by abscission of old parts) and meteorological influences [33]. Thus, the study of metal concentration of both roadside soil, grass and caesalpinia reflects the extent of aerial contamination of the roadside environment. The penetration of heavy metals into the food chains due to vehicular emissions may cause a long- range ecological and health hazard.

The results of our study indicate that the concentration of heavy metals such as Pb, Cd, Mn, Ni, Cu and Cr from the traffic area is an indicative of anthropogenic pollution.

4. CONCLUSION

It was concluded that with an increase in the amount of heavy metals in soil and their uptake by plants also increase. The mobility of heavy metals are highly translocated from soil to plant leaves in all the sampling sites. According to our study grass shows slight greater accumulation potentiality than caesalpinia. High metal concentrations in plants are contained in urban and highway roadsides due to the anthropogenic activities in addition to the traffic density. The heavy metal concentration was maximum in the study area along the roadside of Bagalkot to Mudhol. Caesalpinia is widely distributed at Bagalkot (Navanagar) is used as roadside ornamental plant and the grass is the good food for grazing animals. In accordance with the data presented here grass and caesalpinia possess all characteristics and are selected as bioindicators.

There is also a need for enforcement of regulations to control environmental pollution. Public participation, non-governmental organization and civic agencies of the government required collective approach towards the solution. Continuous air monitoring for one or more pollutants is an absolute necessity for completing a diagnosis of pollutant level in the air, water and soil environment. Most importantly, a network of monitoring stations throughout the country would be helpful to measure the current pollutant level. Accumulation of air monitoring data will provide the criteria needed for establishing air quality standards.

ACKNOWLEDGEMENT S

The authors are thankful to the Principal, BLDEA's Degree College, Jamkhandi (Karnataka), Principal, JSS College Dharwad and to the Chairman, PG Department Botany, Karnatak University, Dharwad (Karnataka) for providing necessary facilities to carry out research work.

COMPETING INTERESTS

Authors have declared that no competing interests exist.

REFERENCES

1. Kelly C. Cadmium theurapeutic agents (Review): The heavy metals concentration along roadside trees of quetta and its effects on public health. Current Pharmaceutical Design. 1999;5:229-240.
2. Dursun A, Aslantas R, Pirlak L. The effect of air pollution on horticultural crops. Journal of Ecology. 1998;27:11-14.
3. Donalas B, Edita B. Transport of heavy metals from soil to *Pinus sylvestris* L. and *Betula pendula* trees. Journal of Ecology. 2007;53(1):29-36.
4. Bakirdere S, Yaman M. Determination of lead, cadmium and copper in roadside soil and plants in Elazig, Turkey. Environmental Monitoring Assessment. 2008;136:401-410.
5. Mortula MM, Rehman MS. Study on waste disposal at DEPZ. Bangladesh Environmental Protection Agency, (BAPA). 2002;2:807-817.
6. Arora M, Kiran B, Rani A, Rani S, Kaur B, Mittal M. Heavy metal accumulation in vegetables irrigated with water from different sources. Food Chemistry. 2008; 111:811-815.
7. Bhuiyan MAH, Suruvi NI, Dampare SB, Islam MA, Qurashi SB, Suzuki S. Investigation of the possible sources of heavy metal contamination in lagoon and canal water in the tannery industrial area in Dhaka, Bangladesh. Environmental Monitoring Assessment. 2011;175:633-649.
8. Viard W, Pihan F, Promeyrat S, Pihan JC. Integrated assessment of heavy metal highway pollution: Bioaccumulation in soil, Graminaceae and land snails. Chemosphere. 2004;55:1349-1359.
9. Othman IM, AlOudat, AlMasri MS. Lead levels in roadside soils and vegetation of Damascus city. Sci. Total Environment. 1997;207:43-48.
10. Hjortenkrans D, Bergback B, Haggerud A. New metal emission patterns in road traffic environments. Environmental Monitoring Assessment. 2006;117:85-98.
11. Grigalaviciene I, Rutkoviene V, Marozas V. The accumulation of heavy metals Pb, Cu and Cd at roadside forest soil. Polish Journal of Environmental Studies. 2005; 14:109-115.
12. Onder S, Dursun S. Air borne heavy metal pollution of *Cedrus libani* (A. Rich.) in city center of Koyna (Turkey). Atmosphere. Environ. 2006;40(6):1122-1133.
13. Aksoy A, Ahin U, Duman F. Hohinia *Pseudo-acacia* L. as a possible biomonitor of heavy metal pollution in Kayseri. Tropical Journal of Botany. 2000a;24:179-284.
14. Fatoki OS. Lead, cadmium and zinc accumulation a long some selected major roads if Eastern cape. International Journal of Environmental Studies. 2003;60:199-204.
15. Aksoy A, Celok A, Ozturk M. Plants as possible indicators of heavy metal pollution in Turkey. Chemia Inzyneria Ekologiczna. 2000b;11:1152-1161.
16. Rashed MN. Monitoring of environmental heavy metals in fish from Nasser Lake. Environmental International. 2001;27:27-33.
17. Aktaruzzaman M, Fakhruddin ANM, Choudhary MAZ, Fardous Z, Alam MK. Accumulation of heavy metals in soil and their transfer to leafy vegetables in the

region of Dhaka Aricha Highways, Savar, Bangladesh. Pakistan Journal of Biological Science. 2013;16(7):332-338.

18. Allen SE, Grimshow HM, Parkinson JA, Quamby C. Chemical analysis of ecological materials. Blackwell Scientific Publications. Osney Mead, Oxford, UK; 1974.
19. Zaidi MI, Astrar A, Mansoor A, Farooqui MA. The heavy metals concentration along roadside trees of Quetta and its effects on public health. J. Appl. Sci. 2005;59(4):708-711.
20. Onder S, Dursun S, Gezgin S, Demirbas A. Determination of heavy metal pollution in grass and soil of city centre green areas (Koyna, Turkey). Pol. J. Environ. Std. 2007;16:145-154.
21. Sharma P, Dubey RS. Lead toxicity in plants. Brazilian Journal of Plant Physiology. 2005;17(1):35-52.
22. Eick MJ, Peak JD, Brady PV, Pesek JD. Kinetics of lead adsorption and desorption of Gooethite: residence time effect. Journal of Soil Science. 1999;164:28-39.
23. Harrison RM, Laxen DPH, Wilson SJ. Chemical association of lead, cadmium, copper and zinc in street dust and roadside soils. Environ. Sci. Technol. 1981;15: 1378-83.
24. Khattak MI, Jana Akhatar, Rehan K. Study of Pb concentration in roadside plants (*Dalbergia sissoo* and *Cannabis sativa*) in region of quetta. Sci. Int. (Lahore). 2013;25(2):347-352.
25. Al-Khashman OA. The investigation of metal concentration in stress dust samples in Aqaba city, Jordan. Environ. Geochemistry & Health. 2007;29:197-207.
26. Saeedi M, Hosseinzadeh M, Jamshidi A, Pajooheshfar SP. Assessment of heavy metals contamination and leaching characteristics in highway side soils, Iran. Environmental Monitoring Assessment. 2009;151(1/4):231-241.
27. Gune A, Alpaslan M, Inal A. Plant growth and fertilizer (J). Ankara University, Agriculture Pub. No. 1539, Ankara, Turkey; 2004.
28. Shaffiq M, Iqbal MZ, Arayne S, Athar M. *Alstonia scolaris* and Cassia siamea as possible biomonitors of Lead and Cadmium in the polluted environment of Karachi city, Pakistan. Journal of Applied Botany and Food Quality. 2011;84:95-101.
29. Barnhart J. Occurrences, uses and properties of chromium. J. Reg. Toxicol. Pharmacol. 1997;26(1Pt2):S37.
30. Sahu KC, Warrier R. Lead, cadmium and copper contamination of soil and vegetation due to vehicular emission along Powai road in North Bombay, India. Indian J Earth Science. 1985;12:50-57.
31. Yassoglou N, Kosomas C, Asimakopoulos J, Kallianou C. Heavy metal contamination of roadside soils in the Greater Athens area. Environmental Pollution. 1987;47: 293-304.
32. Thambavani SD, Vathana VM. Caesalpinia as indicator for heavy metals pollution. Elixir Polllution. 2013;57:14335-1434.
33. Kabata-Pendias A, Pendias H. Trace elements in soils and Plants, third ed. CRC Press, Boca Raton; 2001.

22

Drainage Basin Morphometric Parameters of River Lamurde: Implication for Hydrologic and Geomorphic Processes

E. D. Oruonye[1*], B. B. Ezekiel[1], H. G. Atiku[1], E. Baba[1] and N. I. Musa[1]

[1]*Department of Geography, Taraba State University, P.M.B. 1167, Jalingo, Taraba State, Nigeria.*

Authors' contributions

This work was done in collaboration among all authors. Author EDO designed the study, performed the statistical analysis, wrote the protocol and first draft of manuscript, did the analysis and report writing. Authors BBE, HGA, EB and NIM contributed in fieldwork observation, measurement, recording and data collection. All authors read and approved the final manuscript.

Editor(s):
(1) Chandra Shekhar Kapoor, Department of Environmental Sciences, University College of Science, Mohan Lal Sukhadia University, India.
Reviewers:
(1) Joel Efiong, University of Calabar, Nigeria.
(2) Riad Hosein, University of the West Indies, Trinidad & Tobago.

ABSTRACT

This study has examined the drainage basin morphometric parameters of River Lamurde and its implication on hydrologic and geomorphic processes. Various morphometric parameters of the drainage basin were calculated based on literatures. Lamurde basin is a fourth stream order with dendritic pattern of drainage. The results of the study indicated that the drainage basin area is 553.9 km^2, basin perimeter 197.5 km, form factor 0.19, circulatory ratio 0.5 and drainage texture is 03. The stream frequency of the study area is 0.11 stream segments per square kilometer, drainage density (Dd) is 0.389 km/km^2, bifurcation ratio 3.85 and length of overland flow is 0.13 km. The drainage density, bifurcation ratio and circulatory ratio values indicate that the basin has a gentle slope, more elongated and highly permeable bedrock. The relief ratio of the basin is characterized by less resistant rocks. Both values of stream frequency and drainage density show that Lamurde basin has a low relief and by implication has a low response to surface runoff. This basin shape parameters indicates that the basin has a flatter peak of flow for longer duration. The

*Corresponding author: Email: eoruonye@gmail.com

morphometric parameters studied using GIS has helped us to understand various terrain characteristics such as the nature of the bedrock, infiltration capacity, runoff etc.

Keywords: Drainage basin; geomorphic; hydrologic; lamurde; morphometric and parameters.

1. INTRODUCTION

Drainage basin morphometry can be regarded as the topographical expression of land by way of area, slope, shape, length among others [1]. Morphometric analysis defines more clearly and precisely the general form of the basin landform as represented on a map and serve as a basis for demonstrating the effect of environmental control on fluvial system and for predicting the basin output variables such as discharge [2]. These morphometric parameters affect catchment stream flow pattern through their influence on concentration time [3].

Studies on drainage basin morphometry have been carried out in many parts of the world [4]. In Nigeria, similar studies have been carried out which include those of [5,6,7,8]. The drainage basin morphometric parameters include stream order, stream length, stream number, basin area, basin shape factor (e.g. circularity ratio, elongation ratio, form factor and compaction ratio), basin perimeter, bifurcation ratios, drainage density, stream frequency and drainage intensity.

Studies have shown that data on stream discharge and morphometric parameters of drainage basins in developing countries are either lacking or grossly inadequate where they exist [9,10]. In Nigeria, a major constrain to hydrological studies and water resources planning and management is lack of reliable hydrological data for use by water resource managers and researchers [2]. Many stream discharge gauging stations which were established in the early 1970s are no longer functional and where they function, are poorly maintained [2]. [10] asserted that this development has made hydrological research almost impossible in the country. [2] observed that the situation has made many researchers in the country to abandon hydrological studies in favor of other aspects of the study. The alternative to this is to resort to measures of appraising and evaluating the natural water resources potential of basins without stream gauge records, using series of generalized regional relationships based on morphometric parameters [2].

Morphological studies of rivers are very important to study the behaviour of a river, its aggradations/degradation, shifting of the river course, erosion of river bank etc. and to plan remedial measure for erosion, flooding and other related problems. Quantitative analysis of drainage basin morphometric parameters is of immense utility in river basin evaluation, watershed prioritization for soil and water conservation and natural resources management at micro level.

This type of study has not been carried out on River Lamurde in Jalingo town of Taraba State Nigeria. No segment of the basin is gauged and information on the drainage system is very scarce as not much research work has been carried out in this direction. Yet the basin is a major source of water in Jalingo metropolis and is characterized by hydrologic and geomorphic problems such as flooding, erosion and mass movement. Greater part of the Lamurde River basin flood annually and this flooding has increased in frequency in recent times with devastating effect on the socio-economic life of the people. The river basin is used for a variety of agricultural activities (rainfed and irrigation farming and livestock grazing) and major source of urban water supply in the town. This study is necessitated by the need to provide information on the drainage basin morphometric parameters of River Lamurde. This information would be used to describe the basin as a landform and a hydrological unit and the implication of the drainage basin morphometry on the hydrologic and geomorphic processes in the area. This will enhance the management techniques which will not only aid in the improvement of river regime by retaining greater volume of rainfall intercepted by the basin on the land, but will also help in solving problem of soil erosion, sediment generation, flooding and urban water supply in the area.

1.1 Description of Study Area

River Lamurde is the main river draining Jalingo town and environ. It forms a confluence with its major tributary, River Mayogwoi in Magami ward in the town. It took its sources from the Yorro Mountain near Gangoro and flows downhill through Yorro, Tazarang, Alkali Gwa, Bassa and

Jalingo. The basin is located between latitudes 8°40' to 9°05'N and longitudes 11°05' to 11°45'E (Fig. 1). The river flows for over 96km westward before emptying into the Benue River system near Tau community [11]. The basin is located in a sub humid climate zone with average rainfall of 850 mm^3. Rainfall starts in May and ends in October. The highest amount of rainfall in the basin is received in the months of July, August and September. Temperature is relatively high throughout the year but lower in the middle of the rainy season and winter months of December and January. Mean temperature is 30°C. The vegetation type is the wooded savanna characterized by short grasses interspersed with short trees. Geologically, the drainage basin consists of cretaceous rocks of the Bima sandstone formation with isolated flat-topped hills [11]. Scattered remnants of highly metamorphosed sedimentary rocks and diverse predominantly granitic and plutonic masses generally known as older granites are found all over the basin. Ferruginous tropical soils derived from crystalline acid rocks of the basement complex are dominant soil types in the basin. River Lamurde is the major source of recharge for underground water in the area with a typical minimum yield of 648,240 m^3 per year from the

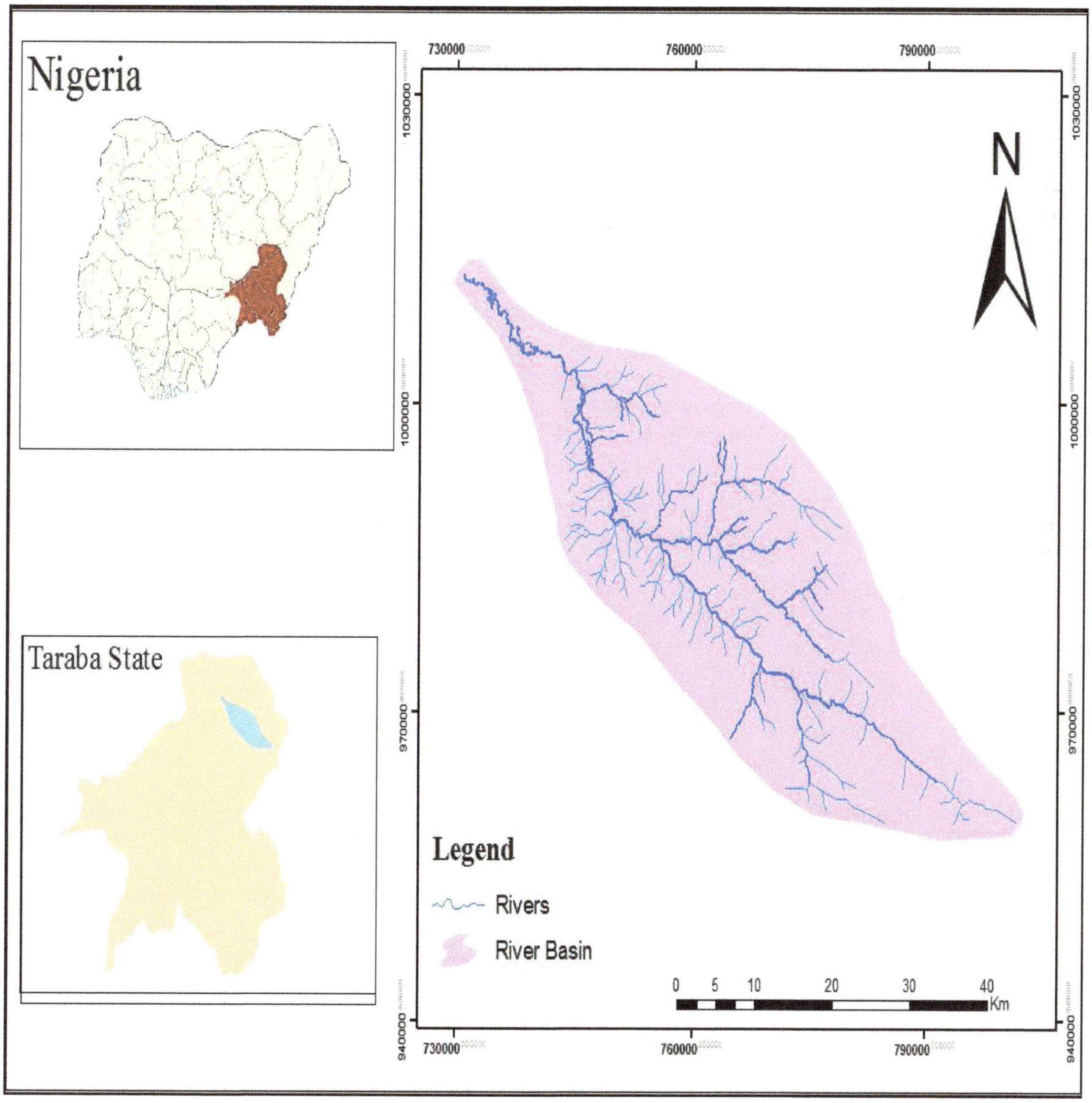

Fig. 1. Location map of the study area

boreholes located within the flood plain of the river [12]. The changes in land uses along the floodplains of the river has been a matter of great concern because of the increase in the devastating effects of flooding in recent times in the area [13,14].

2. MATERIALS AND METHODS

The data used in this study comprises of ETM+ (2015) and Shuttle Radar Topographic Mission (SRTM) digital elevation model of the basin obtained from the United State Geological Survey (USGS) online open resources. The map of the study area was obtained from the Taraba State Bureau of Land and Survey. Satellite images of Landsat 8 were used as base map for the GIS technique. The images were acquired on 15th March 2015 and the projection type is UTM and spheroid is WGS 1984. The datum is the Minna datum, UTM zone 32N and the pixel size is 15 meter. Morphometric analysis and prioritization of basin was based on the integrated use of remote sensing and GIS technique. The drainage map of the study area was prepared using geo-coded data and verified with survey of Jalingo toposheets no. 193, 194, 195, 214, 215, 216, 235, 236 and 237 of the area at 1:50000. The toposheets and satellite data were geometrically rectified and georeferenced to World space coordinate system using ERDAS IMAGINE 9.2 software package.

Morphometric analysis comprises a series of sequential steps. The drainage layer was converted to digital format through on-screen digitization from available topographic maps using ArcGIS 9.1, in the scale of 1:50000 and the attributes were assigned to create the digital database. All measurements were directly computed from the vector data that was extracted from the topographic maps. The entire drainage segments were digitized as lines separately for each order [15]. Digitization work was carried out using Arc GIS 9.2 software. The map showing the drainage pattern was prepared after detailed ground checks (fieldwork) with GPS survey on channel network. Image interpretation comprises data derivation from map analysis, map work, computation from topographical maps and satellite imageries pertaining to different components of the drainage basin morphometric parameters in the study area. The Area (A) and Perimeter (P) of the drainage basin was calculated using measurement tool option of ArcGIS version 9.2. Various morphometric parameters such as linear aspects of the drainage network: stream order (Nu), bifurcation ratio (Rb), stream length (Lu), and areal aspects of the drainage basin: drainage density (D), stream frequency (Fs), texture ratio (T), elongation ratio (Re), circulatory ratio (Rc), form factor (Rf) of the basin were computed. This morphometric parameters of the basin were calculated based on literature [16,17,15,18,19,20,21].

3. RESULTS OF THE FINDINGS

Fig. 2 shows that the basin exhibits a dendritic type of drainage network which indicate the homogeneity in texture and lack of structural control in the study area.

3.1 Stream Order (Nu)

The designation of stream order is usually seen as the first step in drainage basin analysis. Stream order has been defined as a measure of the position of a stream in the hierarchy of tributaries [22]. Hierarchical ordering of streams is necessary to assess hydrodynamic character of a drainage basin. In the present study, the various stream segments of the drainage basin have been ranked according to Strahler's stream ordering system. According to [19], the smallest fingertip tributaries are designated as first stream order. Where two first stream order meets, a second stream order is formed; where two second stream order meets, third stream order is formed and so on. The stream order of River Lamurde is presented in Table 1.

The total number of stream segments is found to decrease as the stream order increases in the basin. The study reveals that the development of 1st order streams is maximum in the basement complex area and minimum in the alluvial plains (Table 1). The number of 1st stream order in a basin of a given size is dependent upon a number of factors which include climatic, geologic and hydrologic. The trunk stream through which all discharge of water and sediment passes is therefore the stream segment of the highest order. River Lamurde, which is the trunk in the study area is of the fourth order (Fig. 2).

3.2 Stream Length (Lu)

The total length of individual stream segments of each order is the stream length of that order. Stream length measures the average (or mean) length of a stream in each orders, and is calculated by dividing the total length of all streams in a particular order by the number of streams in that order. The stream length of the various stream orders in River Lamurde is presented in Table 1. Stream length is one of the most important hydrological features of the basin as it reveals the surface runoff characteristics. Streams of relatively smaller lengths are characteristics of areas with larger slope and finer textures. Streams with longer lengths are generally the characteristics of flatter surface with low gradients. Usually, the total length of stream segments is highest in first stream orders and decreases as the stream order increases. The number of streams of various orders in the basin was counted and their lengths measured from the mouth to the divide of the drainage basin. Stream length is a revelation of the chronological developments of the stream segments including interlude of tectonic disturbances.

3.3 Mean Stream Length

The mean stream length is calculated by dividing the total stream length of given order by number of stream of that order as shown in Table 1. The value for any given order is greater than that of the lower order and less than that of its next higher order in the whole drainage basin. The mean stream length of the study area is 3.65 km.

Table 1. Stream orders, stream number, stream length in the Lamurde river basin

S_no	Stream order	Stream numbers	Total stream length (km)	Mean stream length (S.L/SN)	Bifurcation ratio
1	1st order	132	382.7	2.9	5.08
2	2nd order	26	105.3	4.1	4.33
3	3rd order	6	53.5	8.9	6
4	4th order	1	60.1		

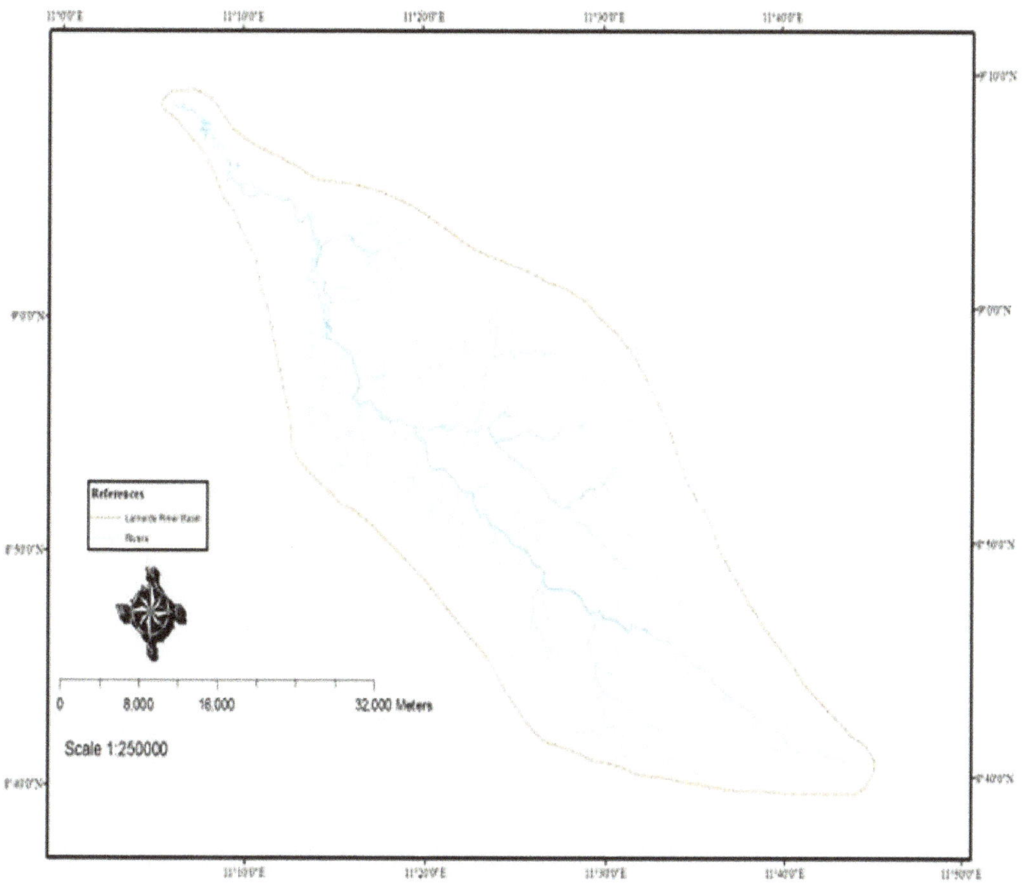

Fig. 2. Lamurde river basin

3.4 Bifurcation Ratio (Rb)

Bifurcation ratio is related to the branching pattern of a drainage network and is defined as the ratio between the total numbers of stream segments of one order to that of the next higher order in a drainage basin [20]. The sum of all the bifurcation ratios in the basin was divided by the number of bifurcation ratios to give the mean bifurcation ratio of the basin. The calculated mean bifurcation ratio for the study area is 3.85, an indication that the study area is a lowland area. This low Rb value indicates less structural disturbance in Lamurde basin. This suggests that the study area has low potentials for discharge compare to those of highland areas with bifurcation ratio of 5.0 [15]. Bifurcation ratios are controlled by basin physiographic factors especially basin relief and drainage density [23]. However, in regions where the network geometry of streams develops without pronounced lithological or structural control, bifurcation ratios between basins of different orders are stable showing little variation from one order basin to another [24,25].

Chorley [26] had noted that the lower the bifurcation ratio, the higher the risk of flooding, particularly of parts and not the entire basin. The low average bifurcation ratio of the Lamurde basin of 3.85 is an indication that parts of its segment are liable to flooding. According to [27], bifurcation ratios ranging from 3 – 5 indicate natural drainage system characterized by homogenous rock. The drainage basin of Lamurde may have been affected by human activities.

3.5 Length of Overland Flow

Length of overland flow is defined as the length of flow path, projected to the horizontal, non channel flow from point on the drainage divide to a point on the adjacent stream channel (Horton, 1945). Horton, for the sake of convenience, had taken it to be roughly equal to half the reciprocal of the drainage density. Overland flow is significantly affected by infiltration (exfiltration) and percolation through the soil, both varying in time and space [28]. In this study, the length of overland flow of the Lamurde drainage basin is 0.13 kilometers, which shows low surface runoff in the study area.

3.6 Areal Aspects

The areal aspects of drainage basin include different morphometric parameters, like drainage density (D), stream frequency (Fs), form factor (R_f), circulatory ratio (Rc), elongation ratio (Re) and length of overland flow (L_g). The values of these parameters were calculated and the results are presented in Table 2.

Table 2. Areal aspects of Lamurde river basin

S_no	Areal parameter	Computed value
1	Area (Sqkm)	1,553.9
2	Perimeter (Km)	197.5
3	Elongation ratio	0.49
4	Drainage density	0.389
5	Stream frequency	0.11
6	Length of overland flow	0.13
7	Form factor	0.19
8	Circulatory ratio	0.5
9	Drainage texture	0.3

3.7 Drainage Area (Au)

The entire area drained by a stream or system of streams such that all streams flow originating in the area is discharged through a single outlet is termed as the drainage area. In other words, the drainage area is a collecting area from which water would go to a river. The boundary of the area is determined by the ridge separating water flowing in opposite directions. The total area of the drainage basin is estimated to be 1,553.9 km^2. Basin area has been identified as the most important of all the morphometric parameters controlling catchment runoff pattern. This is because, the larger the basin, the greater the volume of rainfall it intercepts, and the higher the peak discharge that result [29,7,30,31,32,33].

3.8 Drainage Density (Dd)

Drainage density can be regarded as the total length of the streams of all orders per drainage area. Dd is expressed as the ratio of the total sum of all channel segments within a basin to the basin area i.e., the length of streams per unit of drainage area. It is a dimension inverse of length [16]. Drainage densities can range from less than 5 km/km^2 when slopes are gentle, rainfall low and bedrock permeable (e.g. sandstones), to much larger values of more than 500 km/km^2 in mountainous areas where rocks are impermeable, slopes are steep and rainfall totals are high [34]. The drainage density (Dd) of the study area is 0.389 km/km^2. Thus, in this study, the drainage density falls less than 5 km/km^2 which indicates that the area has a gentle slope, low rainfall and permeable bedrock. Low drainage densities are often associated with widely spaced streams due to the presence of less resistant surface materials (lithologies or rock types), or those with high infiltration capacities. Drainage basin with high Dd values indicates that a large proportion of the precipitation runs off. On the other hand, a low drainage density indicates that most rainfall infiltrates into the ground and few channels are required to carry the runoff. In general, low drainage density leads to coarse texture while high drainage density leads to fine texture [19].

3.9 Stream Frequency (Fs)

The stream frequency (Fs) or channel frequency or drainage frequency of a basin may be defined as the total number of stream segments within the basin per unit area [17]. It mainly depends on the lithology of the basin and reflects the texture of the drainage network. The basins of the structural hills have higher stream frequency, drainage density while the basins of alluvial has minimum. The stream frequency of the study area is 0.11 stream segments per square kilometer. The lower value indicates that the basin possess low relief and less number of streams. The existence of less number of streams in a basin indicates matured topography, while the presence of large number of streams indicates that the stream is youthful and still undergoing erosion. Both values of stream frequency and drainage density show that Lamurde basin has a low relief and by implication has a low response to surface runoff.

3.10 Drainage Texture (Rt)

Horton [17] defined drainage texture as the total number of stream segments of all order in a basin per perimeter of the basin. The drainage texture of the Lamurde basin is 0.84. This shows that the basin has a very coarse texture. Smith have classified five different drainage textures related to various drainage densities as very coarse (below 2), coarse (2 - 4), moderate (4 - 6), fine (6 - 8) and very fine (8 and above). The higher the drainage texture, the more the dissection and erosion in the basin. Drainage texture depends on a number of natural factors such as climate, rainfall, vegetation, rock and soil types, infiltration capacity, relief and stage of development. Weak rocks devoid of vegetative cover produce fine texture, while rocks which are hard and with vegetative cover produce coarse texture.

3.11 Basin Shape (Rf)

This is the measure of the elongation of a basin. As elongation increases for a given area, Rf decreases. Three parameters viz. Elongation Ratio (Re), Circulatory Ratio (Rc) and Form Factor (Rf) are used for characterizing drainage basin shape, which is an important parameter from hydrological point of view. The shape of the basin mainly governs the rate at which the water is supplied to the main channel. The basin has an elongation ratio (0.49), circularity ratio (0.5) and form factor (0.19) which indicates that the shape of the basin tends towards elongation. The elongation shape indicates that the basin will have a flatter peak of flow for longer duration. Basin shape reflects the way that runoff will bunch up at the outlet. An elongated basin having the outlet at one end of the major axis and having the same area as the circular basin would cause runoff to be spread out over time, thus producing a smaller flood peak than that of the circular basin.

3.12 Basin Perimeter (P)

Basin perimeter is the outer boundary of the drainage basin that encloses its area. It is measured along the divides between basins and may be used as an indicator of basin size and shape [20]. The basin perimeter of Lamurde basin is 197.5 kilometers.

3.13 Elongation Ratio (Re)

Elongation ratio (Re) is defined as the ratio of diameter of a circle of the same area as the basin to the maximum basin length [20]. Elongation ratio determines the shape of the basin and can be classified based on these values as circular (0.9 - 1), oval (0.8 – 0.9), less elongated (0.7 – 0.8), elongated (0.5 – 0.7), more elongated (< 0.5). Regions with elongation ratios are susceptible to more erosion whereas regions with high values correspond to high infiltration capacity and low runoff. The elongation ratio of the drainage basin is 0.49 which indicates more elongation and more prone to erosion with less infiltration capacity. Circular drainage basins are more efficient in the discharge of runoff. They are at greater risk from flood hazard because they have a very short lag time and high peak flows than the elongated basins. Elongated drainage basins have low side flow for shorter duration and high main flow for longer duration and are less susceptible to flood hazard.

3.14 Circularity Ratio (Rc)

The circularity ratio is a similar measure as elongation ratio, originally defined by Miller [21] as the ratio of the area of the basin to the area of the circle having same circumference as the basin perimeter. The calculated Rc value for the study area is 0.5 which indicate that the drainage basin is between elongated and circular and is characterized by medium to low relief. The value of circularity ratio varies from 0 (in line) to 1 (in a circle). The circularity ratio (Rc) has been used as a quantitative measure for visualizing the shape of the basin and is expressed as the ratio of basin area (A) to the area of a circle (Ac) having the same perimeter as the basin [21,19]. It is affected by the lithological character of the basin. The ratio is more influenced by length, frequency (Fs), and gradient of streams of various orders rather than slope conditions and drainage pattern of the basin. It is a significant ratio, which indicates the dendritic stage of a basin. Its low, medium and high values are indicative of the youth, mature and old stages of the life cycle of the tributary basins.

3.15 Form Factor (Rf)

Form factor is the numerical index commonly used to represent different basin shapes [16]. The value of form factor ranges between 0.1 to 0.8. The form factor of the study area is 0.19. This shows that the basin is very elongated and thus has low peak flow of longer duration. Consequently, the flood flow of this type of basin is easier to manage than the circular basin [35]. The Smaller the value of form factor, the more elongated will be the basin. The basins with high form factors (0.8), have high peak flows of shorter duration, whereas, elongated drainage basin with low form factors have lower peak flow of longer duration.

3.16 Relief Aspects of the Study Area

Basin relief is an important factor in understanding the denudational characteristics of the basin. Relief is the difference between the maximum and minimum elevations in the basin. The maximum height of the Lamurde basin is 685m and the lowest is 118 m. The relief of the basin is 567 m (Fig. 3). The basin length, relative relief and relief ratio of the study area is presented in Table 3.

3.17 Relief Ratio (Rh)

This is a dimensionless measure of the overall gradient across a basin. It is calculated by dividing the relief (H) of a basin by its length (L). The maximum basin relief is obtained from the elevation difference between highest point on the basin perimeter and its confluence with the trunk stream. The maximum basin relief of the Lamurde drainage area is 567 m (685 m – 118 m). The Rh normally increases with decreasing drainage area and size of sub-watersheds of a given drainage basin [36,37]. The relief ratio of the Lamurde basin is 0.032. The relief ratio of the basin is low, which is characteristics features of less resistant rocks. The lower values may indicate the presence of basement complex rocks that are exposed in the form of small ridges and mounds with lower degree of slope [38]. Low relief ratios also indicate that the recharge capabilities of the basin are low and chances of ground water potential are good [39]. The study area lies within hilly terrain (Yorro hills) in the initial stages and further down stream in the gently or flat sloping terrain (Fig. 3).

Table 3. Relief aspects of river Lamurde

S_no	Morphometric parameters	Result
1	Basin length	89.6 km
2	Relative relief	2.78
3	Relief ratio	0.032

3.18 Relative Relief

Relative relief is defined as the differences in height between the highest and the lowest points (height) in a unit area. It is an important morphometric parameter used in the overall assessment of morphological characteristics of terrain [40]. [41] suggested to calculate relative relief by dividing the difference of height between the highest and lowest points in the basin (H) with basin perimeter (P), thus relative relief = H/P. The relative relief of the study area is 2.87. Relative relief is calculated on the basis of highest and lowest elevations and the data of relative relief so derived are tabulated and classified into three categories viz. (i) low relative relief = 0 m – 100 m, (ii) moderately relative relief 100 m – 300 m and (iii) high relative relief = above 300 m.

3.19 Basin Length (Lb)

Basin length is the longest dimension of a basin to its principal drainage channel. The longer the length of a basins, the lower the chances that such a basin will be flooded when compared with a more compact basin. Basin length (Lb) has

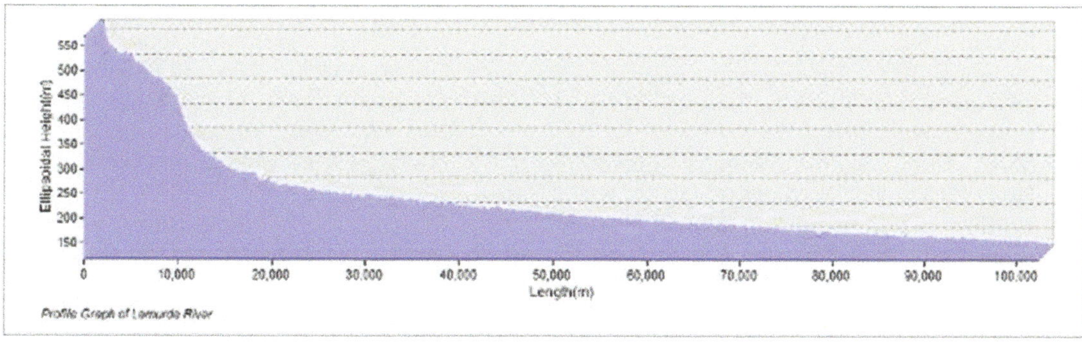

Fig. 3. Profile of river Lamurde

been given different meanings by different workers [20,42,43,44]. The Lb is the longest length of the basin, from the catchment to the point of confluence [42]. The Lamurde stream meets River Benue at the point of confluence in the north-western part of the study area. The length of the Lamurde river basin is 89.6 kilometers.

4. CONCLUSION

This study has examined the morphometric characteristics of River Lamurde basin using remote sensing image and GIS technique. These morphometric parameters studied using GIS has helped us to understand various terrain characteristics such as the nature of the bedrock, infiltration capacity, runoff etc of the study area. The results of the findings has revealed that the basin has a low relief, coarse texture, gentle slope and permeable bedrock. The elongated nature of the basin implies a high main flow for longer duration and less susceptibility to flood hazard. The result of the findings suggest that the study area will be more prone to erosion with less infiltration capacity. The findings revealed that the recharge capabilities of the basin are low and chances of ground water potentials are good. The implication of the findings of the study is that there will hardly be enough water supplies to support large scale water development project, particularly hydropower development projects except with intensive development of land and water resources of the basin.

COMPETING INTERESTS

Authors have declared that no competing interests exist.

REFERENCES

1. Ajibade LT, Ifabiyi IP, Iroye KA, Ogunteru S. Morphometric analysis of Ogunpa and Ogbere Drainage Basins, Ibadan, Nigeria. Ethiopian Journal of Environmental Studies and Management. 2010; 3(1):13 –19.
2. Ezemonye MN, Emeribe CN. Appraisal of the hydrological potential of ungauged basin using morphometric parameters. Ethiopian Journal of Environmental Studies and Management. 2013;6(4):376–380.
3. Jones JAA. Global hydrology: Processes, resources and environmental management. London. 1999;399.
4. Eze EB, Efiong J. Morphometric parameters of the Calabar river basin: Implication for hydrologic processes. Journal of Geography and Geology. 2010; 2(1):18–26.
5. Okechukwu GC. Fluvial – Geomorphic interrelationships in some river catchments in the Nigerian Precambrian basement complex. Unpublished Ph.D Thesis, Department of Geography, University of Ibadan, Nigeria; 1974.
6. Ebisemiju FS. The structure of the inter–relationship of Drainage basin characteristics. Unpublished Ph.D. Thesis, University of Ibadan, Ibadan. 1976;321.
7. Faniran A, Ojo O. Man's physical environment. Heinemann Educational Books, London. 1980;404.
8. Anyadike RNG, Phil-Eze RC. (). Runoff response to basin parameters in South – Western Nigeria. Geographical Annals. 1989;71A(1&2):75–84.
9. Abiodun AA. Water resources projects in Nigeria and the hydrological data employed in their planning and development'. Proceedings Symposium on Design of Water Resources Projects with Inadequate Data. Madrid. 1973;21-32.
10. Ayoade JO. Hydrology and water resources. In Areola O, Okafor SI. (eds.), Fifty years of geography in Nigeria: The

Ibadan story. Ibadan University Press, Ibadan. 1998;331–347.
11. Oruonye ED. Geographical aspects of Jalingo region. Phimela Concept Printing Press, Jalingo; 2014.
12. Musa Buba Siam. Technical feasibility study for the improvement of Jalingo water supply scheme in Taraba State of Nigeria. A Thesis submitted to the School of Graduate Studies, Kwame Nkrumah University of Science and Technology, Kumasi in partial fulfillment of the requirements of Masters of Science; 2002.
13. Oruonye ED. Socio-economic impact assessment of flash flood in Jalingo metropolis, Taraba State, Nigeria. International Journal of Environmental Sciences. 2012a;1(3):135–140.
14. Oruonye ED. The challenges of urban FLOOD disaster management in Nigeria: A case study of Jalingo metropolis, Taraba State Nigeria. Recent Scientific Journal. 2012b;37-42.
15. Strahler AN. Dynamics basis of geomorphology. Geophysical Society of American Bulletin. 1952a;63(9):923-938.
16. Horton RE. Drainage basin characteristics. Transaction of American Geophysics Union. 1932;13:350–361.
17. Horton RE. Erosional development of streams and their drainage basins: Hydrophysical approach to quantitative morphology. Bulletin of Geological Society of America. 1945;5:275-370.
18. Strahler AN. Hypsometric (area-altitude) analysis of erosional topography. Bull Geol, Soc. Am. 1952b;63.
19. Strahler AN. Quantitative geomorphology of drainage basins and channel networks. In: Chow VT (ed) Handbook of Applied Hydrology. McGraw-Hill, New York. 1964; 439–476.
20. Schumm SA. Evolution of drainage systems and slopes in Badlands at Perth Amboy, New Jersey. Bull. Geol. Soc. Amer. 1956;67:597-646.
21. Miller N. A quantitative geomorphic study of drainage basins characteristics in clinish mountain area, Verginia and Tenssesse. Columbia University. 1953;30. (Technical Report No. 3)
22. Leopold LB, Wolman MG, Miller JP. Fluvial processes in geomorphology. Freeman and Co., San Francisco; 1964.
23. Milton LE. The geomorphic irrelevance of some drainage network laws. Australian Geographical Studies. 1966;4:89-95.
24. Strahler AN. Quantitative analysis of watershed geomorphology. Trans Am Geophys Union. 1957;38:913–920.
25. Chorley RJ, Donald EG, Malm Pogorzelski HA. A new standard for estimating drainage basin shape. American Journal of Science. 1957;255:138-141.
26. Chorley RJ. Introduction to physical hydrology. Methuen and Co. Ltd., Suffolk. 1969;211.
27. Kale VS, Gupta A. Introduction to geomorphology. Orient Blackswan Private Limited. 2001;84-86.
28. Schmid BH. Critical rainfall duration for overland flow an infiltrating plane surface. Journal of Hydrology. 1997;193:45-60.
29. Morisawa ME. Relation of morphometric properties to runoff in the little mill creek, Ohio, drainage basin, Technical Report 17, Columbia University, Department of Geology, ONR, New York; 1959.
30. Pitlick J. Relations between peakflows, precipitation and physiography for five mountainous regions in Western U.S.A. Journal of Hydrology. 1994;158:219-240.
31. Jain V, Sinha R. Evaluation of geomorphic control on flood hazard through geomorphic instantaneous unit hydrograph. Current Science. 2003;85(11): 26-32.
32. Okoko EE, Olujinmi JAB. The role of geomorphic features in urban flooding: The case of Ala river in Akure, Nigeria, Int. Journal of Environmental Issues. 2003; 1(1):192-201.
33. Nabegu AB. Determinant of sediment yield in river basins. Implications for reservoir management. International Journal of Environmental Issues. 2005;3(2):40-48.
34. Hugget RJ. Fundamentals of geomorphology. London, New York, Routledge; 2003.
35. Onosemuode C, Adetimirin OI, Aboderin SO. Hydrological analysis of onitsha North East drainage basin using geoinformatic techniques. World Applied Sciences Journal. 2010;11(10):1297-1302.
36. Guttschalk LC. Reservoir sedimentation. In: Chow VT (ed.), Handbook of applied hydrology, McGraw Hill, Book Company, Newyork, Sections. 1964;4-11.
37. Subodh CP, Gopal CD. Morphometric analysis and associated land use study of a part of the Dwarkeswar watershed. International Journal of Geomatics and Geosciences. 2012;3(2):351–363.

38. Datt R, Ramanathan NL. Environmental monitoring'. Proceedings of Seminar on the Status of Environmental Studies in India, Thiruvanathapuram. 1981;284-287.
39. Parveen R, Kumar U, Singh VK. Geomorphometric characterization of upper South Koel basin, Jharkhand: A remote sensing and GIS approach. Journal of Water Resource and Protection. 2012;4: 1042–1050.
40. Nongkyrin JM, Hussain Z. Morphometric analysis of the Manads river basin using earth observation data and geographical information system. International Journal of Geomatics and Geosciences. 2011; 2(2):647-654.
41. Melton MA. Correlation structures of morphometric properties of drainage systems and their controlling agents. Journal of Geology. 1958;66:442-460.
42. Gregory KJ, Walling DE. Drainage basin form and process: A geomorphological approach. Arnold, London; 1973.
43. Gardener V. Drainage basin morphometry. In Goudie A. (Ed.), Geomorphological Techniques. London: George Allen and Unwin. 1990;47-55.
44. Cannon JP. Generation of explicit parameters for a quantitative geomorphic study of the mill creek drainage basin. Oklahoma Geology Notes. 1976;36(1):13–17.

23

Effect of Soil Physico-chemical Properties on Agriculture: A Study in Tangail District, Bangladesh

M. Rahman[1] and M. M. Mian[1,2*]

[1]School of Environmental Science, Bangladesh Agricultural University, Mymensingh, 2202, Bangladesh.
[2]School of Urban Construction and Environmental Engineering, Chongqing University, 174 Shazhengjie, Shapingba, Chongqing, 400044, China.

Authors' contributions

The research was carried out equally by both authors. Author MR conceived the main idea and plan of the study. The literature study, statistical analysis and manuscript writing were conducted by authors MMM and MR. Both authors read and approved the final manuscript.

Editor(s):
(1) Abdel-Tawab H. Mossa, Environmental Toxicology Research Unit (ETRU), Pesticide Chemistry Department, National Research Centre, Egypt.
Reviewers:
(1) Mohammed Suleiman, Umaru Musa Yar'adua University, Katsina, Nigeria.
(2) Evans Ntim Amedor, Institute of Tropical Agriculture, Ghana.

ABSTRACT

Aims: The economic development of Bangladesh is mainly influenced by agriculture sector, but lacks of available information regarding soil characteristics hindering the process of appropriate agricultural practice. This study tries to examine the physico-chemical characteristics of soil in Tangail district, Bangladesh, and evaluate the scope of agriculture.
Place and Duration of Study: Entire soil sample was collected from three major zones (Bhuapur, Gopalpur and Tangail Sadar Upazilla) of Tangail district, Bangladesh during the period of September, 2013 to July, 2014.
Methodology: Random quadrate methods were used to collect soil samples. Collected samples were dried at room temperature, mixed carefully, crushed and sieved by 20 mesh net and analyzed soil texture, bulk density, soil pH, organic matter, total N, available P, K, and S. Entire analysis was conducted twice and average value of results was considered for calculation.
Results: It is revealed that, the soil texture varied largely In the three study zones, Tangail Sadar soil is mainly sandy clay loam, Bhuapur and Gopalpur Upazilla are silt loam and loam texture

**Corresponding author: Email: manikmbstu@gmail.com*

respectively. The mean bulk density is lower in Tangail Sadar (0.093) and higher in Bhuapur (5.019) and Gopalpur (4.706), which indicate soil compactness. However the pH of the three study zones were 5.70, 6.22 and 6.28, which is suitable for some crop cultivation, but the entire study area lacks soil macronutrients.

Conclusion: Adequate fertilizers are required to cultivate crops in these zones. Beside chemical fertilizer, organic fertilizer should also be used for retention of soil organic contents.

Keywords: Soil properties; nutrients; scope; agriculture; fertilizer.

1. INTRODUCTION

The economic development of Bangladesh largely depends on agricultural development. Total GDP of this country is dominated by manufacturing 17.78%, agriculture 16.11% and trade 13.41% sectors [1]. Agriculture is the second GDP earning sector, and majority of the people are associated with its practice. However, the agriculture of this country is suffering from various problems such as nutrient deficiency, toxicity in soil, natural calamities, insect and disease hazards, improper soil and crop management. For proper land use, agricultural planning and better crop production, basic data on the physical and chemical properties of soil are very important. Both climatic and physico-chemical characteristics of an area are very important to determine the type of crops to be grown [2].

Bangladesh has a wider range and greater complexity of lands. Earlier soils of Bangladesh were divided into 7 tracts, then into 20 general soil types and 530 soil series. At present, the soils of Bangladesh are divided into 30 Agro-Ecological Zones (AEZ). The soil series are classified on the basis of the characteristics of soils such as soil texture, structure, soil color, organic matter content, soil pH etc. in different layer of soils. Moreover, the chemical characteristics of soils are almost same under the same soil series and the soils are constituted by same parent materials. Evaluations of physical and chemical characteristics of soil are helpful for economic use of fertilizer and irrigation water supply and, thereby, they can be used to maximize crop production. Total area of the country is about 147570 km^2 of which about 11.3 million (79%) hectare is floodplain, 1.2 million (8.3%) hectare is terrace land and 1.8 million (12.6%) is hill area. Hilly areas occur in the north, northeast and east parts of the country [3].

Chemical characteristics of soils represent the nature of genetic process, its development and present nutrient status. For evaluating the characteristics of soil profile and present fertility status of soil, chemical analysis has got much importance. Proper fertilizer use depends on the nutrient content of the soil. As a consequence of anthropogenic activities over a long period of time, the vegetation as well as soil properties are also changed. The quality of soil refers not only its lack of degradation or contamination, but also to its overall fitness, or effectiveness for supporting plant growth, managing water and responding to environmental stresses [4]. Hence, a person dealing with soil should be acquainted with the chemical properties to decide whether the soil is suitable for plant growth.

Nowadays, the farmers of the country are desperately trying to increase the crop yield by applying more and more inorganic chemicals, mainly nitrogenous fertilizers, pesticides and by increasing cropping intensity. For these reasons, reserve of nutrient becomes depleted and available nutrient becomes unavailable. Again, excessive amount of applied fertilizers hinders the availability of other nutrients too. Our farmer do not know the inherent nutrient status of the soils and they use fertilizers blindly and uneconomically. In order to minimize these problems, it is very pertinent to evaluate the fertility status of our soils at the farmers' level [5-7].

The rigidity and supporting powers, freedom of drainage, moisture storage capacity, and ease of penetration by roots, aeration and retention of plant nutrients are all intimately related to physical and chemical condition of soil. If a soil has good physical and chemical properties, then it will be considered as an appropriate soil for agriculture [8].

Tangail is a district located in the central region of Bangladesh. Agriculture is the main occupation of the Tangail district. About 49.53% people are involved with agricultural activities. The main agricultural products are paddy, potato, jute, sugarcane, sesame, linseed, wheat, mustard seed and pulse. About 2800 km^2 cultivable lands are available in Tangail. The main fruit products are mangos, jackfruit, bananas, litchis, and pineapples. It produces about 847309 tons of crops every year. Among 12 Upazillas of Tangail district Bhuapur, Gopalpur and Tangail Sadar are three important Upazillas in respect of agricultural production. Tangail Sadar, Bhuapur and Gopalpur Upazilla have 61594, 40125 and 47295 acres of cultivable lands respectively [9]. For successful crop production in this area, knowledge of physical and chemical properties of soil is very important. But data on physical and chemical properties in this area is not sufficient. This study tries to examine the physical and chemical properties of soil of three upazillas of Tangail district, Bangladesh, and analyze the scope of agricultural practice in these areas.

2. MATERIALS AND METHODS

2.1 Materials

Entire soil samples were collected from, Bhuapur, Gopalpur and Tangail Sadar Upazilla of Tangail district. Among 30 Agro-ecological Zones (AEZ) of Bangladesh, Bhuapur and Tangail Sadar Upazilla were located in the Old Brahmaputra Floodplain and Gopalpur Upazilla was located in Young Brahmaputra-Jamuna Floodplain. A total of 87 samples was collected from 29 locations of 3 Upazilla, particularly 27 samples were collected from Bhuapur Upazilla, 24 samples from Gopalpur Upazilla and 36 samples from Tangail Sadar Upazilla of Tangail district (Fig. 1). All the samples were collected from (0 – 30) cm depth of soil. The collected samples were dried at room temperature, mixed thoroughly, crushed, sieved with a 20-mesh sieve and preserved in plastic bags for subsequent laboratory analysis.

2.2 Methods

2.2.1 Soil texture

Soil textures were analyzed by using hydrometer method [10]. Fifty grams of oven dry soil from each sampling depth was taken separately in dispersion cup and 10 ml of 5% calgon solution was added to the sample and allowed to soak for 15 minutes. 90 ml tap water was added to the

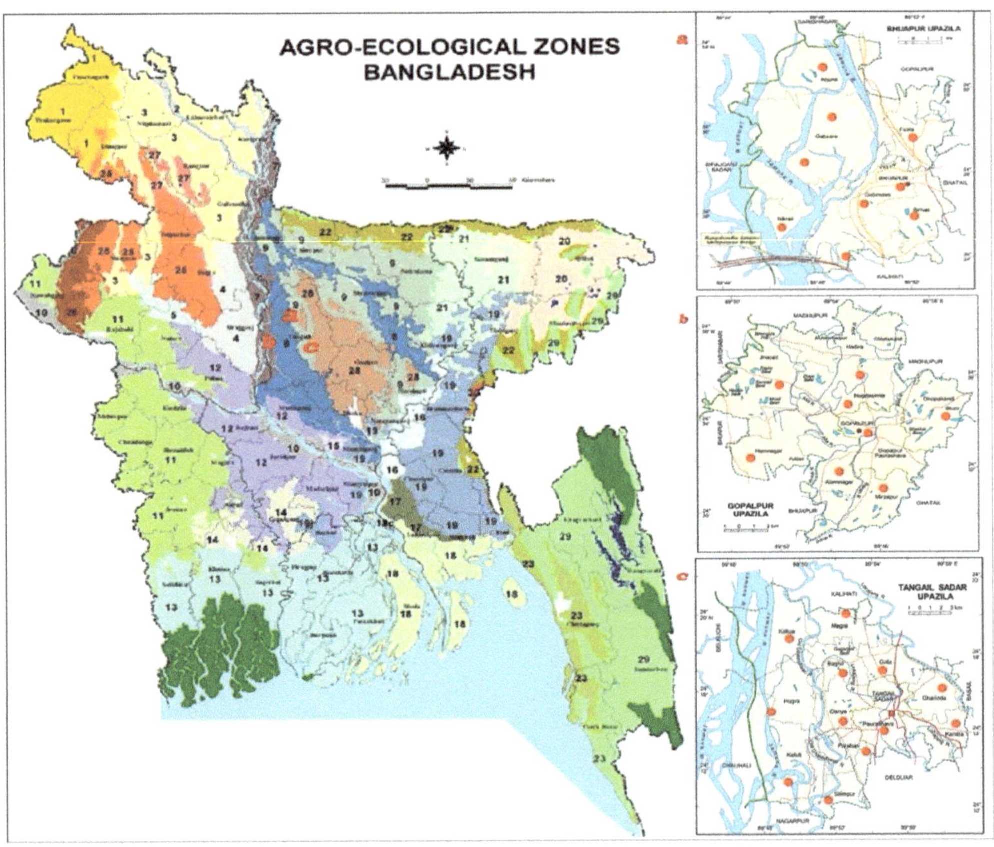

Fig. 1. Location of study area: (a) Bhuapur, (b) Gopalpur and (c) Tangail Sadar Upazilla of Tangail District, Bangladesh

dispersion cup. The suspension was then stirred with an electric stirrer for 15 minutes. The content of the dispersion cup was then transferred to one liter sedimentation cylinder and tap water was added to make the volume up to mark. A cork was placed in the mouth of the cylinder and the cylinder was inverted several times until the whole soil mass appeared in the suspension. The cylinder was set up right and the hydrometer readings were taken at 40 second and 2 hours of sedimentation. The temperature of suspension was recorded with a thermometer at 40 second and 2 hours of sedimentation. The corrections of hydrometer readings were made as the hydrometer was calibrated at 68℉.

The percentage of sand, silt and clay were calculated as follows:

% (Silt + Clay) = (Corrected hydrometer reading after 40 seconds)/(Oven dry weight of soil)×100
% (Clay) = (Corrected hydrometer reading after 2 hours)/(Oven dry weight of soil)×100
% Sand = 100 – % (Silt + Clay)
% Silt = % (Silt + Clay) – % Clay

The textural class of soil for different depths of each series was determined by plotting the results (sand, silt and clay) on triangular diagram (Fig. 2) designed by Marshall [11].

2.2.2 Bulk Density

Bulk density is a measure of a soil mass per unit volume of soil. It is used as a measure of soil wetness, volumetric water content, and porosity. Factors that influence the measurement include; organic matter content, the porosity of the soil, and the soil structure these factors will intern control hydraulic conductivity.

Bulk Density (p) = Mass of oven dried soil / Total volume

2.2.3 Soil Ph

Soil pH was determined by glass electrode pH meter [12]. Twenty grams of air dried soil from each sampling depth was taken in 50 ml beaker separately and 50 ml of distilled water added to each beaker. The suspension was stirred well for several times during the next 30 minutes and allowed to stand for about an hour. The position of the electrode was adjusted in the clamp of the electrode holder. Then the electrode was immersed into partly settled soil suspension and pH was measured. The result was reported as "soil pH measured in water" (Soil: water ratio being 1:2.5).

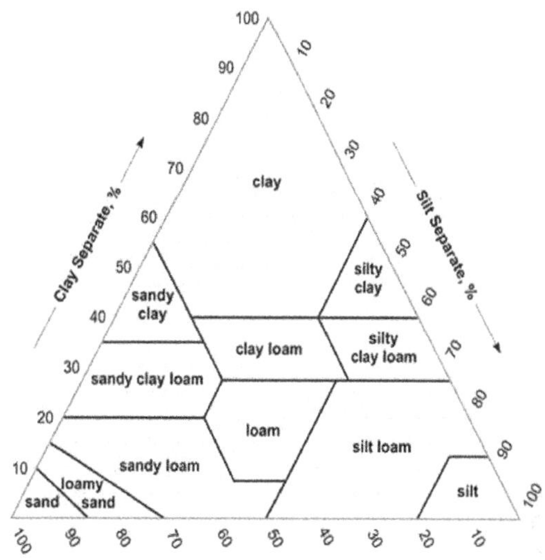

Fig. 2. Triangular diagram of textural classes of soil

2.2.4 Organic matter

The percentage of organic carbon was determined titrimetrically using Walkley and Black method with oxidation of organic matter using potassium dichromate ($K_2Cr_2O_7$). Following the procedure of previous study [13], 2g soil along with 10 ml 1N $K_2Cr_2O_7$ solution and 20 ml of Conc. H_2SO_4 was taken in a 500 ml conical flask. After 30 minutes rest, about 200 ml distilled water and 10 ml concentrated H_3PO_4 was added. 40 drops of diphenylamine indicator ($C_{12}H_{11}N$) was added and titrated against 0.2 N ferrous sulphate ($FeSO_4$. $7H_2O$) until purplish blue color was turned to green color. The organic matter was obtained by multiplying the content of organic carbon by Van Bemmelen factor of 1.73 [14].

2.2.5 Total nitrogen

Total nitrogen content in soil was determined by Semi-Micro Kjeldahl method, following concentrated sulphuric acid digestion and distillation with 40% sodium hydroxide. The ammonia evolved was collected in boric acid indicator and was titrated with 0.01 N sulphuric acid [15]. The % N in soil was calculated by the following formula:

$\% N = [\{(R-B) \times 1.4\}/ W] \times S$

Where, R= Reading of the sample in burette (H_2SO_4). S= Strength of H_2SO_4., B= Value of blank reading., W = Weight of the soil in mg., 1.4= Correction factor.

2.2.6 Available phosphorus

Available phosphorus present in soil was determined by Olsen's method colorimetrically, where $SnCl_2$ was used as a reductant. Extracted with 0.5 M sodium bicarbonate solution [16] and developed blue color by $SnCl_2$ reduction and measured the color colorimetrically [15], with the help of spectrophotometer at 660 nm wave length [12].

2.2.7 Available sulphur

Available sulphur (S) was determined by extracting the soil samples by calcium chloride solution (0.15%) [14]. The amount of S content in the extract was estimated turbidimetrically by spectrophotometer at 420 nm wave length.

2.2.8 Available potassium

Available potassium was determined by ammonium acetate extraction method. Soil samples were saturated with ammonium acetate (NH_4OAC) solution and the supernatant clear solution was separated and collected. Then NH_4OAC solution was added again to make the final volume up to 100 ml for each sample. For the above cation the Flame photometer was separated, adjusted and different standard curves were prepared to final out the concentration of the cation but the extracting solution used for each sample was same. The reading was recorded by flame photometer. From each reading the concentration of the ion was obtained from its standard curves.

2.3 Quality Control of Experiment

In order to ensure accurate and reliable results each experiment was done twice and the average result was taken for calculation, and a further experiment was carried out, if the difference of two results were more than 5%. Entire data were calculated using MS office 2010 and SPSS 2012.

3. RESULTS AND DISCUSSION

3.1 Physical Characteristics of Soil

3.1.1 Soil texture

The percentage of sand, silt and clay content in the three Upazilla were varied largely. The particle size distribution of soil of Tangail Sadar, Bhuapur and Gopalpur are presented in Table 1. The mechanical analysis was done for the determination of the relative percentage of the sand, silt and clay in the soils and ultimately to find out the textural classes.

In Tangail Sadar Upazilla, the soil texture was sandy loam. The sand, silt and clay fraction of the soil were varied from (36.8% - 64.43%), (9.34% – 28.08%) and (25.23% – 44.33%) respectively. The surface layer had a higher percentage of sand compare to a deeper layer and the deeper layer was dominated by high percentages of silt and clay. The soil texture of Bhuapur Upazilla was silt loam, the fraction of sand, silt and clay were varied from (9.23% - 30.64%), (58% - 76%) and (0.25% - 20.36%) respectively. In Gopalpur Upazilla, the soil textural classes were mostly loam, and some silt loam also found. The fraction of sand, silt and clay were (11% - 38.64%), (45% - 76%) and (0.25% - 14%) respectively.

Soil like silt loam has higher agricultural values being less susceptible to become loose and open [17]. Silt loam soil can easily be kept in a state good tilt, favorable for germination of seed, easy for root penetration and has considerable water holding capacity. This soil may be highly productive if managed properly.

Table 1. Particle size fractions and textural classes of selected soils

Location	Particle size fractions			Textural class
	Sand (%)	Silt (%)	Clay (%)	
Tangail Sadar	54.831	17.337	28.782	Sandy clay loam
Bhuapur	17.475	69.417	11.825	Silt loam
Gopalpur	29.271	57.124	10.958	Loam

3.1.2 Bulk density

The bulk densities of three Upazillas were quite different, the average bulk density of three Upazilla's were 0.09 g/cm^3, 5.02 g/cm^3 and 4.71 g/cm^3 respectively (Table 2). However, the bulk densities of Tangail Sadar and Bhuapur upazilla are relatively similar type, but in Gopalpur Upazilla show a wide variation throughout the area.

The average bulk density 4.71 g/cm^3 in Gopalpur Upazilla is quite high i.e. soils of this region are compacted. Hence, there would be restricted seed emergence and root growth which will affect the total plant growth and yield. The use of tractors will directly affect the soils bulk density, causing extreme compaction especially if the soil is wet. High bulk density also means that there is little pore spaces in soil to be occupied by water and air.

3.2 Chemical Characteristics of Soil

3.2.1 Soil pH

The soil pH of the three Upazilla is slightly acidic, Tangail Sadar Upazilla are mostly acidic (5.70) soil compared to other to Bhuapur Upazilla and Gopalpur Upazilla. The data on soil pH for the three Upazilla are presented in Table 3. The pH of Tangail Sadar Upazilla varied within that area, whereas Gopalpur Upazilla shows higher variation of pH throughout the area.

Table 2. Bulk density of selected soils

	Location	Mean	SD.
Bulk density (gcm-3)	Tangail Sadar	0.093	0.001
	Bhuapur Upazilla	5.019	0.436
	Gopalpur Upazilla	4.706	1.013

The pH value of this area is similar to the results of previous studies [18]. Hence it can be speculated that soil of this zone is slightly acidic in nature. So, rice, potato, tomato, rye, peanut, corn, tobacco, carrot, cauliflower and cucumber, can be cultivated in these three zones. However, the soil pH depends on kinds of basic rock or parent materials. Rainfall is also a great factor to increase soil acidity. The use of nitrogenous fertilizers without the addition of lime contributed to this decline in soil pH. Due to the acidic nature, these zones are not most nutritious soil for all kinds of plants. Fig. 3, illustrate the suitable pH range for a variety of crops [19] and mean pH of study areas.

The soil of Bhuapur Upazilla and Gopalpur Upazilla have suitable pH range for a wide variety of crops, whereas lower soil pH in Tangail Sadar is not suitable for wheat, onion, pea, soybean, sugarcane and asparagus.

Table 3. Soil reaction (pH) of the three Upazilla's of Tangail

	Location	Mean	SD.
pH	Tangail Sadar	5.70	0.322
	Bhuapur Upazilla	6.22	0.588
	Gopalpur Upazilla	6.28	0.930

3.2.2 Macronutrients

The macronutrients N, organic C %, S, P and K ppm of the study area are illustrated in Fig. 4. Among the three Upazillas of Tangail district, the total N content was highest (0.73%) at Tangail Sadar Upazilla. The amount of total N in maximum locations was very low to low. Perhaps the accelerated decomposition of organic matter, addition of less organic matter, changed in cropping systems and variations in soil characteristics affect N content of these study areas. Entire soil samples contained N below critical level. The total N content in Bhuapur and Gopalpur Upazilla were close to the findings of previous studies [5,20]. However, the N status can easily be increased by the addition of nitrogenous fertilizers for better crop production.

The number of organic carbon content in three different regions varied a little, and the percentage of organic carbons shows approximately similar pattern as N. The available S contents of soils showed wide variation among different locations. The S content was lower in Tangail Sadar Upazilla, due to oxidation and leaching or exhausted by continuous cultivation affecting lower S content of this region, which is similar to the result of previous study [21]. The highest available P was found to be 25.08 ppm in Gopalpur Upazilla, due to the addition of phosphatic fertilizer applied to previous crops. The available P in Bhuapur Upazilla was medium and in Tangail Sadar Upazilla was low. Available P is mainly affected by soil characteristics and farming practices. It might be decreased because of inadequate P fertilization and phosphate fixation in acidic terrace soils. The K is higher in Bhuapur and Gopalpur Upazilla, and very lower in Tangail Sadar Upazilla. Perhaps the presence of more primary and secondary K bearing minerals was the reason of such variation. Soils in temperate regions that supply adequate K for crop needs often contain considerable trioctahedral mica (biotite) in their clay fractions [6]. Potassic fertilizer should be applied to maintain the K status above critical level for optimum yield. Continual K fertilization is necessary for K deficient soils [6].

3.3 Soil Assessment

The correlations of soil physical and chemical properties are well supported by the study findings (Table 4). The bulk density of soil has significant negative correlation with N and Organic C, and positive correlation with P, K, S and soil pH. The Sand and Clay soil has availability of N compounds, whereas silt dominating soil has higher P, K and S ratio. The soil pH is positively correlated with silt, comparatively higher pH can be found in Gopalpur Upazilla, where it recorded percentage of (69.417) silt.

Though the pH level of the study area is favorable for some crop, but overall macronutrients are low to very low. Fig. 5, illustrate the recommended nutrient level for a variety of plants [22].

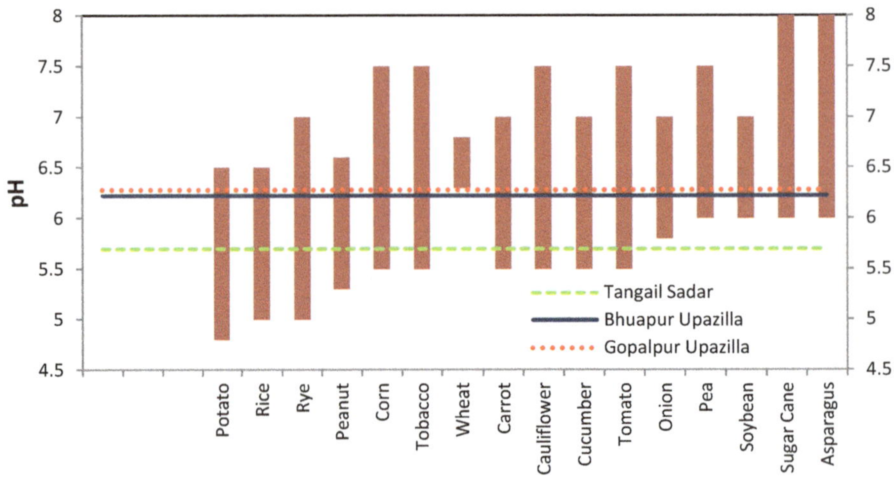

Fig. 3. The pH of the study areas and suitable range for various crops cultivation

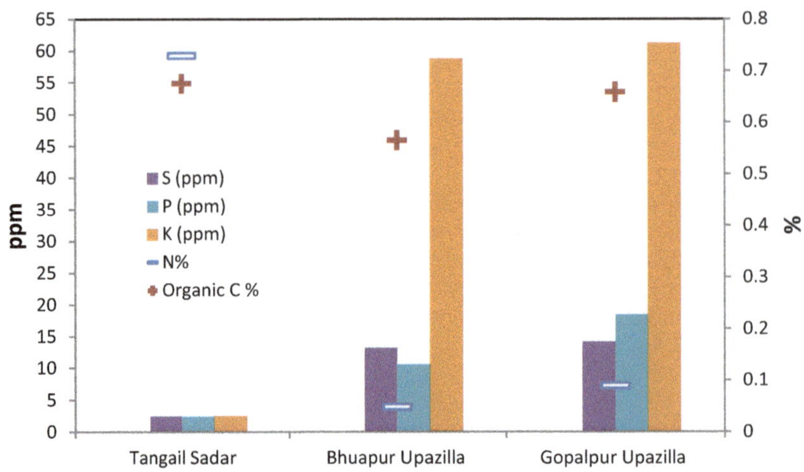

Fig. 4. Macronutrient contents in study area

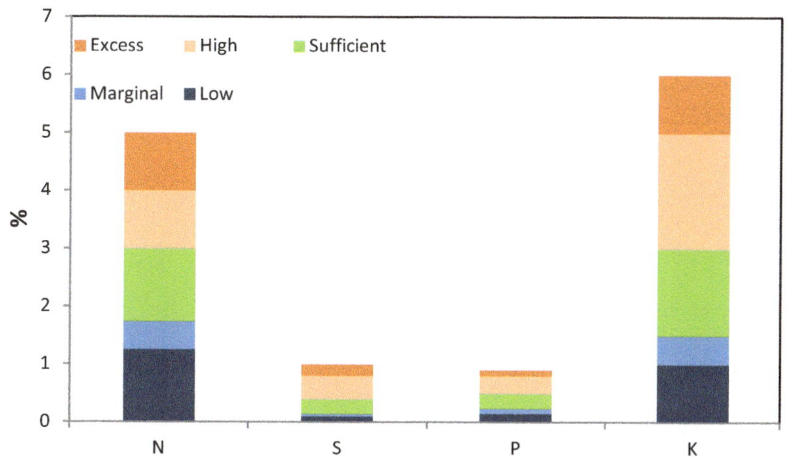

Fig. 5. The recommended level of nutrients for a variety of plants

Table 4. Correlation of soil physical and chemical properties

	N	Organic C	P	K	S	pH
Bulk density	-0.954 **	-0.177	0.600 **	0.474 **	0.698 **	0.431 **
Sand	0.843 **	0.215	-0.364 *	-0.234	-0.517 **	-0.412 *
Silt	-0.920 **	-0.171	0.462 **	0.320	0.600 **	0.485 **
Clay	0.841 **	0.067	-0.611 **	-0.446 **	-0.675 **	-0.532 **

*** Correlation is significant at the 0.01 level (2-tailed); * Correlation is significant at the 0.05 level (2-tailed)*

Three zones of the study area have very lower percentages of N, S, P and K. Additional fertilizer is required for the cultivation of these areas. Beside chemical fertilizer, organic fertilizer is also required for the organic retention of the soil.

4. CONCLUSION

The physical and chemical properties of soil in three major zones (Tangail Sadar, Bhuapur and Gopalpur) of Tangail District, Bangladesh were successfully investigated. The soil of Tangail Sadar is sandy clay loam and comparatively less compact, and Bhuapur and Gopalpur Upazilla soils are silt loam and loam respectively, and more compact soil. The pH of the study area is slightly acidic and sufficient for some crop cultivation, but selective areas have lack of soil micronutrients, appropriate fertilizer is required to cultivate crops in these zones.

COMPETING INTERESTS

Authors have declared that no competing interests exist.

REFERENCES

1. BBS. GDP of Bangladesh; 2013-2014. (Base: 2005-06). Bangladesh Bureau of Statistics, Bangladesh; 2014.
2. Hossain MM, Sattar MA. Physical and chemical properties of some selected soils of Bangladesh. MS Thesis Department of Soil Science BAU. Mymensingh. 2002;1-58.
3. SRDI. Land and soil resources utilization guide (in Bengali). Thana nirdeshika series- tangail sadar thana, soil resources development institute. Dhaka, Bangladesh. 2000;51-61.
4. Lewandowski A, Zumwinkle M, Fish A. Assessing the soil system: Arevies of soil quality literature. Minnesota department of agriculture, energy and sustainable agriculture program. St Paul; 1999.
5. Bhuiya ZH, Sattar MA, Islam MS. Physical and chemical studies of some soils of Bangladesh. Asiatic Society of Bangladesh. 1974;19.
6. Bhuiyan NI. Co-ordinated project on potassium studies Progress Report (1987-88). BARRI, Joydevpur, Gazipur. 1988; 1-45.
7. Ahsan E, Karim Z. Soil and management research on upland soils of Bangladesh. In: Proceeding of the international conference on the management and fertilization of upland soils in the Tropics and Sub-tropics. 1988;247-51.

8. Foth HD. Fundamentals of Soil Science. John Wiley & Sons. New York; 1990.
9. Zilla Tyatha Batayan. Available:http://www.dctangail.gov.bd/ (Accessed; 2012)
10. Bouyoucos GJ. The hydrometer as a method for the mechanical analysis of soils. Soil Science. 1927;23:343-53.
11. Marshall TJ. Mechanical composition of soils in relation to field description of texture. Common wealth of Australia. Council for Scientific and Industrial Research. Melbourne, Australia. Bull. 1947;227:20.
12. Jackson ML. Soil chemical analysis. Constable and Co Ltd. London; 1988.
13. Gosh AB, Bajaj JC, Hasan R, Singh D. Soil and soil water testing method. A laboratory manual, division of Soil Science And Agricultural Chemistry. IARI. New Delhi-110012. 1983;1-45.
14. Page AL. Methods of soil analysis. American society of agronomy and soil science society of amecica publisher. USA Part II, chemical and microbial properties, 2nd edition; 1982.
15. Black CA. Methods of soil analysis. Amer Soc Agron Inc Pub Madison, Wiscosin, USA, Part I and Part II; 1965.
16. Olsen SR, Cole CV, Watanabe FS, Dean LA. Estimation of available phosphorus in soils by extraction with sodium bicarbonate. US Department Agriculture 1954;929.
17. Weir WW. Soil science, its Principles and Practices. JB Lippincott CO, Philadelphis, Pa, USA; 1949.
18. Hannan MA. Structure and related physical properties of some soils of Bangladesh. M.Sc. thesis. Department of soil science. University of Dhaka. Bangladesh. 1995; 138-52.
19. SC H. Soil reaction preferences of plants. Mich. Agr. Exp. Sta. Spec. Bull. 306. East Lansing; 1941.
20. Islam MS. Soil Management in Agricultural Research in Bangladesh. BARC, Dhaka. 1983;105-9.
21. Beaton JD, Wanger RE. Sulphur a vital component of maximum economic yields systems. Sulphur in Agriculture. 1985;9: 2-7.
22. Dr. Ross McKenzie. Soil fertility/ crop nutrition alberta agriculture. Food and Rural Development Lethbridge; 1998.

24

A Study of Water Quality Status of Mangrove Vegetation in Pichavaram Estuary

V. E. Nethaji Mariappan[1*], A. Hari Nivas[1], T. Kanmani[2] and S. Parthiban[1]

[1]Centre for Remote Sensing and Geoinformatics, Sathyabama University, Rajiv Gandhi Road, Jeppiaar Nagar, Chennai – 600 119, India.
[2]M. E. Environmental Engineering, Sathyabama University, Rajiv Gandhi Road, Jeppiaar Nagar, Chennai – 600 119, India.

Authors' contributions

This work was carried out in collaboration between all authors. All authors read and approved the final manuscript.

Editor(s):
(1) Worku L. Mulat, Civil and Environmental Engineering Department, University of Connecticut, USA And Ethiopian Institute of Resilience & Climate Change Adaptation, USA.
(2) Maria Panitsa, Department of Environmental and Natural Resource Management, University of Patras, Greece.
Reviewers:
(1) Pavlina Simeonova, Institute of Solid State Physics, Bulgaria.
(2) Thokozani Simelane, Africa Institute of South Africa, South Africa.
(3) Anonymous, Instituto Nacional de Investigaciones Forestales Agrícolas y Pecuarias, Mexico.

ABSTRACT

Mangrove ecosystem is unique in nature due to the combination of Halophytic plants, animals, birds and aquatic organisms. Such wetland ecosystems possess rich nutrition status that contributes rich bio resources to aquatic flora and fauna. Brackish water within the ambit of Pichavaram plays a pivotal role on growth and development of mangrove community. A study was undertaken to understand the surface water dynamics quality parameters in Pichavaram Estuary. Analysis of water quality parameters was assessed for the period of November 2012. Adequate care was taken to cover all tidal regimes (i.e.) (i) Inlet: the point where the principal feeder opens into the estuary; (ii) Center: the point that gives the general water quality of the estuary; (iii) Outlet: the place where the overflow occurs. Study sites from which samples were collected, were identified adjoining the region of Parangipettai Vellar Estuary and Pichavaram mangrove forest. Ten surface water samples were collected. These were analyzed for physico-chemical parameters such as temperature, pH, salinity, dissolved oxygen and total dissolved solids vary significantly

**Corresponding author: Email: nethajim@gmail.com*

among the three regions. All studied parameters were spatially interpolated in GIS environment to understand the distribution of individual nutrient in the ecosystem. The results of the statistical analysis were subjected to cluster analysis for grouping parameters based on similarity and dissimilarities. Individual parameter values were overlaid on the LANDSAT imagery to understand spatial representation of water quality with respect to bio-physical value of the mangrove ecosystem.

Keywords: Water quality parameters; pichavaram; mangroves; spatial interpolation; cluster analysis.

1. INTRODUCTION

Estuaries consist of a complex mixture of many distinctive habitat types that do not exist in isolation [1]. These can typically contain mangrove forests, salt marshes, sea grass meadows, oyster reefs and non-vegetated areas. It is widely accepted that estuaries provide habitat for numerous fauna that include fish and invertebrate species, many of which are economically important. Any significant loss of habitats that are provided by estuaries could drastically affect the local ecological systems and associated fauna [2].

Mangrove occur in diverse environmental settings of geophysical (i.e. climate, tides and sea level) and geomorphological (dynamic history of the land surface and contemporary processes and biological components). Mangroves and its associated ecosystems as biologically most productive, socioeconomically important, and aesthetically attractive while providing food and shelter for many vital biotic species, some that are commercially important [3,4]. Mangroves support a wide variety of ecosystem goods (fuel wood, medicine, construction materials) and services (shoreline stabilization, fisheries breeding and nursery grounds, sedimentation trapping, uptake of nutrients and heavy metals-phytoremediation). They have immense value to ecological processes. Pichavaram Mangrove Forest is said to be the world's second largest mangrove forest which is located in cuddalore district, southern part of Tamil nadu, India. Pichavaram mangroves are surrounded by rivers and lakes and brackish water that changes with seasons and geographic domain. Water quality parameters provide basic scientific information about spatial quality parameters and ecologically relevant toxicological threshold values. These are important to understand critical physical and chemical parameters influencing the aquatic environment. Physical and chemical variables like temperature, rainfall, pH, salinity, dissolved oxygen and carbon di oxide influence growth and vigour of mangroves. Other parameters such as total suspended and dissolved solids, total alkalinity, acidity and heavy metal contaminants are limiting factors for aquatic organisms.

Temperature is a limiting factor in the mangrove environment [5-7]. Water temperature is probably the most important environmental variable for the sustenance of aquatic flora and fauna. pH is one of the vital environmental variable that determines the survival, metabolism, physiology and growth of aquatic organisms. Salinity is a dynamic indicator for the growth of selective species of mangroves in pichavaram where cation exchange system act as chelates in binding salts in ion exchange system. It is expressed as the total concentration of electrically charged ions (cations) in water in part per thousand (‰). The anions include CO_3^-, SO_4^{2-} , Cl^-, $HCO3^-$, NO_3^-, NH_4^{2-}, PO_4^{2-}. Alkalinity of a water body is a measure of its capacity to neutralize acids to a designated pH [8,9].

The tides in Pichavaram mangrove forest are semi-diurnal and vary in amplitude from about 15 to 100 cm in different regions, reaches maximum during monsoon and post-monsoon and a minimum during summer [10]. The rise and fall of the tidal water is through a direct connection with the sea at the Chinnavaikal mouth and also through the two adjacent estuaries. The depth of the water-ways ranges from 0.3 to 3 m [10] which indicates that shallow area predominate in Pichavaram brackish water.

Somjai Sremongkontip et al. [11] describes the paper on Detecting Changes In The Mangrove Forests Of Southern Thailand Using Remotely Sensed Data And GIS. Remote sensing and GIS were used to analyze the changes in the mangrove forest. Mangrove forests have been used by different user groups. Changing of government policies on mangrove forest utilization and management caused the occurrence of three major activities; mangrove forest concession, tin-mining activities and intensive shrimp farming. These activities were considered as the major reasons for changes of mangrove forest areas and led to the degradation of mangrove forest, the loss of mangrove forest and ecological disturbance. Furthermore, it affected the mangrove forest dependence-dwellers.

Marília Cunha-Lignon et al. [12] presented a paper analysis of Mangrove Forest Succession, Using Sediment Cores: A Case Study in the Cananéia –Iguape Coastal System. Sediment cores are an essential tool for the analysis of the dynamics of mangrove succession. Coring was used to correlate changes in depositional environments and lateral sedimentary facies with discrete stages of forest succession at the Cananéia-Iguape Coastal System in south eastern Brazil. A local level successional pattern was examined based on four core series T1) a sediment bank; T2) a smooth cordgrass Spartina alterniflora bank; T3) an active mangrove progradation fringe and; T4) a mature mangrove forest. Cores were macroscopically described in terms of color, texture, sedimentary structure and organic components. The base of all cores exhibited a similar pattern suggesting common vertical progressive changes in depositional conditions and subsequent successional colonization pattern throughout the forest. The progradation zone is an exposed bank, As the substrate consolidates attains the greatest structural development in the mature forest. Cores provide a reliable approach to describe local-level successional sequences in dynamic settings subject to drivers operating on multiple temporal and spatial scales where spatial heterogeneity can lead to multiple equilibria and where similar successional end-points may be reached through convergent paths.

Giri et al. [13] had done a work on Status and distribution of mangrove forests of the world using earth observation satellite data. Our scientific understanding of the extent and distribution of mangrove forests of the world is inadequate. The available global mangrove databases, compiled using disparate geospatial data sources and national statistics, need to be improved. Here, we mapped the status and distributions of global mangroves using recently available Global Land Survey (GLS) data and the Landsat archive. 1000 Landsat scenes were interpreted using hybrid supervised and unsupervised digital image classification techniques. Each image was normalized for variation in solar angle and earth–sun distance by converting the digital number values to the top-of-the-atmosphere reflectance. Ground truth data and existing maps and databases were used to select training samples and also for iterative labelling. Results were validated using existing GIS data and the published literature to map 'true mangroves'. The total area of mangroves in the year 2000 was 137,760 km2 in 118 countries and territories in the tropical and subtropical regions of the world. Approximately 75% of world's mangroves are found in just 15 countries, and only 6.9% are protected under the existing protected areas network (IUCN I-IV). This study confirms earlier findings that the biogeographic distribution of mangroves is generally confined to the tropical and subtropical regions and the largest percentage of mangroves is found between 5° N and 5° S latitude.

Claudia Kuenzer et al. [8] had taken a review on Remote Sensing of Mangrove Ecosystems. Mangrove ecosystems dominate the coastal wetlands of tropical and subtropical regions throughout the world. They provide various ecological and economical ecosystem services contributing to coastal erosion protection, water filtration, provision of areas for fish and shrimp breeding, provision of building material and medicinal ingredients, and the attraction of tourists, amongt many other factors. Mangrove mapping is one of the most demanding tasks in remote sensing, because the remotely sensed signal from mangrove ecosystems is composed of several components and is influenced by a large number of other parameters. In optical data, the spectrum of a pixel containing mangrove is usually influenced by pixel fractions of mangrove leaves, stems, and branches; underlying mudflats; soils; and water surfaces. All of these components differ depending on mangrove species, vigor, age, and season, as well as soil type and water turbidity and quality, among others. Other parameters influencing the spectral signal include plant and leaf geometry, LAI, stand density, and atmospheric conditions, to name a few. Furthermore, the spectral signal, its —mixing, ‖ and its distinctiveness in optical data vary, depending on the spatial and spectral resolution of the sensors used, ranging from aerial photography (pixels in centimeter to meter range) to highest-resolution spaceborne multispectral data (pixel size in meter range) to multispectral data of medium resolution (10–30 m) to airborne or spaceborne hyperspectral data (pixel size in 1–30-m range, but up to 200 spectral bands). The goal of all remote-sensing–based mangrove mapping and monitoring activities should be the protection of these unique ecosystems, whose value cannot be overestimated. This is especially true with respect to climate change-related sea level rise scenarios globally. Sea level rise would have a severe impact on coastal communities in the tropics and subtropics. Natural mangrove ecosystems are a productive, extremely valuable shield against this threat.

Chandra and Joseph [14] had done a work on Mangrove Forest Distributions and Dynamics in Madagascar(1975–2005). Mangrove forests of Madagascar are declining, albeit at a much slower rate than the global average. The forests are declining due to conversion to other land uses and forest degradation. time-series satellite data of 1975, 1990, 2000, and 2005 were interpreted using a hybrid supervised and unsupervised classification approach. Landsat data were geometrically corrected to an accuracy of ± one-half pixel, an accuracy necessary for change analysis. The post classification change detection approach had been used. The results showed that Madagascar lost 7% of mangrove forests from 1975 to 2005, to a present extent of ~2,797 km2. Deforestation rates and causes varied both spatially and temporally. The forests increased by 5.6% (212 km2) from 1975 to 1990, decreased by 14.3% (455 km2) from 1990 to 2000, and decreased by 2.6% (73 km2) from 2000 to 2005. Similarly, major changes occurred in Bombekota Bay, Mahajamba Bay, the coast of Ambanja, the Tsiribihina River, and Cap St Vincent. The main factors responsible for mangrove deforestation include conversion to agriculture (35%), logging (16%), conversion to aquaculture (3%), and urban development (1%). The analyses show the potential for producing consistent and timely mangrove forest databases of Madagascar using the historical archive of Landsat data. The full potential of remote sensing technology for identifying mangrove forests, measuring the biophysical properties, and detecting forest cover changes can only be realized through a robust and operational mangrove forest assessment and monitoring program. Future research is needed to map very small mangrove areas that were not able to discern with the Landsat resolution.

Marwa and Evan [15] had done a thesis on The Economic Value of Mangroves: A Meta-Analysis. This paper presents a synthesis of the mangrove ecosystem valuation literature through a meta-regression analysis. The main contribution of this study is that it is the first meta-analysis focusing solely on mangrove forests, whereas previous studies have included different types of wetlands. The number of studies included in the regression analysis is 44 for a total of 145 observations. several regressions are included with the objective of addressing outliers in the data as well as the possible correlations between observations of the same study. The possible interaction effects also investigated between type of service and GDP per capita. The findings indicate that mangroves exhibit decreasing returns to scale, that GDP per capita has a positive effect on mangrove values and that using the replacement cost and contingent valuation methods produce higher estimates than do other methods and also there are statistically significant interaction effects that influence the data. Finally, the results indicate that employing weighted regressions provide a better fit than others. However, in terms of forecast performance we find that all the estimated models performed similarly and were not able to conclude decisively that one outperforms the other.

2. METHODOLOGY

2.1 Description of Study Area

The study area, Pichavaram mangrove forest lies in (Lat. 11° 29' N; Long. 79°47' E) is located between the Vellar and Coleroon estuaries. The forest occurs on 51 islets, ranging in size from 10 sq km to 2 sq km, separated by intricate waterways that connect the Vellar and Coleroon estuaries [16]. The southern part near the Coleroon estuary is predominantly covered by mangrove vegetation, while the northern part near the Vellar estuary is dominated by mud-flats. The Vellar estuary opens into the Bay of Bengal at Parangipettai and links with the Coleroon River, which is a tributary to the river Cauvery. The mangrove ecosystem is influenced by the mixing of nearby water types such as the neritic, estuarine and freshwater. The mangroves cover an area of about 1100 ha, 40% being covered by water- which 50% by forest, ways and the remaining filled by sand-flats and mud-flats (Fig. 1 and Table 2).

2.2 Satellite Data Analysis

LANDSAT ETM data was downloaded from GLCF website. Satellite data was composed of six bands that area imported as TIFF file format for visualization. They are then imported to (img) format LANDASAT image is represented as

Digital number as of 8 bit multi spectral continuous data products were re-projected WGS84 datum to UTM and WGS84 datum in ERDAS Imagine software Fig. 2.

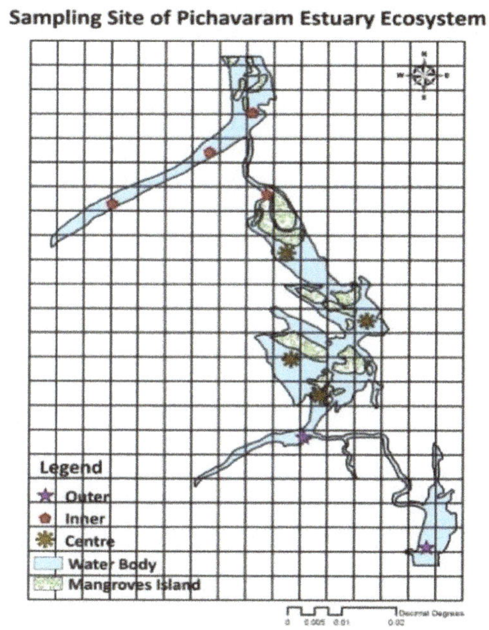

Fig. 1. Sampling points in the study area

Fig. 2. Landsat ETM satellite image of Pichavaram study area

2.2.1 Sampling

Generally three sampling sites were selected for monitoring. They are (i) Inlet: the point where the principal feeder opens into the estuary. (ii) Center: the point that gives the general water quality of the estuary. (iii) Outlet: the place where the overflow occurs. A total of ten locations covering four sample locations in the inlet and centre and two sample location in outer region in Fig. 1 and table 1. Surface water samples were collected during monsoon period of the year 2012. Adequate care was taken to collect samples during low tide conditions. Samples were collected in clean plastics bottles. Collected samples were stored in laboratory as per standard procedure till the analysis is over.

The following physical parameters namely temperature, turbidity, pH, Electrical Conductivity (E.C). Chemical parameters like Total Solids (TS), Total Dissolved Solids (TDS), Total Suspended Solids (TSS), Total Hardness, Calcium Hardness, Magnesium Hardness, Nitrates, Phosphates, Sulphate, Chlorides, Fluorides, Potassium and Sodium were analyzed for this study.

2.2.2 Cluster analysis

Cluster Analysis was used in this study to examine the differences between the monitoring stations during the four seasons of the year. It shows the complexity of the system and the multitude of contributing sources. Cluster analysis of all sampling station was performed by Bray-Curtis method using Biodiversity Pro ver. 2.0 a open source software. Clustering analysis involve combining of homogenous parameters within the specified samples. The heterogeneous are sorted out based on the distance from the nearest ones. Cluster was split at 40 per cent level at all the sampling points. They were later grouped as a single zone except sampling station 5 pertaining to Chinnavaikal. Similarly clustering was performed for physio-chemical parameters for all sampling station.

Based on the cluster analysis, the concentration of sulphide and organic matter in cluster 3 (i.e. Stations 3 and 4) were high compared to clusters 1 (Stations 1 and 2) and 2 (all sites during the monsoon season).The results yielded that at 10 per cent clustering distance two zones were identified they area zone were in the order of i) K, Mg, So4, Ca, Cl, Na, Total Hardness, TDS and EC zone ii) Nitrite, Tidy test, Free Ammonia, Fluoride, Fe, Po4, Si, Nitrate, pH and Turbidity. Hierarchical Cluster Analysis is the only way to observe how homogeneous groups of variables are formed and K-Means Cluster Analysis only supports classifying observations.

Table 1. Sampling site representing water quality parameters and its range

S.No	Longitude	Latitude	Date of Collection	Sampling Station	Turbidity NTU	pH	EC (micro mho/cm)
1	79.7780	11.4966	16.11.2012	Vellor Mouth	6.2	7.29	42864
2	79.7701	11.4885	16.11.2012	Opp.CAS	3.8	7.10	26950
3	79.7521	11.4778	16.11.2012	New Bridge	5.1	7.08	16739
4	79.7843	11.4677	16.11.2012	Fish Market	4.3	7.09	9974
5	79.7991	11.4537	16.11.2012	Chinnavaikal	5	7.11	3701
6	79.7852	11.4458	16.11.2012	Vaikal suthal	8.6	7.26	47165
7	79.7876	11.4299	16.11.2012	Killai Tourist	4.9	7.32	43509
8	79.8103	11.4073	16.11.2012	Kollidam	6.8	7.46	46735
9	79.7807	11.4797	16.11.2012	Mudasalodai	8.2	7.22	18011
10	79.7907	11.4384	16.11.2012	MGR thittu	6.2	7.24	18488

K-Means Cluster Analysis is used to classify observations through K number of clusters. The idea is to minimize the distance between the data and the corresponding cluster centroid. K-means analysis is based on one of the simplest algorithms for solving the cluster problem, and is therefore much faster than hierarchical cluster analysis. Users should typically consider K-means analysis when the sample size is larger than 100. However, that K-means cluster analysis assumes the user already knows the centroid of the observations, or, at least, the number of groups to be clustered.

For a matrix *X* with *n* observations by *p* variables, initial cluster centres can be specified with a *K*-by-*p* matrix, or chosen from the matrix *X* with a predefined number of clusters.

- Select Initial Cluster Centres from Observations
- Allocate Observations to the Closest Cluster

3. RESULTS

The results exhibited variability in water quality within and between the sample site collections. All the observations were significantly higher at outer region followed by inner and middle region as shown in table 2 and 3. Higher concentration of nutrients at the outer region was attributed to certain environmental and meteorological factors exert remarkable influence in this mangrove biotype. Pichavaram mangrove is saline in nature. This was due to mixing up of sea water with estuarine water. pH was highest (7.46) at Kollidam region and minimum at New Bridge (7.08) (Table 2). The physic chemical variables of mangrove water such as turbidity, pH, EC, TDS, total hardness, calcium, magnesium, sodium, potassium, free ammonia, iron, nitrite, nitrate, chloride, fluoride, sulphate, phosphate and silica vary significantly among the three regions (Table 3).

Electrical Conductivity (3701 - 47165 micro mho/cm) and Total dissolved solids (330.16-32714 mg/l) were more pronounced in mangrove ecosystem. This again was due to mixing of sea water with river water. There was a large variation in chloride concentration (891 mg/L) to (15,345 mg/L) at Vaikal Sudal. Higher value seems to be more pronounced in outer region and centre region and less levels in the inner region. Calcium and magnesium are responsible of increase in salt concentration towards salinity of water. Both the variables have the lowest concentration in inner region, and their level are moderate in Centre region and highest in the outer region.

Total hardness is attributed to sodium concentration in the surface water of the pichavaram region. It is understood that sodium induces hardness in water quality. A linear regression was developed between total hardness and sodium concentration as identified by a positive relationship (r^2-0.8) in Fig. 3. Potassium concentration was lower than sodium concentration invariably at all the sampling points. Mangrove region is possessed higher levels of nitrate and phosphate in Table 3.

4. DISCUSSION

Pichavaram region falls in a tidal region where river water Confluence Sea. Tides action seems to be high in new moon and full moon day's average in other days. Water samples are muddy, due to intermixing of water in soil that makes colloidal system. Colloidal system is rich in nutrients as it contains chelates that combine most of the cautions.

Results from Water samples collected from centre and outer region if the estuary covering the entire spectrum of wetland ecosystem showed that all three regions have its own soil water status, ecological diversity of flora and fauna. The salinity acts as a limiting factor in the distribution of living organisms. Its variation caused by dilute ion and evaporation is most likely to influence the fauna in the intertidal zone [17]. A reflected by the pH and EC values that were observed to be at invariably higher concentrations at all the sampling points.

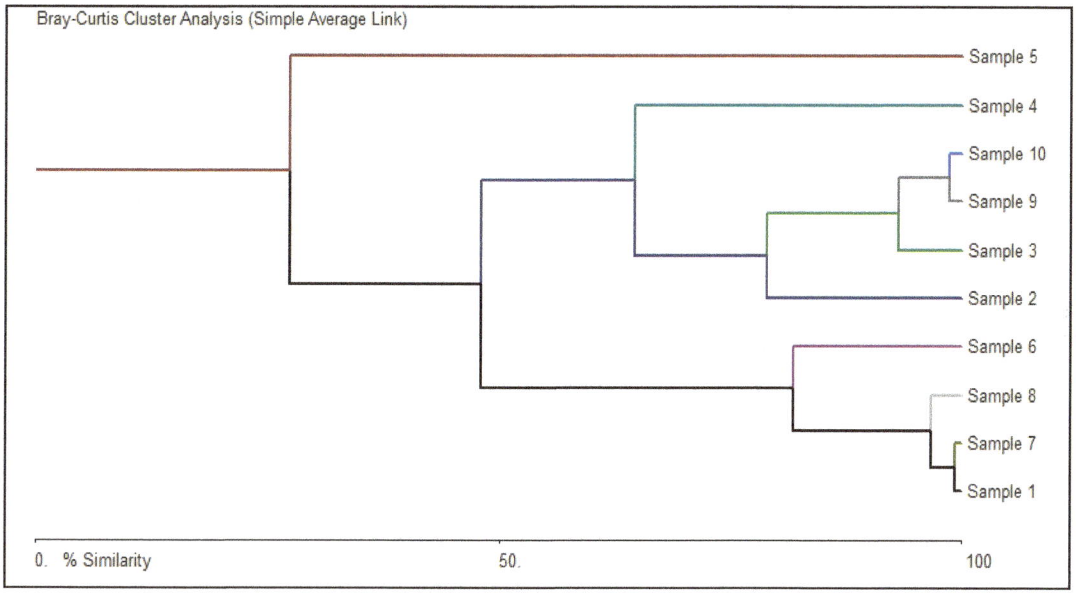

Fig. 3. Cluster analysis for sampling station

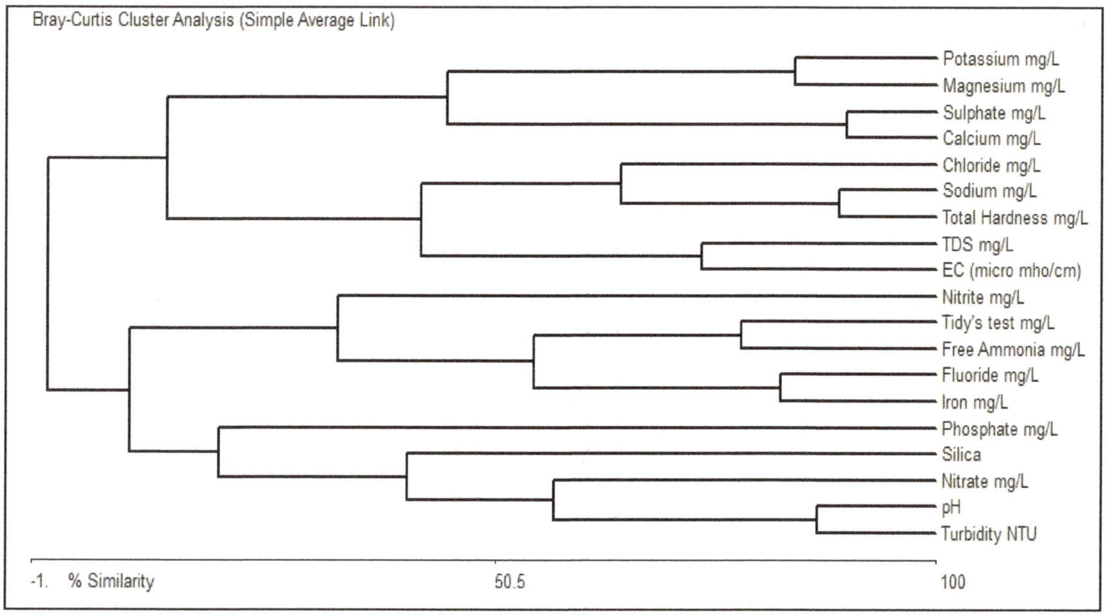

Fig. 4. Cluster analysis for Physico- chemical parameter

Principal component analysis creates variables that are linear combinations of the original variables. The new variables have the property that the variables are all orthogonal. The principal components can be used to find clusters in a set of data. PCA is a variance-focused approach seeking to reproduce the total variable variance, in which components reflect both common and unique variance of the variable. PCA is generally preferred for purposes of data reduction (i.e., translating variable space into optimal factor space) but not when the goal is to detect the latent construct or factors.

The values of sodium, chlorides and sulphate were at higher concentration in the study area that might be due to geomorphology prevailing in that area. Higher EC values were observed in outer region and centre region and less levels in the inner region.

Calcium and magnesium are responsible for increase in salt concentration towards salinity of water. Both variables have the lowest concentration in inner region, and their level are moderate in Centre region and highest in the outer region. Such higher concentration of sodium might be due to absorption of iron and silica as silicate minerals. Higher concentration of nitrates and phosphates are due to leaching of urea and di-ammonium phosphate, ammonium chloride fertilizers from the soil to wetland region through agricultural activities that are carried upon in the adjoining mangrove ecosystem as depicted in Fig. 1. [18] found very clear seasonal fluctuation in the concentrations of Fe, Mn, Cu and Zn in the environmental compartments such as dissolved, particulate and sediment fractions and found the sediments of mangrove which is normally rich in organic matter as a prime sink for these elements that are being brought in by freshwater.

Table 2. Range values of selected variables and its sampling station

Parameter	Range (max/min)		Sampling Site
pH	Maximum	7.46	Kollidam
	Minimum	7.08	New Bridge
Turbidity	Maximum	8.6	Vaaikal suthal
	Minimum	3.8	Opp.CAS
Sulphate (mg/L)	Maximum	2024	Vaaikal suthal
	Minimum	252	Chinnavaaikal
Calcium (mg/L)	Maximum	1840	Vaaikal suthal
	Minimum	168	Chinnavaaikal
Sodium (mg/L)	Maximum	7400	Vaaikal suthal
	Minimum	520	Chinnavaaikal
Total Hardness (mg/L)	Maximum	6800	Vaaikal suthal
	Minimum	680	Chinnavaaikal

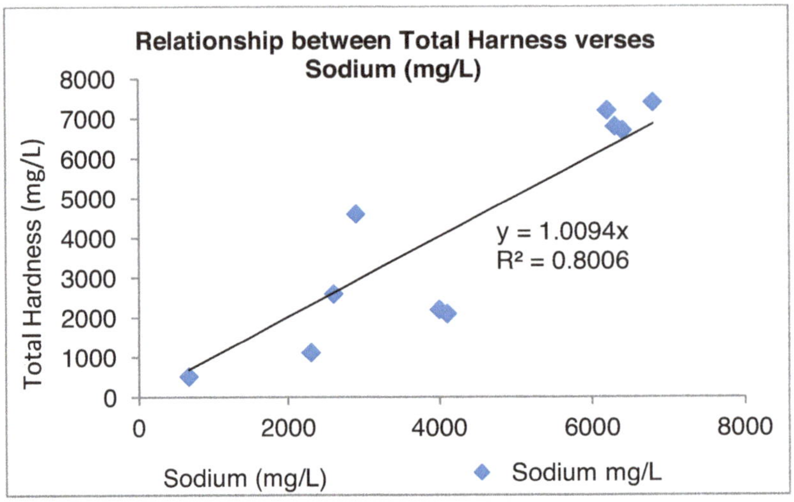

Fig. 5. Depicts a good linear relationship of sodium and total hardness

Table 3. Water sample parameters and its values in (mg/L)

TDS	Total Hardness	Calcium	Magnesium	Sodium	Potassium	Iron	Free Ammonia	Nitrite	Nitrate	Chloride	Fluoride	Sulphate	Phosphate	Tidy's test	Silica
30004	6400	1680	528	6700	500	0.4	0.22	0.03	2	14602	0.42	1786	0.1	0.2	16.41
18865	2900	180	216	4600	400	0.5	0.2	0.05	3	7376	0.29	1420	0.4	0.1	16.76
11715	2600	640	240	2600	200	0.5	0.19	2.01	4	4950	0.32	726	0.4	0.1	13.91
6958	2300	600	192	1120	140	0.4	0.22	0.82	4	2970	0.28	520	0.5	0.2	15.3
2264	680	168	62	520	30	0.4	0.04	0.04	4	891	0.31	252	34	0.2	19.52
330.2	6800	1840	528	7400	460	0.5	0.14	0.009	1	15345	0.28	2024	0.1	0.1	16.11
30456	6300	1680	504	6800	400	0.3	0.12	0.05	1	14603	0.26	1826	0.2	0.1	17.8
32714	6200	1640	504	7200	450	0.5	0.16	0.06	1	15097	0.25	1932	0.1	0.1	19.96
12608	4000	880	432	2200	200	0.4	0.13	1.00	3	6039	0.31	832	0.3	0.1	17.29
12942	4100	840	480	2100	200	0.4	0.11	0.79	3	5890	0.42	846	0.3	0.1	18.71

5. CONCLUSION

The study of surface water samples on biologically rich wetland ecosystem exhibited spatial variability of parameters between inner, centre and outer region. Most of the variables showed higher variability in outer region rather than inner and centre region. Salinity influences water quality parameters that promote growth of aquatic flora and fauna in the ecosystem. Intensification of agriculture promotes an increase in nitrate and phosphate content in soil water is detrimental growth of mangroves in pichavaram region if it is excessive.

ACKNOWLEDGEMENTS

We thank Principal Chief Conservator of Forest, Govt. of Tamil Nadu, Divisional Forest Officer, Parengipettai Division for granting permission to collect water samples from Pichavaram region. Special thanks are due to Dr. Joshua Amarnath, Prof. & Head, Department of Chemical Engineering for Student allotment for this research work. We also thank our management for providing infrastructure and moral support extended for this research work.

COMPETING INTERESTS

Authors have declared that no competing interests exist.

REFERENCES

1. Prince Jeyaseelan MJ. Studies on Ichthyo fauna of the mangrove of Pichavaram (South India). Ph.D. Thesis, Annamalai University, Parangipettai, India; 1983.
2. Bulthuis DA, Axelrad DM, Mickelson MJ. Growth of the seagrass *Heterozostera tasmanica* limited by nitrogen in Port Phillip Bay, Australia. Mar Ecol Prog Ser. 1992;89:269–277.
3. Naylor RL, Goldburg RJ, Primavera JH, Kautsky N, Beveridge MCM, et al. Effect of aquaculture on world fish supplies. Nature. 2000;405:1017-1024.
4. Zhang Q, Shuzhen S. The mangrove wetland resources and their conservation in China. Journal of Natural Resources. 2001;16:28-36.
5. Odum EP. Fundamentals of Ecology. 3rd ed. W.B. Saunders: Philadelphia; 1971.
6. Pihl L, Cattrijsse A, Codling I, Mathieson S, McLusky DS, Roberts C. Habitat use by fishes; 2002.
7. Boyd CE, Water quality in warm water fish ponds. University Press: Alabama, USA; 1979.
8. Claudia Kuenzer, Andrea Bluemel, Steffen Gebhardt, Tuan Vo Quoc, Stefan Dech. Remote sensing of mangrove ecosystems: A review. Remote Sens. 2011:3(5):878-928.
DOI: 10.3390/rs3050878
9. Edokpayi CA. Variation of chemical constituents of a brackish water prawn habitat in southern Nigeria. Acta Satech. 2005;2(1):11-18.
10. Muniyandi K. Studies on mangroves of Pichavaram (Southeast coast of India). Ph.D. Thesis. Annamalai University, Parangipettai, India; 1986.
11. Somjai Sremongkontip, Yousif Ali Hussin, Liza Groenindijk. Detecting changes in the mangrove forests of Southern Thailand Using Remotely Sensed Data and GIS. In. International Archives of Photogrammetry and Remote Sensing. 2000:12. Part B7. Amsterdam
12. Marília Cunha-Lignon, Michel M, Mahiques, Yara Schaeffer-Novelli, Marcelo Rodrigues, Daniel A. Klein, Samara C. Goya, Ricardo P. Menghini, et al., Analysis of mangrove forest succession, using sediment cores: A case study in the Cananéia –Iguape Coastal System, São Paulo-Brazil. Brazilian Journal of Oceanography. 2009;57(3):161-174
13. Giri C, Ochieng E, Tieszen LL, Zhu Z, Singh A, Loveland T, Masek J, Duke N. Status and distribution of mangrove forests of the world using earth observation satellite data. Global Ecology and Biogeography. 2011;20:154–159.
14. Chandra Giri, Joseph Muhlhausen. Mangrove forest distributions and dynamics in Madagascar (1975-2005). Sensors. 2008;8(4):2104-2117.
DOI: 10.3390/s8042104
15. Marwa E Salem, Evan Mercer D. The economic value of mangroves: A meta-analysis: Sustainability. 2012;4(3):359-383.
DOI: 10.3390/su4030359
16. Kathiresan K. A review of studies on Pichavaram mangrove, Southeast India. Hydrobiologia. 2000;430:185-205.

17. Gibson RN. Recent studies on the biology of intertidal fishes. Oceanogr. Mar. Biol. Ann. Rev. In estuaries and other brackish areas. In: Elliott M, Hemingway KL, ed. Fishes in Estuaries. Blackwell Science. 1982;20:363-414.

18. Subramanian V, Ramanathan AL, Vaidhyanathan P. Heavy metal distribution and fractionation in the Cauvery estuary. Mar Pollut Bull. 1989;21(7):324-340.

25

The Spread, Impact and Control of *Chromolaena odorata* (L.) R.M. King and H. Robinson in Grassland Area

Muhammad Rusdy[1*]

[1]*Department of Forage Crops and Grassland Management, Faculty of Animal Science, Hasanuddin University, Makassar, Indonesia.*

Author's contribution

The sole author designed, analyzed and interpreted and prepared the manuscript.

Editor(s):
(1) Ahmed Esmat Abdel Moneim, Department of Zoology, Helwan University, Egypt And Institute of Biomedical Research Center, University of Granada, Spain.
Reviewers:
(1) Anonymous, Universidade do Estado do Rio de Janeiro, Brazil.
(2) Anonymous, University of Sciences and Technology Houari Boumediene, Algeria.

ABSTRACT

Chromolaena odorata is native to tropical America and has been reported as one of the world's most invasive species in the humid tropics and subtropics. As an attempt to generate information useful for preventing and controlling the weed in grassland area, the spreads, impacts and controls of the weed are reviewed. Some attributes that contribute to its success as invader weed were described. Its impacts, both negative and positive on grassland ecosystem were presented. Possible solutions to prevent the spread of the weed and its control in grassland area were also discussed.

Keywords: Chromolaena odorata; dispersal; impact; control; grassland.

1. INTRODUCTION

A major constraint to livestock production in many tropical and developing countries is the scarcity and fluctuating quantity and quality of year around forage supply. During the rainy season, tropical forage species grow at very fast rates, with forage yield often exceeding animal's

**Corresponding author: Email: muhrusdy79@yahoo.co.id*

requirements. If not grazed or fed by animals, such forage becomes fibrous and lack of most essential nutrients which required for improved microbial fermentation and performance of animals. Conversely, during the dry season, pasture plant growth becomes slower or stop and the effect of inadequate feeding on livestock productivity may be serious.

Since the last four decades, forage supply from grassland area in many tropical countries has been continuously decreasing, mainly due to conversion of grassland to crop cultivation and invasion of noxious exotic plants. Invasion of world's grassland by shrub plants is among the dominant changes in the earth's vegetation during the last two centuries [1]. In Indonesia, invasion the exotic invasive plant *Chromolaena odorata* (Fig. 1) has transformed grassland ecosystems into worse conditions. Its fast growing and highly competitive ability traits making this weed lowering productivity and reducing grassland area for livestock. In the pasture area owned by Faculty of Animal Science Hasanuddin University in Enrekang regency, Indonesia, the weed has covered more than 50% of grassland area. Lacks of forage because of reducing carrying capacity of invaded grassland generally occurs during the dry season and during the season, many cattle are dead because of starvation. Furthermore, this weed is avoided by livestock as its leaves have offensive odor and bitter tasting. If it is not controlled, the change from grassland to almost pure stand of Chromolaena as reported by [2] can happen rapidly. To avoid this possibility becomes real, understanding in depth about biology and ecology of this weed is needed. Many investigations relating to the spread, impact and control of the weed have been reported and it seems appropriate to review and discuss these findings to reduce invasiveness of the weed in grassland area.

2. SPREADS

Chromolaena odorata (L.) R.M. King and H. Robinson, commonly known as siam weed, devil weed, bunga semak putih (Indonesia) and several other names in different countries (hereafter is called Chromolaena), is native to tropical America and has become a major invasive weed of grasslands, plantation crops and forests in the Old World tropics; from west and south Africa to Southeast, South Asia and the Pacific region [3,4]. The initial spread of

Fig. 1. *Chromolaena odorata* plant

the weed in Asia took place in the early 1880's when it was introduced as ornamental plant to the Botanical Garden in Calcutta India, then propagated throughout South East Asia and parts of Oceania [5,6]. In Indonesia, it was introduced to Deli in the east coast of North Sumatera in the beginning of commercial cultivation of tobacco in the turn of twentieth century [7] and in eastern Indonesia, the weed was first sighted in 1970's [8]. Transmigration program, with the movement of people, vehicles and equipments from Java was responsible for the spread of the weed to eastern Indonesia [4]. The weed has been introduced either intentionally or unintentionally to a new area far from its native areas. It was intentionally introduced to Ivory Coast in 1952 to control *Imperata cylindrica* other coarse grasses growth [9], however the major causes of its spread is through unintentional seed contamination [4].

The weed can spread through both sexual and vegetative means but the main spread method is through sexual propagation. Sexual propagation starts when the weed forms flower at one year old and increased until the stand is ten years old [10]. Flowering is initiated by a decrease in both day length and rainfall and peaked in December-January in northern hemisphere and in June-July in southern hemisphere [11], and close to the equator, the flowering may lose synchrony and generally occurs at the start of dry season [4]. Flowering is often prolific, but fertile seed is produced without pollination, as the weed is apomictic [12]. Flowers are capable of producing huge quantities of seeds, with estimated from 93,000 [13] to 1,600,000 seeds per plant [14]. Seed viability varies; most seeds didn't viable for more than 1 year; the remaining 1.4% seedlings survived into the second year [9]. Seeds tend to degrade in high moisture and temperature environments [10]. The light, narrow seeds, topped with a turf are easily dispersed by wind over a short distance and can spread over a long distance by attaching to clothing, fur animals, machinery, etc. [4], but the primary long-distance vector responsible for its spread is human activity. During the World War II, contamination of the weed seeds with moved military equipment and personnel were the major source of long-distance spread in the Asia and Pacific regions [6].

Seeds which fall under dense vegetation may lie dormant until the land is disturbed by clearing, like burning, cultivation and overgrazing [4], because light is needed for germination and growth [15]. In grassland area, animals will avoid land invaded by the weed and subsequently tend to overgraze on non-invaded lands. If water is sufficient, overgrazing which leaves patches of bare soil may be sufficient to stimulate germination. Once it germinated and established, *Chromolaena* that has a high competitive ability and allelopathic trait making this weed posses the ability to suppress the growth of forage plants and other shrub.

Besides by seeds, cut branches of roots and stem of *Chromolaena* can form roots and shoot, and these may spread during cultivation and other soil moving activities. Stem cutting could produce adventitious roots from the internodes within eight days under field conditions [15]. Under moist conditions, branches trailing on the ground may form roots [11]. The underground organ also can aid its survival in case of fire, drought or mechanical damage through coppicing. But vegetative propagation does not contribute significantly to its spread [11].

3. IMPACTS

Once established, *Chromoleena* grows rapidly. It has extremely fast growth rate (up to 20 mm/day) and often forms a dense thicket of height 1,5 – 2.5 m in open areas that can grow through and over the existing vegetation. In grassland area, the weed exerted positive and negative impacts. Table 1 below listed some impacts of the weed on grassland ecosystem.

4. CONTROLS

Studies have been carried out all over the world to control the weed. The best form of invasive species management is prevention. Preventing the spread of the weed will reduce future problems. If prevention is no longer possible, it is best to treat the weed infestations when they are small to prevent them from establishing. Investigations have shown that the weed can be controlled mechanically, chemically, culturally or biologically Mechanical controls have been the predominant method to control the weed. Hand weeding such as slashing, uprooting and digging out of the young plant is the most common method used in many countries. The use of tractor drawn equipments are limited to accessible area. Slashing reduces the standing weed biomass but it is not a solution because the

Table 1. Impact of *Chromolaena* on grassland ecosystem

No.	Negative impacts	Authors
1.	It has allelopathic effects on growth of other plant species	[16]
2.	Dense thicket of the weed prevents the free movement of livestock	[11]
3.	Shades out the growing forage, reducing the growth of grasses and legumes and plant biodiversity, preventing establishment of other plant species, interfering natural ecosystem process and altering integrity of natural plant community.	[17]
4.	Reduced carrying capacity of grassland from approximately 6 ha/head of cattle to more than 15 ha/head (in South Africa).	[18]
5.	Reduced cattle population from 2000 to 800 heads at Magoga Sisal ranch, completely closed down of Nkhlashhane Sisal ranch in Swaziland	[2]
6.	Harms the economy of small holder farmers in Indonesia because of lack of forage for their livestock.	[19]
7.	Reduced the population of banteng (*Bos javanicus*) and deer (*Cervus timorensis*) as caused by severe reduction of herbage yield in East Java, Indonesia.	[20]
8.	Threatening the endangered black rhinoceros (*Diceros bicornis*) as resulted from reducing its natural feed supply (in Maputaland, Africa).	[21]
9.	Reduces ecotourism potential and game-farming areas as reduced game areas and visibility.	[21]
10.	Consumption of the weed flowers may cause pyrrolizidine toxicity that can destroy the liver of cattle and goat.	[22]
11.	Consumption of the weed leaves may cause nitrate poisoning in animals.	[23]
12	Invaded grassland can be a harbor introduced wild pigs, rodents and insects	[24]
13.	The weed harbors of grasshopper, *Zonocerus variegates* that breeds in the weed and from there, moves to cassava, feeding on the leaves and causing yield losses.	[25]
14.	Reduced water run-off in water catchment as the greater biomass and higher rates of transpiration of *Chromolaena* compared to indigenous vegetation	[21]
No.	**Positive impacts**	
1.	Soil under *Chromolaena* had higher minerals and better physical soil properties.	[26]
2.	Increased organic carbon (27.9%), total nitrogen (36.7%), total phosphorus (56.8%), extractable calcium (68.3%) and magnesium (140.3%) relative to savannah soils.	[27]
3.	As surface mulching, it increased soil available N, P and K by 11.92, 36.65 and 51.64%, respectively.	[28]
4.	Can carry on atmospheric nitrogen fixation with the help of free living bacteria in its roots.	[15]

plant coppices profusely from the stumps after slashing. To kill the weed, it must be slashed in short term interval to exhaust reserve carbohydrates and [29] recommended slashing four times a year in cocoa plantation, but [30] and [2] reported that slashing is not effective to control the weed. The most effective means of control for small infestations is to remove the entire plant including roots by hand weeding, but follow-up clearance every 2 – 3 month is needed because of rapid of new infestation from the seeds.

The negative effect of slashing and uprooting is that it opens the soil to the sun, which promote the exposure of seeds to light and this facilitate seed germination and creates ideal sites for recolonization by the weed [31]. Mechanical control is labor intensive and can be effective if infestations are small and accessible. A combination with other methods such as cultural and chemical control has been found to be economical and effective.

Controlled burning was the best control method of the weed [2]. Fire was found to be very effective at eliminating seeds, small and medium size of the weed [32], but [30] reported that in South Sulawesi Indonesia, burning without followed by planting any vegetation or mulch application resulted in infestation of hard seed plant of *Mimosa pudica*. Wu R et al. [33] reported that burning followed by planting signal grass (*Brachiaria decumbens*) effectively suppressed the weed resulting complete eradication after three years. The cost of planting with grass species is expensive, therefore, this approach, if successful, is only likely to be used where intensive livestock production is conducted. Castillo and Moog [34] reported that planting of *Leucaena leucocepahala* in pastures reduced the weed population in Philippines. Torres and Paller [35] suggested planting legumes like *Calopogonium muconoides* and *Centrosema pubescens* and grass, *Setaria* sp. immediately after burning to reduce the weed infestation.

Chemical control using herbicides applied at seedling stage on early growth has given encouraging results. Pre-emergence herbicides such as diuron at dose 1.5 kg/ha and atrazine 2.0 kg/ha were effective to control of germination and establishment of the weed. Paraquat, glyphosate, and 2,4-D were the post emergence herbicides that resulted in in drastic reduction in the density of the weed [36]. In South Africa, the use of triclopyr, 2,4,5-T and tebuthiuron were consistently effective [37]. Triclopyr was more effective than glyphosate and that an acceptable level of control could be obtained with 2.5 dm^3/ha triclopyr and between 3.5 and 4.3 dm^3/ha of glyphosate was required for effective control [38]. In grassland area, triclopyr was more suitable to be used than glyphosate because triclopyr selectively kill broad-leaves plants like Chromolaena but did not kill grasses [39]. Efficacy of triclopyr can be enhanced when it mixed with asulam as opposed to individual products was applied. A 95% reduction of regrowth was achieved using a mixture of asulam and triclopyr, while over 89% reduction was achieved using glyphosate alone, however, glyphosate applied alone was more efficacious than when it mixed with asulam [40]. Tjitrosemito et al. [41] found 2,4-D, triclopyr and picloram are effective against Chromolaena in rubber plantations. Although widely used, efficacy of triclopyr is grassland area was not lasting, as at 90 days after spraying, Chromolaena became dominant plant again [39].

Although many herbicides have been reported to give effective results, but these generally fail to give long time of control, as reinfestation of cleared areas invariably occurs. Also, those herbicides require repeated applications with high dosages. Therefore most herbicide based control methods are not cost effective under most economical conditions in developing countries. Biological control, would therefore is expected as a means to reduce the vigor and reproductive capacity of the weed to a level at which other means of control become cost effective.

There are many natural enemies of this weed. McFadyen [42] identified 207 insect and 2 mite species that attacked the weed in America. Of these, about a quarter is specific to Chromolaena [43]. The moth *Parachutes pseudoinsulata* was the first agent to be deliberately introduced, from Trinidad to many countries. In the early 1980s, it was reported established in Sri Lanka and in the mid 1980s it had been established in Sabah Malaysia, but apparently did not causing much damage to the weed [6]. Introduction after mid 1980s resulted in establishment in some countries like Guam [44], Indonesia [6] and Ghana [45]. Attempts to introduce and establish *Parachutes pseudoinsulata* in some countries like Nigeria, South Africa, India Thailand and Ivory Coast were failed. The failures mostly caused by predatory pressures [5]. The moth was introduced into Indonesia in 1992 and has been established successfully in North Sumatera, but in Java and other Indonesia islands, it has not been reported to establish [20,6].

The second insect most widely introduced is gall fly (*Cecidochares connexa*). This fly was successfully reduced growth of the weed in Guam [44] and in India [46]. In 1995 the fly was released in Indonesia where it readily established and now it has established on all the major islands of Indonesia [47]. In Timor Leste, the gall fly established at most of the release sites and caused a visible reduction in plant height and density, however, control of the weed by the gall fly is limited by severe dry season and the widespread use of fire in clearing land for agriculture [8]. In South Sulawesi, the gall fly doesn't give good control. Although the fly is well established, *Chromolaena* still growing well.

5. CONCLUSION

The huge amounts of light and persistent seeds of Chromolaena making the weed can be dispersed easily far from its native places. In grassland areas, seeds germination and establishment are enhanced by open soil conditions during the wet season. The main negative impact of the weed in grassland ecosystem is reduced forage production for livestock. Although Chromolaena spread in grassland area has positive impacts, however, its negative impacts are far exceeding its positive impacts. Control of the weed is very difficult and even after many years of research, problems of getting effective and cheap control remains unsolved.

COMPETING INTERESTS

Author has declared that no competing interests exist.

REFERENCES

1. Polley HW, Wisley BJ, Derner JD. Do species evenness and plant density influence the magnitude of selection and complementary effects in annual plant mixtures? Ecology Letters. 2003;6:248–257.
2. Ossom E, Lupupa B, Mhlongo S, Khumalo L. Implication of weed control methods on Sandanezwe (*Chromolaena odorata*) in Swaziland. World. J. Agric. Sci. 2007;3(6): 704–713.
3. McFadyen RE, Skarrat. B. Potential distribution of *Chromolaena odorata* (Siam Weed) in Australia, Africa and Oceania. Agric. Ecosyst. Environ. 1996;59:89–96.
4. McFadyen REC. *Chromolaena odorata* in South East Asia and the Pacific. In: Agriculture, New Directions for New Nation. Ed.by H.da Costa, C. Piggin, C.J.da Cruz, J.J. Fox. ACIAR Proceedings. 2003, No. 113.
5. Muniappan R, Bamba J. Biological control of *Chromolaena odorata*: Successes and failures. In: Proceeding of the X International Symposium on Biological Control of Weeds 4 – 14 July 1944. Montana State University, Montana, USA; 2000.
6. CrutwellMcFadyen RE. *Chromolaena* in Asia and the Pacific: Spreads continues but control prospects improve. Proceeding of the Fifth International Workshop on Biological Control and management of *Chromolaena odorata*, Durban, South Africa, 23 – 15 October 2000. Zachariades, C., R. Muniappan and L.W. Strathie (eds). APC-PPRI. 2002;13–18.
7. Sipayung A, Desmier de Chenon R, Sudharto PS. Observation on *Chromolaena odorata* (L.) RM King and H Robinson in Indonesia. Biotrop Spec. Publicaion. 1991;44:43–49.
8. Day MD, Brito AA, da Costa Gutteres, da Costa Alves AP, Paul T, Wilson CG. Biocontrol of *Chromolaena odorata* in Timor Leste. Proceedings of 8th International Workshop on Biological Control and Management of *Chromolaena odorata* and other Eupatorieae, Nairobi, Kenya, 1 – 2 November 2010. Zachariades C, Strathie L.W., M.D. Day, R. Muniappan (eds) ARC-PPRI, Pretoria. 2010;134 -140.
9. Binggeli P. *Chromolaena odorata* (L.) King and H. Robinson (Asteraceae); 1999. Available:http://members.tripodco.uk/Woody Plant Ecology/docs/web-sp4htm
10. Witowski EFT, Wilson M. Changes in density, biomass, seed production and soil seed banks for the non-native invasive plant, *Chromolaena odorata*, along a 15 year chronosequence. Plant Ecol. 2001; 152:13–27.
11. Zachariades C, Day M, Muniappan R. Reddy GVP. *Chromolaena odorata* (L.) King and H. Robinson (Asteraceae). In: Biological Control of Tropical Weeds using Arthropods. Ed. R. Muniappan, GVP Reddy and A. Raman. Published by Cambridge University Press; 2009.
12. Rambuda TD, Johnson SD. Breeding system of alien plants in South Africa: Does Baker's rule apply?. Diversity and Distribution. 2004;10:409–416.
13. Weerakoon L. Studies on biology and control of *Eupatorium odoratum* (L.). M.Sc. Thesis, Vidyodaya Campus, University of Ceylon; 1972.
14. Wilson M. Autecology of invasive alien plant, *Chromolaena odorata* in the Greater St. Lucia Wetland Park. M.Sc. Thesis. University of Witwatersrand, Johannesburg; 1995.
15. Ambika SR. Ecological adaptation of *Chromolaena odorata* (L.) King and Robinson. Department of Botany, Bangalore University, India. Available:www.ehs.cdu.edu.au/chromolaena/proceedings/fourth/ambika1.html (Accessed on October 22, 2014)
16. Ambika SR, Poornima. Allelochemicals from *Chromolaena odorata*(L.) King and Robinson for increasing crop productivity. In: Chromolaena in the Asia Pacific region. Ed by Technical Reports No. 55. (Printed version published in 2004).
17. Goodal JM, Zacharias PJK. Managing *Chromolaena odorata* in subtropical grasslands in Kwazulu-Natal, South Africa.

Proceedings the Fifth International Workshop of Biological Control and management of *Chromolaena odorata*, Durban, South Africa, 23 – 15 October, 2000. Zachariades C, Muniappan R, Strathie LW (eds). 2002;120–127.

18. Anonymous. *Chromolaena odorata* (Chromolaena). Available:www.keys.lucidcentralorg/keys/v3/eafrinet/weeds/key/weeds/Media/htm (Accessed on March 24, 2014)
19. Wilson CG, Widiyanto EB. The biological control programme against *Chromolaena odorata* in eastern Indonesia. In: Proceeding of the fifth International Workshop on Biological Control and Management of *Chromolaena odorata*, Durban, South Africa, October 2000, eds. Costas Zachariades, Muniappan R, Strathie LW. 2002;34–39.
20. Tjitrosemito S. Integrated management of *Chromolaena odorata* emphasizing the classical biological control. Biotropia. 1998; 11:9–21.
21. Zachariades C, Goodal JM. Distribution, impact and management of *Chromolaena odorata* in southern Africa. In: Proceeding of the Fifth International Workshop on Biological Control of *Chromolaena odorata*, Durban, South Africa, October, 2002. Eds. Zachariades C, Muniappan R, Strathie LW. 2002;34–39.
22. Pancho JV, Plucknett DL. *Chromolaena odorata* (L.) R.M. King and H. Robinson, - A new record of a noxious weed in the Philippnes. Philippine Journal of Animal Science. 1971;8:143–149.
23. Sajise PE, Palis PK, Lales NV, Lales JS. Flowering behavior, pattern of growth and nitrate metabolism of *Chromolaena odorata*. Phil. Weed Sci. Bull. 1974;1:17–24.
24. Parsons WT, Cuthbertson EG. Noxious weed of Australia. CSIRO Publishing, Australia; 2001.
25. Boopre M. A non-nutritional relationship of *Zonocerus* (Orthopera) to *Chromolaena* (Asteraceae) and general implications for weed management. Ecology of *Chromolaena odorata*. Biotrop Special Publication. 1991;44:153–161.
26. Ojeniyi SO, Odenina SA, Agdebe TM. Soil productivity improving attributes of Mexican sunflower (*Tithonia diversifolia*) and siam weed (*Chromolaena odorata*). Emirates J. Food. Agric. 2012;24(3):243–247.
27. Tondoh JE, Kone AW, N'Dri JK, Tamene L, Brunet D. Changes in soil quality after subsequent of *Chromolaena odorata* fallows in humid savannah, Ivory Coast. Catena. 2013;101:99–107.
28. Manjappa K, Jowkin V, Channabasappa KS, Kelaginamani SV. Use of eupatorium as surface mulch in cashew nut plantations. Karanataka. J. Agric. Sci. 2010;23(2):337–338.
29. Are LA, Folarin JO. Cultural weed control experiment. Ann. Rep. Cocoa Res. Inst. Nigeria. 1970;19:68–69.
30. Rusdy M, Sjahril R, Riadi M, Budiman B. Integration of mechanical and cultural control to manage invasive shrub *Chromolaena odorata* and other weeds under drought conditions in pasture area. Indonesian J. Indonesian Trop. Anim. Agric. 2013;38(1):65–71.
31. Roux P. Use of chemicals for the control of *Chromolaena odorata* in South Africa. Natal Parks Board, Pietrmaritzburg; 2014.
32. Wessels MF. The role of fire and mechanical clearing in the management of *Chromoaena odorata* Master Thesis, School of Biological and Conservation Sciences. Faculty of Science and Agriculture, University of Kwazulu-Natal, South Africa; 2014.
33. Wu R, Xu X. Cultural control of feijicao (*Chromolaena odorata* (L.) R.M. King and H. Robinson by planting signalgrass (*Brachiaria decumbens* Stapf) in southern Yunnan, The People's Republic of China. J. Biotrop. Special Publication. 1991;44: 83–89.
34. Castillo AC, Moog FA. Introduction of ipil-ipil in "goony" infested pastures. Philippines J. Animal Industry. 1977;32:1–10.
35. Torres DO, Paller EC. The devil weed (*Chromolaena odorata* (L.) R. King and H. Robinson) and its management. SEAWIC Weed Leaflet. 1989;4:1–5.
36. Abraham T, Thomas CG, Joseph PA. Herbicides for control of *Chromolaena odorata*. Available:www.ehs.cdu.edu.au/choromolaena /proceedings/ fourth/abra.htm (Accessed on December 2014)
37. Erasmus DJ, Van Staden J. Screening for candidate herbicides in field trials for the

chemical control of *Chromolaena odorata*. S. Afr. J. Plant Soil. 1986;3(2):66–70.
38. Van Staden J. Chemical control of *Chromoaena odorata*: Efficacy of triclopyr and glyphosate applied to regrowth. Applied Plant Science. 1987;1(1):39-42.
39. Rusdy M. Integrating mechanical and chemical control treatments to manage invasive weed *Chromolaena* odorata in grassland area. Am. J. Exp. Agric. 2015; 6(3):134–139.
40. Utulu SN. Controlling regrowth of *Chrmolaena odorata* (L.) King and H. Robinson, using herbicide mixtures in young oil palm plantation in Nigeria. Nigerian Institute for Oil Palm Research, Benin City, Nigeria. Available:www.chs.cdu.au/chromolaena/proceedings/third/3ut.html (Accessed on April 25, 2014)
41. Tjitrosemito S, Sastroutomo S, Utomo IH. Weed management in young rubber plantation in Indonesia. Weed Watcher. 1986;1(supplement):4.
42. McFadyen C. Phytophagous insect recorded from *Chromolaen odorata*. Chromolaena Newsletter. 1988;2:5–33.
43. Waterhouse DF. Biological control of weeds. Southeast Asian Prospects. Australian Centre for International Centre for International Agricultural Research. Canberra. 1994;302.
44. Reddy GVP, Kikuchi RS, Muniappan R. The impact of *Cecidochares connexa* on *Chromolaena odorata* in Guam. Proceedings of the Eight International Workshop and Management of *Chromolaena odorata* and other Eupatoriae, Nairobi, Kenya. 1 -2 November 2010. Zachariades C, Strathie LW, Day MD, Muniappan R (eds) ARC-PPRI, Pretoria. 2003;128–133.
45. Uyi US, Wilson DD, Zachariades C. The impact of *Pareuchaetes pseudoinsulata* (Lepidoptera: Arctidae) on the growth rate of *Chromolaena odorata* (L.) in southern Ghana. J. Appl. Environ. Manage. 2009; 13(1):59–65.
46. Bhumannavar BS. Ramani S. Introduction of *Cecidochares connexa* (Macquat) (Diptera: Tephritidae) into India for the biological control of *Chromolaena odorata*. Available:www.cabi.org/isc/FullTextPdf/2012/20123159075.pdf (Accessed on February 25, 2015)
47. Day MD, McFadyen R. *Chromolaena odorata* (L.) King and Robinson – chromolaena – In: Julien MH, McFadyen REC, Cullen JM (eds). Biological control of weeds in Australia, CSIRO Publishing, Cellingwood, Australia. 2012;162–169.

26

The Study on Some of Ecological Specifications of *Gentiana olivieri* a Medicinal Species in Rangelands of Fars Province, Iran

Damoun Razmjoue[1*] and Zahra Zarei[2]

[1]Department of Range and Watershed Management, Faculty of Natural Resources, Behbahan Khatam Alanbia University of Technology, Behbahan, Iran.
[2]The Range management, Behbahan Khatam Alanbia University of Technology, Behbahan, Iran.

Authors' contributions

This work was developed in collaboration by the both authors, who contributed equally to the literature review and writing of the manuscript. Both authors read and approved the final manuscript.

Editor(s):
(1) Małgorzata Pietrowska-Borek, Department of Biochemistry and Biotechnology, Poznań University of Life Sciences, Poznań, Poland.
Reviewers:
(1) Sholeh Ghollasimod, Natural resources and Environment, University of Birjand, Iran.
(2) Anonymous, Kocaeli University, Turkey.

ABSTRACT

According to the importance of medicinal plants and the necessity of comprehensive understanding them for exploitaion; some ecological aspects of *Gentiana olivieri* plant that is from Gentianaceae family were reviewed. This study was done in 5 counties that were as the habitat of this plant. In first stage in each site physiognomy units were separated, and in each unit, one reagent zone was specified; and in each site, by random systematic method, 4 transect with 100 m length and along each of transect, 5 plots 2×2 were selected. And the data related to vegetation and soil were gathered. The obtained results showed that the factors such as: vegetation percent, density, abundance and the height of plant in different sites are affected by the factors related to soil. As the areas that had the most percent of sand and minimum percent silt, clay and EC had the minimum vegetation percent; and the areas having the most amount of K,OC and EC, had the most plant height and the areas having the maximum clay amount and pH had the minimum density and abundance and plant height. Meanwhile, the habitats of this plant were from 1600 to 2000 m from

**Corresponding author: Email: d.razmjoue@gmail.com, d.razmjoue@bkatu.ac.ir*

sea level and were as the relatively hillside lands with the slop less that 20 percent and in all geographical directions.

Keywords: Ecological; medicinal plant; Gentiana olivieri; Fars province; rangelands.

1. INTRODUCTION

The plants as one of the main components of ecosystem, has a significant role in the life of other living things and protection and balance of ecosystem. According to the significance of plant species in rangelands, it is necessary to proceed for understanding the ecological specification of them [1]. The study on ecological specification of rangeland species, provides the necessary knowledge for proper usage of this species in modifying of rangeland ecosystem and its scientific management [2]. A key prerequisite of a rangeland management, is understanding the ecological relationship between its components, when in an environment a species plant grows, it affects on the soil specifications and also soil will affect on vegetation. The obtained results of studies have showed that soil specifications are as one of the effective factors on vegetation [3]. Many studies have done in the field of reviewing the interaction of vegetation and soil up to now. Kansaran et al. [4] by reviewing the ecological specifications of *Erisimum amasianum* species found that presence of this species causes increase of nitrogen, potassium and calcium in the soil. According to the obtained results of study on the relation between soil specifications with vegetation factors done by Akbarlou et al. [5] it was found that the percent of vegetation canopy and species diversity in comparison with the changes of soil factors is more affected by the changes in the soil factors. And the amount of clay percent and electrical conduction are as the effective factors on the vegetation factors. Moradi et al. [6] to study on some of the ecological specifications of plant communities of Baghe-Shad found that scattering and establishing the plant communities are related with the depth and texture of soil. And the other ecological factors such as pH, phosphorus, potassium, nitrogen and the percent of organic materials has less importance in dispersion of region plant communities. Jafari et al. [7] to study on the relation of vegetation with some of soil specifications in Nadooshan Rngeland of Yazd province got to this result that there is an especial relation between vegetation dispersal and soil specification; as the soil texture, gypsum, potassium salts, lime and electrical conduction has a significant effect on regional vegetation. Fu et al. [8] in a study on relation between soil specifications and plant diversity in China found that the organic materials of soil and the total amount of nitrogen, have the most effect on the dispersion of plants communities. Medicinal plant of *Gentiana olivieri* form Gentianaceae family has dispersion in Turkey, Iran and Iraq. This plant is as a perennial herbaceous plant, having blue- violet flowers, oval and brown seeds, with the length of 0.8 to 1 mm, Stems erect, slender, simple, glabrous. Inflorescences terminal corymbose cymes of 3-5 flowers. Stamens inserted just below middle of corolla tube; filaments 6-8 mm; anthers linear, 2.5-3 mm Testa is thin reticulate and grows in calcareous soils, clay, maren and the rangeland covered by grass and in height of 350- 2300 m. Height of this plant is 10-30 cm from a basal rosette. Root stock is sheated with a fibrous collar at apex [9]. This plant because of having flavonoids and alkaloid compounds, is used for diabetes, depression, dyspepsia, gastrointestinal diseases and colds. Also is useful as a analgesic and anti-inflammatory [10,11,12]. According to the importance of *Gentiana olivieri* plant and as the analysis of vegetation and soil status provide the most data for classification of habitat and availability of food, the main purpose of this study is studing on some ecological factors of this species and determining the most important effective soil factors on the mentioned species so that, by cognition of the relation and generalize the obtained results in similar areas, by spending minimum time and cost, propose the managerial strategies to improve and development of rangelands.

2. MATERIALS AND METHODS

2.1 Introduction of Studied Sites

- Khoh Pardis site of Laar county

This site is located 50 km from this county between longitude of 53°33 ′ 17 ″ to 53°22 ′ 56 ″and latitude of 27°29 ′ 19 ″ to 27°25 ′ 32 ″ ; the average temperature is between 25-30°C and rainfall mean is annual 80-210 mm (Fig. 1).

Fig. 1. Study sites of *Gentiana olivieri* in Fars province

- The site of Abadeh-Tashk county

This site is located 100 km from Neyriz county that is between longitude of 53°45′ 58″ to 53°39′ 48″ and latitude of 29°49′ 22″ to 29°45′ 17″ and the mean temperature is between 20-25 ^{0}C and rainfall average is annual 250- 350 mm (Fig. 1).

- The site of Jahroom county

This site is located 160 km from south of Shiraz that is between longitude of 53° and 32° eastern and latitude of 28° and 29° northern. Rainfall average is 230 mm and the annual mean temperature is 23°C (Fig. 1).

- The site of Darab county

This site is located 180 km from Shiraz that is between longitude 33° and 54′ eastern, and latitude of 28° and 45′ northern; that the rainfall average in this site is 180 mm and the annual mean temperature is 19°C (Fig. 1).

- The Site of Fasa county

This site is located 145 km from south of Shiraz that is between longitude 53° and 39′ eastern and latitude 28° and 55′ northern. The annual rainfall average of this site is about 300 mm and annual average temperature is 18.5°C (Fig. 1).

2.2 Vegetation Sampling

In 2013 by desert reviews and librarian studies, the main habitats of mentioned plant were identified; and by using aerial photographs scale of 1: 20000 and topographic map with a scale of 1: 50000, it was preceded to separate the physiognomy units.

In each physiognomy unit the reagent zone was selected and in each site by using random systematic method, 4 transects with 100 m length (according to the distribution, density and vegetation) and along each transect, 5 plots of 2×2 S^2 were selected. The size of sampling unit in each type by minimum area method and also the number of plot by statistical method was specified. After installing transects and plots, the data related to vegetation include percentage of vegetation canopy, density, abundant, height of plant (In the months of May and June) and plant phenology (In all months) were gathered and measured.

2.3 Soil Sampling

In relation to the Soil (agrology) studies, according to the habitat characteristics in terms of size and homogeneity in geology and topography, in order to investigate the characteristics of the soil 3 profiles along transects were dug in each site at the foot of *Gentiana olivieri* (totally 15 profiles in two depth). Likewise in each profile two depth, one affected by the organic matter (10-0 cm) and the other not affected by the organic matter (50-10 cm) were taken.

From each level and each sample approximately two kilograms were harvested and to evaluate the physical and chemical properties such as texture (clay, silt and sand) with a hydrometer, organic carbon (OC) using black linen and nitrogen (N) using crude, phosphorus (P) by Olson and colleagues, and potassium (K) measured by flame (flame photometry), PH and EC Soil were transferred to the laboratory.

Soil samples transferred to the laboratory were flattened on thick papers and rough parts, pebbles, roots, leaves and other inseparable organic residues were separated. After drying the samples in air, we ground them in a mortar and screened them in two mm sieve John.

3. RESULTS

3.1 Study on Vegetation

According to the obtained results of floristic study, 160 plant species were identified in 5 considered sites that the most species are belonging to families (Laminaceae, Papilionaceae, Poaceae) in sites of Abadeh Tashk and laar and (Astreaceae, Poaceae, Apiaceae) were found in Darab, Jahroom and Fasa sites.

3.2 Phenology Study on *Gentiana olivieri* Plant

The study on phenologic stages showed that the start time of this species growh according to the climatic conditions in all sites is mid February and early March and in late April, vegetative growth completely terminate according to the climate and physiographic specifications of the zone.

From early May, flowering stage of this plant starts and in late June ends. Its seeding stage starts completely from late July and ends to the late August; and then aerial parts of plant become dry and from early October is seen as a dried and inactive plant (Fig. 2).

3.3 Vegetation Specifications

To determine the difference between vegetation specifications of table analysis, the one side variance was formed; and it was determined that the percent of vegetation in level of %5 and density and plant height in level of %1 have significant difference (Table 1).

By using Duncan multi range comparison test (Table 2) it is seen that the most vegetation percent is related to Abadeh Tashk site and their minimum is related to Laar site , although there is not any significant difference between these four sites (Laar, Jahroom, Darab, Fasa) (Fig. 3).

In terms of density, the minimum is belongs to two sites of Darab and Laar and the most of them is related to Fasa an Abadeh Tashk (Fig. 4). In terms of abundance factor, the maximum abundance is related to Fasa, and the minimum is related to Darab (Fig. 5). And finally Jahroom and Abadeh Tashk allocated the most rates of plant height to themselves; while the minimum plant height was seen in Darab (Fig. 6).

3.4 Study on Soil Specifications

General comparison of variance analysis table of sites in both deeps of (1-10 and 10 -50) shows that only in K amount, there is a significant difference in level of %5 (Table 3).

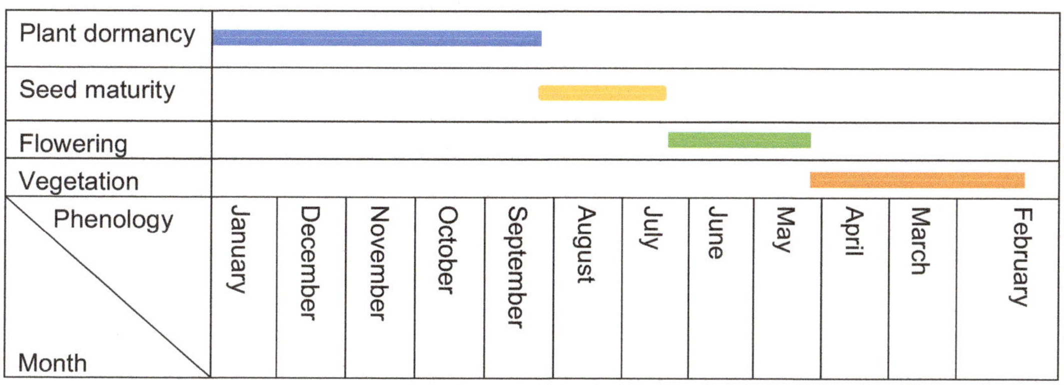

Fig. 2. Phenology stages of *Gentiana olivieri* in Fars province

Table 1. The results of variance analysis and significant levels of sites in vegetation specifications

	Significance level	F amount	Mean squares	Freedom degree	Sum of squares		
*	0.049	2.59	53.148	4	212.592	Sight	Vegetation percent (%)
			20.549	47	965.814	error	
				51	1178.406	total	
**	0.000	7.854	3.179	4	12.716	Sight	Density
			0.405	46	18.618	error	
				50	31.333	total	
ns	0.164	1.668	0.41	4	1.64	Sight	Abundance
			0.246	95	23.35	error	
				99	24.99	total	
**	0.000	11.895	664.326	4	2657.304	Sight	Plant height (cm)
			55.849	42	2345.675	error	
				46	5002.979	total	

*In each column, the averages having common letters don't have significant difference with each other in level of 5 percent. * Significant at the 5% level; ** Significant at the 1% level; NS- Not significant*

Table 2. The comparison results of vegetation specifications average in different sites by Duncan test in level of 5 %

Height (cm)		Frequency		Density		Cover (%)		Site
B	32.40	AB	0.50	B	1.80	B	0.140	Lar
A	48.90	AB	0.55	A	2.55	B	0.460	Jahrom
B	28.50	B	0.30	B	1.33	B	0.147	Darab
B	34.00	A	0.70	A	2.71	B	0.164	Fasa
A	45.13	AB	0.50	A	2.70	A	3.531	Abade Tashk

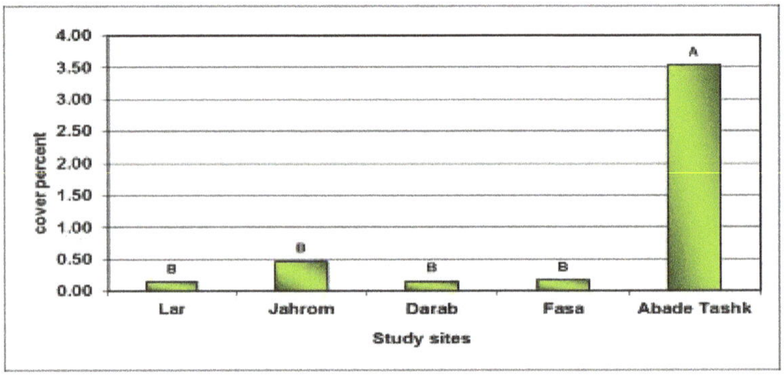

Fig. 3. Cover percent of *Gentiana olivieri* in studied sites

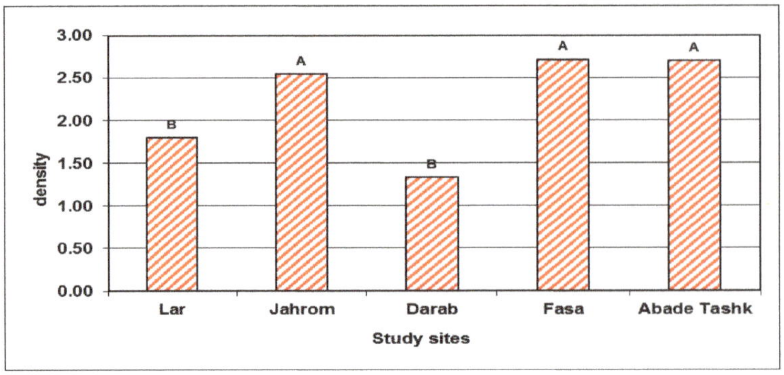

Fig. 4. Density of *Gentiana olivieri* in studied sites

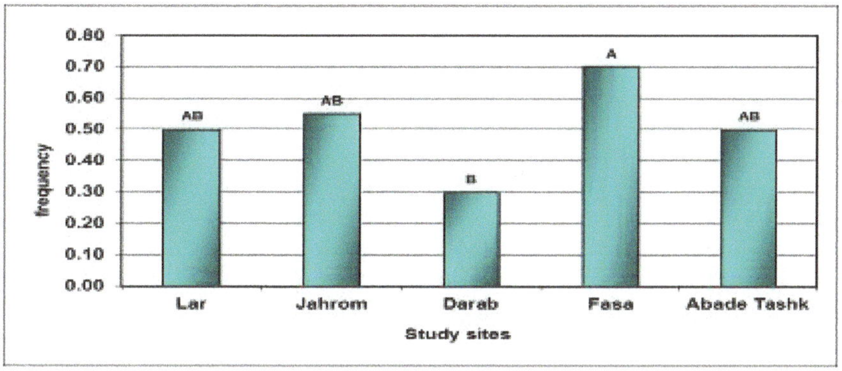

Fig. 5. Abundance of *Gentiana olivieri* in studied sites

Table 3. The variance analysis results of pedological specifications in 0-10 (cm) and 10-50 (cm) deeps

	Significant level	Famount	Average of squares	Free degree	Sum of squares		
ns	0.229	1.513	0.234	4	0.934	Site	EC
			0.154	25	3.858	Error	
				29	4.792	Total	
ns	0.105	2.142	0.258	4	1.033	Site	pH
			0.121	25	3.014	Error	
				29	4.048	Total	
ns	0.251	1.438	0.751	4	3.003	Site	OC
			0.522	25	13.05	Error	
				29	16.053	Total	
ns	0.594	0.708	0.009	4	0.034	Site	N1
			0.012	25	0.3	Error	
				29	0.334	Total	
ns	0.255	1.426	36.533	4	146.133	Site	P1
			25.628	25	640.708	Error	
				29	786.842	Total	
*	0.05	2.763	13417.8	4	53671.2	Site	K1
			4855.92	25	121398	Error	
				29	175069.2	Total	
ns	0.858	0.325	11.617	4	46.467	Site	Clay
			35.72	25	893	Error	
				29	939.467	Total	
ns	0.104	2.15	77.783	4	311.133	Site	Silt
			36.173	25	904.333	Error	
				29	1215.467	Total	
ns	0.583	0.726	57.167	4	228.667	Site	Sand
			78.78	25	1969.5	Error	
				29	2198.167	Total	

** Significant at the 5% level; ** Significant at the 1% level; ns Not significant*

Also it can be concluded from comparison of sites average (Table 4) related to both deeps that the most amount of OC, EC and K has seen in Jahroom, while the most percent of clay and pH is related to Darab site. Laar also has the most sand percent amount and the minimum percent amount of silt, clay and EC. The difference of sand, clay and silt percent is probably related to different textures of studied sites (Fig. 7).

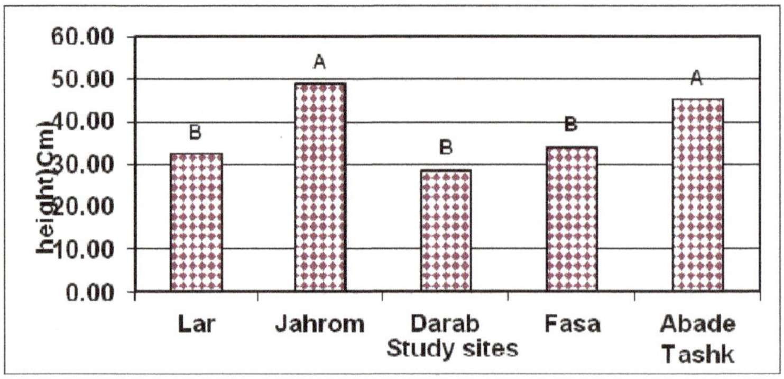

Fig. 6. The height of *Gentiana olivieri* in studied sites

Table 4. The comparison results of pedological specifications average in 0-10 (cm) and 10-50 (cm)

Site	EC		pH		OC		N		P		K		Clay		Silt		Sand	
Lar	0.585	B	7.833	AB	0.675	A	0.082	A	2.583	A	84.500	B	22.833	A	34.167	A	41.500	A
Jahroom	1.107	A	7.800	AB	1.510	A	0.143	A	8.417	A	192.670	A	25.833	A	41.500	A	35.667	A
Darab	0.783	AB	7.997	A	1.443	A	0.108	A	7.583	A	147.170	AB	26.000	A	37.333	A	37.333	A
Fasa	0.680	AB	7.900	A	1.250	A	0.182	A	7.250	A	150.170	AB	23.667	A	41.667	A	41.333	A
Abadeh Tashk	0.828	AB	7.450	B	0.927	A	0.123	A	8.583	A	82.500	B	24.000	A	34.667	A	35.000	A

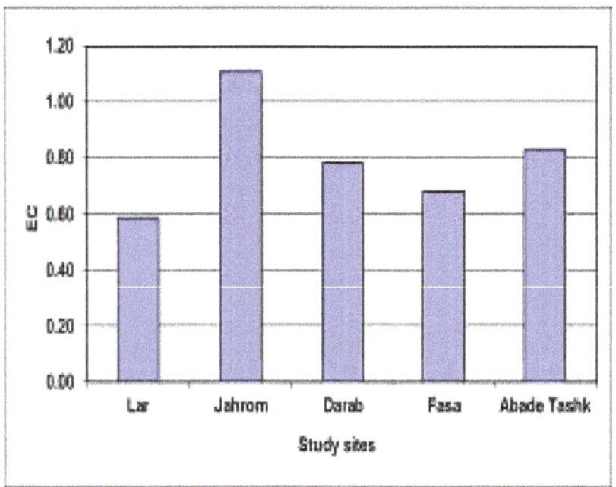

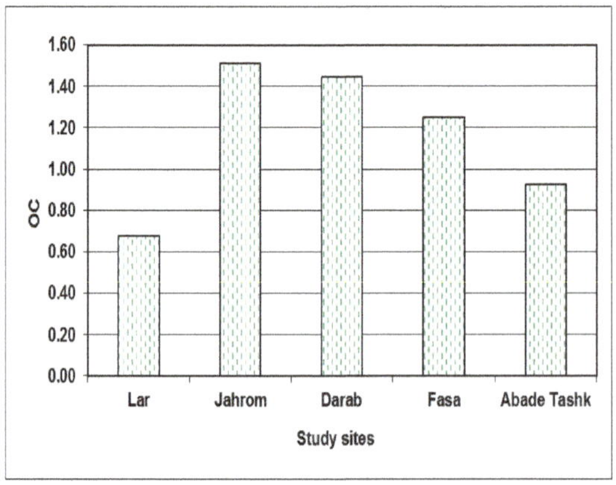

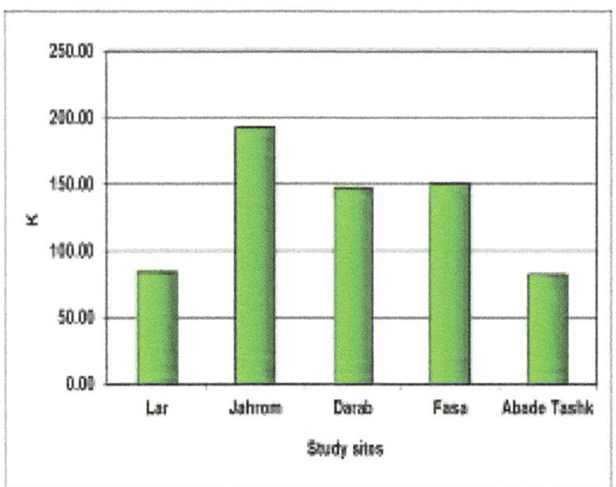

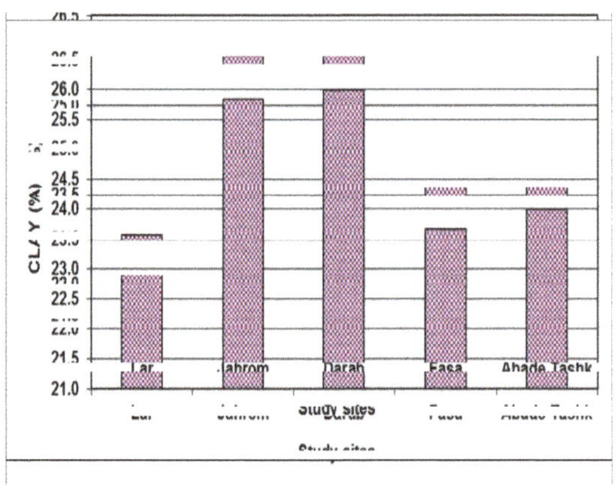

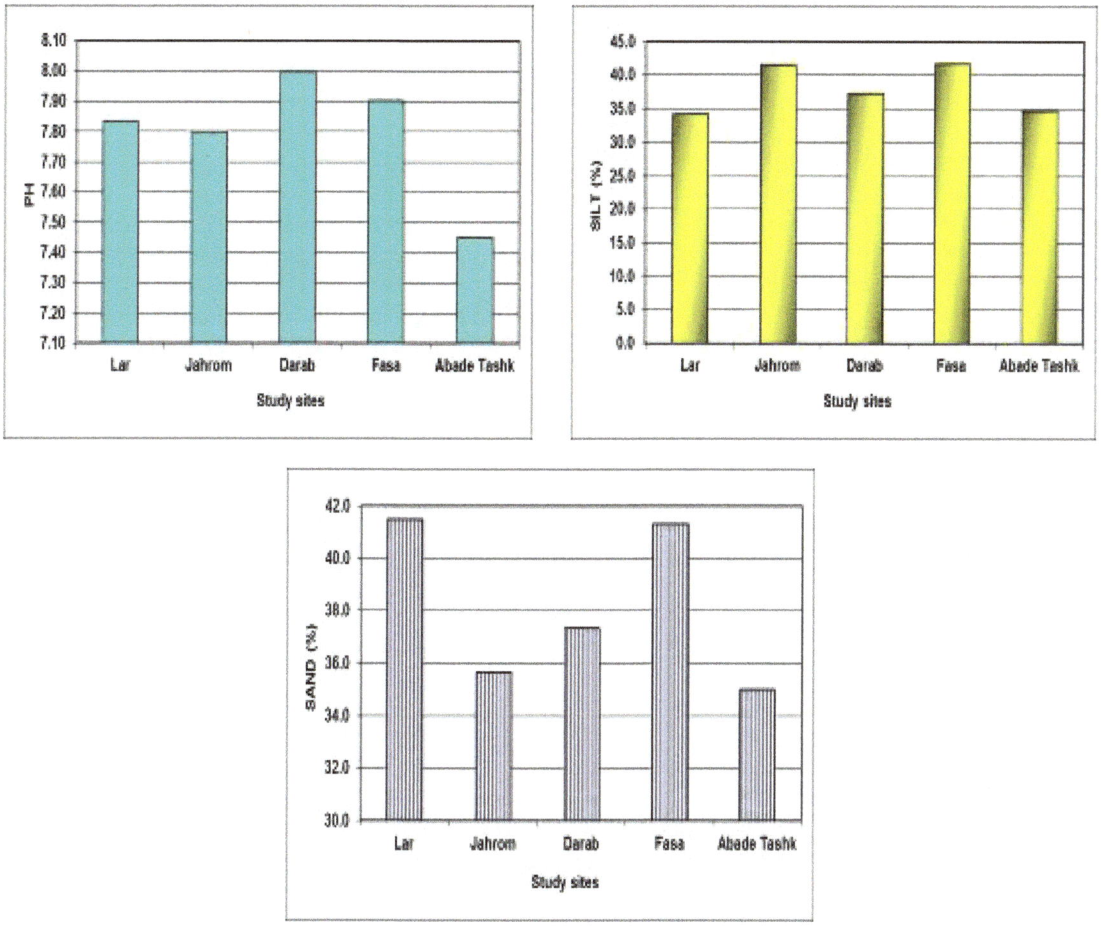

Fig. 7. The amount of soil components in studied sites

4. DISCUSSION AND CONCLUSION

Gentiana olivieri species in rangelands of Fars province have height range of 1600-2000 m. The long term rainfall average of this habitat is annual about 230 mm and annual average temperature of these zones is 20°C.

This species is commonly seen in flat zones toward hillsides and in low slops. Phenologic process of *Gentiana olivieri* plant starts always from mid February and to late April its vegetative growth ends.

From early May the plant flowering stage starts and to late July ends; and then the plant starts its seeding stage and after producing seed, it is seen as a dried plant. Some differences in the phenology time of this plant in studied sites were seen but this difference is negligible and is not significant; so this principle confirms the Ahrar's theory [13].

According to the obtained results of botany field studies of this research, in each site about 50 species and totally 160 species were identified. It is worth mentioning that there is remarkable number of another species in the ecological habitat of *Gentiana olivieri* species; and it is hoped that this ecosystems are sustainable; and is as principle of competitive exclusion or ball [14].

According to obtained results of studies on the soils of studied zone it can be concluded that the zones with most percent sand and minimum silt, clay and EC, has the minimum percent of vegetation; this results are comply with the results of study done by Akbarlou et al. [5] that said the percent of vegetation canopy and species diversity are affected by changes of soil factors such as amount and percent of clay and electronic conductivity.

The plants in the regions having the most amount of K, OC and EC, have the maximum plant height and the regions having the most amounts of clay and pH, have the minimum density and abundance and plant height. That the obtained results of this study in quite comply with the results of studies done by Jafari et al [7] and Moradi et al. [6] and Fu et al. [8] that knew soil chemical and physical factors as the factor of plants dispersion and also considered it effective on vegetation factors. This mentioned items totally are caused by climatic status and are more affected by soils of studied regions [15].

Najafi [16] in a study on the relation of ecological factors with dispersion of plant communities of protected Geno zone, has mentioned organic carbon as one of the effective factors on separation of plants communities. In another study, Boer and Sargeant [17] also in the east of Saudi Arabia proved the relation between vegetation and texture; in the mentioned study, this matter is proved as well.

Anyway, in general, this study showed that soil factors is as one of the effective ecological factors on dispersion of this plant in 5 considered sites; and because *Gentiana olivieri* plant is as one of the medicinal plant of this regions, so, by recognition of other regions soils, some effective steps can be taken toward modification and development of this plant.

COMPETING INTERESTS

Authors have declared that no competing interests exist.

REFERENCES

1. Ahmadi A, Shahmardi A. At ecology of species *Agropyron cristatum* in Western Azarbijan province. Iran Natural Resource Journal. 2005;58(3):691-701.
2. Azarnivand H, Dastmalchi VH. Phenology of four species of Kashan Desert regions plants. Biaban Jounal. 2005;2(5):15-25.
3. Mesdaghi M. Description and analysis of vegetation (first print). Mashhad Jahad Daneshgahi. 2001.
4. Kansaran A, Ergenakcin O, Kandemir N. A study on the morphology, anatomy and autecology of *Erysimum amasianum* Hausskn. & Born. (Brassicaceae) distributed in centeral black sea region (Amasya- Turkey). International Journal of science & technology. 2007;2(1):13-24.
5. Akbarlou M, Yar S, Mohammadesmaeili M. Study on the relation of some physical and chemical specifications of soil with factors of vegetation (Case Study: Ghareh Tapeh Saveh). Pazhoheshhaye Hefazate Aab va Khak. 2012;19(2).
6. Moradi H, Asri Y, Kashi Pazha AH. Study on some of plant society ecological specification of Baghe Shad Zone. 2008;2(3):225-236.
7. Jafari M, Javadi SA, Bagherpourzarchii MA, Tahmoreth M. The study on vegetation relation with some of soil specifications in rangelands of Nadooshan of Yazd province. Rangeland. 2009;3(1):29-40.
8. Fu BJ, Liu SL, Ma KM, Zhu YG. Relationships between soil characteristic, topography and plant diversity in a heterogeneous broad-leaved forest near Beijing China. J. Plant and Soil. 2003;261: 47-54.
9. Davis PH. Edinburgh university press. Flora of Turkey and east Eagean islands. 1987;311.
10. Mansoor A. Pakistan: PhD. Thesis, Institute of Biochemistry, University of Balochistan, Quetta. Entomological and biochemical studies on the etiology of malaria, malaria studies – I; 1996.
11. Aslan M. Turkey: PhD Thesis, Faculty of Pharmacy, Gazi University, Ankara, seker hastaligina karsi halk ilaci olarak kullanilan bitkiler uzerinde farmakognozik arastirmalar; 2000.
12. Orhan DD, Aslan M, Aktay G, Ergun E, Yesilada E, Ergun F. Evaluation of hepatoprotective effect of *Gentiana olivieri* herbs on subacute adminstration and isolation of active principle. Life Sci. 2003; 72:2273-83.
13. Ahrar M. The Plant of *Prangos ferulaceae* and its importance in breeding. Reports Series of Second Conference of Country Reference Expert, Shiraz. Iran Forest and Rangeland and Watershed Management Organization; 2012.
14. Vahabzadeh E. The foundation of environment (Translation) first print. Atrack Mashhad Publications. 2010;145.

15. Arzani H. Forage quality and daily requirement of grazing livestock from Rangeland. Tehran University Publications. 2009;354.

16. Najafitirehshabankareh K, Jalilii A, Khorasanni N, Jamzad Z, Asrii Y. Research- science quarterly of Tahghighate Martaa va Biabane Iran. 2008;15(2):179-199.

17. Boer B, Sargeant D. Desert perennials as plant and soil indicator in Eastern Arabia. Plant Soil. 1998;199:261-266.

27

Effect of Solid Waste Slurry from Biogas Plant on Soil Parameters and Yield of Spinach (*Spinacia oleracea L.*)

Md. Rabiul Islam[1], Md. Nur Hossain[2*], Md. Fakruddin[2], Khusnade Rabbi[3] and Md. Abdul Baten[1]

[1]Department of Environmental Science, Bangladesh Agricultural University, Mymensingh-2202, Bangladesh.
[2]Institute of Food Science and Technology (IFST), BCSIR, Dhaka-1205, Bangladesh.
[3]Ulipur Government College, Ulipur, Kurigram, Bangladesh.

Authors' contributions

This work was carried out in collaboration between all authors. Authors MNH, MF and MNH designed the study, wrote the protocol and wrote the first draft of the manuscript. Authors MRI, MNH and KR performed the experiments and analyzed the data. Authors MF and MAB managed the literature searches and discuss the conclusion. All authors read and approved the final manuscript.

Editor(s):
(1) Chandra Shekhar Kapoor, Department of Environmental Sciences, University College of Science, Mohan Lal Sukhadia University, India.
Reviewers:
(1) Petigrosso, Lucas Ricardo, National University of Mar del Plata, Argentina.
(2) E. U. Eteng, Michael Okpara University of Agriculture, Nigeria.
(3) Raúl Leonel Grijalva Contreras, Instituto Nacional de Investigaciones Forestales Agrícolas y Pecuarias, Mexico.

ABSTRACT

Aims: To determine the effect of Solid Waste (MSW) slurry from biogas plant on soil micro-parameters and yield of spinach.
Study Design: A completely randomized design was used in this study.
Place and Duration of Study: The study was conducted at the field laboratory of Bangladesh Institute of Nuclear Agriculture, Mymensingh during Rabi-1 season (October-December, 2012).
Methodology: Different combination of MSW and regular fertilizer was applied to experimental fields and effect of MSW on soil micro-parameters and yield of spinach was determined.

**Corresponding author: Email: fakruddinmurad@gmail.com*

Results: Application of MSW slurry showed significant effect on soil moisture, organic matter, total nitrogen, available phosphorus, exchangeable potassium and available sulphur in post plant soil except soil pH over control. For ensuring of organic farming, improving soil parameters and soil health, 20 ton/ha slurry + 50% of the recommended fertilizer dose was found the best. Application of MSW slurry had significant positive effect on plant height, leaves per plant and individual plant weight over control. Rate of photosynthesis did not vary significantly due to the application of MSW slurry.

Conclusion: MSW slurry showed a positive impact on growth and yield contributing characters of spinach and soil microenvironment which help improving the fertility status of soil.

Keywords: B solid waste; biogas; soil; spinach; yield.

1. INTRODUCTION

Bangladesh is a highly populated developing country with 0.15 acres lands per capita [1]. Her agriculture plays a vital role in national economy as about 85% of the total population depends directly or indirectly on agriculture for their livelihood. Rapid growth of urbanization and uncontrolled urban expansion, severely degrade environment and consequently undetermined equitable and sustainable development [2]. With the rapid urbanization and industrialization in our country, the average production of waste is increasing day by day. Being a densely populated country, huge amount of wastes are produced in Bangladesh every day especially solid wastes in urban areas. About 1.5 ton per day and about 2 kg per household solid waste were generated in Dhaka [3]. Islam et al. [4] found that the average waste from household waste, kitchen waste and ashes were 0.517, 0.055 and 1.290 ton/year respectively in Mymensingh. The situation in other cities or towns prevails more or less same.

Solid waste management (SWM), is a important environmental health service, and is an integral part of basic urban services. From the earliest primitive human society there have been attempts to safely dispose of solid waste [5]. In the early days, disposal did not pose difficulty as habitations were sparse and land was plentiful. Disposal became problematic with the rise of towns and cities where large numbers of people started to congregate in relatively small areas in pursuit of livelihoods. On one hand, the density of population increased in these centers of congregation and therefore wastes generated per unit area also increased [6]. On the other hand, available land for disposal of waste decreased in proportion. SWM thus emerged as an essential, specialized sector for keeping cities healthy and livable [7]. Karim [8] more than 50% of cultivated soils of Bangladesh have organic matter content below the critical level (1.5%). Deficiencies of sulphur in almost 9.92 million acres (45%) and zinc in 4.96 million acres (23%) of arable lands that limits to attain the expected yield [9]. In addition to that about 100-200 kg/acres of nutrients are removed each year through nutrient mining by the intensive rice cultivation. Most of the soils of Bangladesh have low organic matter content, usually less than 2% [10]. Under such concerns, new agronomic and soil amendment practices are necessary for sustainable agricultural development [11]. Municipal waste based organic farming is expected to be such an effort of the environment friendly agricultural production system as it conserves natural resources and does not cause the pollution of the atmosphere and water bodies [12].

MSW slurry is an anaerobically digested organic material released as byproduct from this biogas plant. The percentage composition of MSW combined from all locations is about 74.4% organic matter, 9.1% paper, 3.5% plastic, 1.9% textile and wood, 0.8% leather and rubber, 1.5% metal, 0.8% glass and 8% other waste. The biodegradable fraction (organic matter) is normally very high as compared to other fractions, due to the use of fresh vegetables and foods, which is common in each city [13]. The biodegradable part of MSW can use as feedstock in biogas plant. It reduces the amount of waste as well as produce biogas and slurry. About 25 to 30% of organic matter is converted into biogas during the anaerobic fermentation process, while the rest becomes available as manure (bio-slurry). This residual manure is normally rich in macro and micro nutrients and is considered as a quality organic fertilizer [14]. The slurry can be used as organic fertilizer most suitable for organic farming of some high value field and horticultural crops including vegetables, fruits, flowers as well as ornamental plants. Beside this yield responses of vegetable crops to bio-slurry manure application have been reported in different crops including okra, maize and cabbage and organic manure can serve as a substitute to mineral fertilizers [15]. Manures supply the required nutrients, improve soil structure, increase microbial population and at the same time maintain the quality of crop produce [16-18].

Spinach *(Spinacia oleracea* L.) a member of the family Chenopodiaceae is an important vegetable crop in the world. Spinach is a rich vegetable from the nutritional point of view, containing appreciable amount of vitamins and minerals. The edible part contains high amount of P-carotene, calcium and iron [19]. The chief nutritive value of spinach lies in their content of P-carotene (Precursor of vitamin A) and vitamin C. It contains carotene (9300 I.U) vitamin C (43 mg), calcium (274 mg), carbohydrate (5.0 g), protein (3.4 g), fat (0.7 g) and calories (30 Kcal.) per 100 g of edible portion [12]. According to the nutritionists, 3% of the total food calories come from vegetables. The availability of vegetables per capita in Bangladesh is only 53 g/day whereas the Japanese consume 425 g/day on an average [4]. Vegetable crops response more to organic fertilizer than any other crops. Spinach is a leafy vegetable and organic matter has profound effect on its quality. Beside this, organically produced crops and fruits are healthy and nutritious, and have better shelf life as well as higher market values. Demand for organically produced crops are increasing day by day in Bangladesh and elsewhere in the world [20].

However, there is very little information available on effect of MSW slurry on soil microenvironment and quality of spinach in the world. In fact, organic farming is not well known in Bangladesh. This study was designed to determine the effect of MSW slurry from biogas plant on the growth and yield of spinach plant and soil micro environment.

2. MATERIALS AND METHODS

The experiment was carried out during *Rabi-1* season from November-December of 2012 at the Field Laboratory of Bangladesh Institute of Nuclear Agriculture (BINA), Mymensingh. The materials used and methods followed in the whole experiment are divided into the following:

2.1 Experimental Site

The experiment was set up at the Field Laboratory of Bangladesh Institute of Nuclear Agriculture (BINA), Mymensingh is situated at 20°28' N latitude and 88°44' longitudes.

2.2 Soil

The soil of the experimental field was medium high, non-calcarious; dark gray flood plain with loamy texture belongs to the Agro Ecological Zone (AEZ) - Old Brahmaputra flood plain. The morphological, physical and chemical characteristics of the soil are shown in the Tables1 and 2.

Table 1. Morphological characteristics of pre plant soil

Location	BINA farm, Mymensingh
General soil type	Old Brahmaputra flood plain
Parent material	Brahmaputra river borne deposits
Soil series	Sonatala series
Vegetation	Cropped with paddy, wheat, pulse etc.
Flood level	Above flood level

2.3 Climate

The climate of the experimental area is characterized by high temperature, humidity and heavy rainfall with occasional gusty winds during *kharif* season (March-September) and scanty rainfall associated with moderately low temperature during the *Rabi* season (October-February).

2.4 Land Preparation

The land was first opened on November 2012 by repeated plowing with a power tiller followed by spading. Laddering which helped breaking clods and leveling the land followed by every plowing. Weeds, stabbles, and crop residues were removed from the land.

2.5 Spinach Crop

"Copi palong" a high yielding variety of spinach was used as test crop in the experiment. The average yield of the variety generally lies between 8-10 t ha^{-1} [12].

2.6 Layout of the Experiment

The experiment was laid out in Randomized Complete Block Design (RCBD), with three replications. Thus total number of unit plots was 15. The unit plot size was 2m × 2m and plots were separated from each other by drains (50 cm). Treatments were randomly distributed within the blocks.

2.7 Treatments

The treatments used for the experiment were as follows-

- T_0 = No slurry (Control)
- T_1 = 10 ton/ha slurry
- T_2 = 20 ton/ha slurry
- T_3 = 10 ton/ha + 50% of the recommended fertilizers
- T_4 = 20 ton/ha + 50% of the recommended fertilizers

2.8 Manures and Fertilizer Application

Nitrogen (N) (100 kg ha^{-1} from urea), phosphorus (P) (14 kg ha^{-1} from triple super phosphate (TSP)), potassium (K) (30.00 kg ha^{-1} from muriate of potash (MP)), sulphur (S) (15 kg ha^{-1} from gypsum) [28] were applied in the treatment (T_0) plot and 50% of all fertilizers were applied in the treatments (T_3 & T_4). Half (50%) of urea and MP, and rest of all were applied as basal dose one day before sowing and rest 50% urea and 50% MP were applied in the plots 21 days after seed sowing (DAS). The MSW slurry (wet basis) was incorporated in the plots as per treatments at 3 days before sowing and after two days they were mixed thoroughly with the soil.

2.9 Seed Sowing

Before sowing seeds were soaked in water for 18 hours. For sowing small (2.5 cm deep) holes were made at requisite distance and 3 seeds were placed in each hole which was then filled with soil. Three seeds per hole were planted in order to ensure uniform stand of the crops in 15th November, 2012. Row to row distance was 20 cm and plant to plant distance was 15 cm. Immediately after sowing watering was done twice a day for ensuring uniform germination.

2.10 Intercultural Operations

Thinning and first weeding was done 21 days after sowing. After thinning one seedling per hill was retained. The growing points of the seedlings to be removed were pinched off. Uprooting was not done since there were possibilities of injuring the adjacent seedlings that were left behind. There were many kinds of weeds but Durba grass (*Cynodon dactylon*) was most troublesome. The plants were irrigated by watering-can initially and as the plants grew older, sprinkler irrigation were given whenever required. During early stage, the rate of growth was rather slow, probably due to infestation of weeds and slow release of nutrient. There was no attack of insects and diseases. Top dressing of urea and muriate of potash was done in the plots after three weeks of seed sowing which was followed by irrigation.

2.11 Harvesting Green Crop (Foliage)

Harvesting was done one time. Harvesting was done at 38 DAS, by uprooting the whole plants. The soil adhering to the roots could be easily thrown away by jerking. Weight of the sample plants and total plants with attached roots were taken from each unit plot immediately after harvesting.

Table 2. Physical and chemical characteristics of pre plant soil

Characteristics	Value	Reference
pH	6.42	Source: Agriversity
Sand (%)	60.00	Humbolt soil laboratory,
Silt (%)	32.00	Department of Soil
Clay (%)	8.00	science, Bangladesh
Textural class	Sandy loam	Agricultural University,
Soil moisture	23.32	Mymensingh-2202,
Total nitrogen (%)	0.11	Bangladesh.
Organic carbon (%)	0.68	
Available phosphorus (ppm)	12	
Exchangeable potassium (m.e. %)	0.10	
Non- Exchangeable potassium (m.e. %)	0.61	
Available sulphur (ppm)	19.00	
Available zinc (ppm)	2.40	

2.12 Data Collection

Data on following parameters were collected and recorded.

2.12.1 Collection and preparation of soil sample

After crop harvest, soil samples were collected from each plot at 0-15 cm depth. After getting weeds and other unwanted things removed, the samples were air dried ground, sieved through a 2 mm sieve and kept for analysis.

2.12.2 Soil moisture

The moisture of soil was determined by gravimetric method and was calculated using following formula [21].

$$\text{Sample moisture Sample moisture (Soil moisture \%)} = \frac{(W - W1)}{W1} \times 100 \text{ (weight basis)}$$

Where,

W = Weight of moist sample in grams
W_1= Weight of air dry sample in grams

2.12.3 Soil Ph

Soil pH was measured with the help of a glass electrode pH meter using soil water suspension of 1:2.5 as described by Black [22].

2.12.4 Organic carbon

Organic carbon was determined by wet oxidation method as described by Black [22] and the organic matter content was calculated by multiplying the percent organic carbon with the Van Bemmelen factor of 1.73 [23].

2.12.5 Total nitrogen

Total nitrogen in soil and treatments were determined by micro-kjeldahl method. Digestion was made with H_2O_2, concentrated H_2SO_4 and catalyst mixture (K_2SO_4:$CuSO_4.5H_2O$: Se in the ratio of 100:10:1). Nitrogen in the digest was estimated by distillation with 40% NaOH followed by titration of the distillate trapped in $H_3BO_3$with 0.01N H_2SO_4 [24].

2.12.6 Available phosphorus

Available phosphorus was extracted from the soil and treatments by shaking with 0.5 M NaHCO3 solutions at pH 8.5 following the method of [25]. The extracted phosphorus was determined by developing blue color by $SnCl_2$ reduction of phosphomolybdate complex and measuring the intensity of color colorimetrically at 660 nm wave length and calibrating the reading to the standard P curve.

2.12.7 Exchangeable potassium

Exchangeable K was extracted with 1.0 N NH40Ac (pH-7) and determined from the extract by flame photometer [22] and calibrated with a standard K curve.

2.12.8 Available sulphur

Available sulphur was determined by extracting the soil and treated samples with CaCl2 solution (0.15%). The extraction method was described [23]. The S content in the extract was estimated turbidimetrically with spectrophotometer at 420 nm [23].

2.13 Physical Quality Analysis and Yield of Spinach

2.13.1 Rate of photosynthesis

To determine the rate of photosynthesis 5 plants were randomly selected from each plot. The rate of photosynthesis was measured by using Portable Photosynthesis System (Model: Li 6400 XT) in μ mol co_2 m^{-2} s^{-1}.

2.13.2 Yield and yield component

To evaluate the effect of MSW slurry on quality of spinach 10 plants were selected at random from each unit plot at harvesting. The following physical quality related characters were studied using 10 plants in each unit of plot and yield per plot (kg $plot^{-1}$) and total yield (t ha^{-1}) was recorded.

2.13.3 Statistical analysis

The collected data were compiled and tabulated in proper form and were subjected to statistical analysis by F-test to examine the treatment effects. The analysis of variance was done following the computer package programme MSTAT. The mean differences were adjudged by Duncan's Multiple Range Test (DMRT) [26] and ranking was indicated by letters.

3. RESULTS AND DISCUSSION

3.1 Soil Parameters

Application of slurry manures caused decreasing effect on the pH of the post-harvest soils (Table 3). All the treatments except T_1 (10 ton/ha slurry) slightly decreased the soil pH as compared to pre-plant soil. The decreasing effect was higher where (T_4) 20 ton/ha slurry + 50% of the recommended fertilizer treatment was applied. This might be due to organic acids released from organic matter through decomposition leading to a decreasing effect on soil pH. The pH value of post-harvest soils ranged from 6.37-6.16 and varied significantly. Yadav et al. [27] observed that the pH of the soil decreased significantly with the addition of organic manure and Gypsum. Similar results have been reported by More [28] and Mian and Eaqub [9].

Application of organic manures and chemical fertilizers caused an increasing effect on soil moisture percentage (Table 3). All the treatments increased the soil moisture as compared to pre-plant soil except T_1 and T_3 (10 ton/ha + 50% of the recommended fertilizers). The increasing effect was higher where 20 ton/ha slurry + 50% of the recommended fertilizers was applied. The ranking of soil moisture was T_4>T_2 (20 ton/ha slurry)>T_3>T_0 (no slurry)>T_1. This might be due to bulky organic matter conserving more water. The soil moisture of post-harvest soils ranged from 22.37 to 25.25% and varied significantly. Sharma et al. [29] observed that integrated application of farmyard manure (FYM) and chemical fertilizer significantly improve water holding capacity of soil. Malewar et al. [30] reported that FYM improved soil physical properties.

The organic matter content of the post-harvest soil was increased in all treated samples compared to the initial soil (Table 3). Soil organic matter content due to treatments ranged from 1.16 - 1.27% after harvest of crops and the treatment T_2 (20 ton/ha slurry) recorded the highest value. Many investigators found that application of organic manures increased the organic matter content in the soil with the chemical caused a decreasing effect [31]. Organic carbon also increased due to application of organic manure as reported by Mathew and Nair [32], Ahmad et al. [33] and Salakinkop and Hunshal [34].

The total N content of post-harvest soils varied significantly among treatments (Table 3). The highest N content was 0.169% obtained in treatment T_4 where 20 ton/ha slurry + 50% of the recommended fertilizers were applied. The lowest value (0.090%) was found in control (T_0) treatment where fertilizers were not used. The data indicated that slurry showed better performance in maintaining total N in soil compared to inorganic source of fertilizer. Application of organic manure increased the total N concentration in soil as reported by Wong et al. [16], Mathew and Nair [32] and Abdel and Hussein [31].

Available P content of post-harvest soil was influenced by different treatments (Table 3). Available P content varied from 10.11 to 19.63 ppm. The P content in pre-plant soil was 12.00 ppm. The highest and lowest available P content were recorded in the treatments T_2 (20 ton/ha slurry) and T_3 (10 ton/ha slurry + 50% of the recommended fertilizers). Soils treated with 20 ton/ha slurry gave higher value of available P compared to other treatments. The release of more available P from the decomposition of slurry might cause higher value in soils treated with slurry manure. Application of organic manures increased the available P in soil as reported by Mathew and Nair [32], Abdel and Hussein [31], Iftikhar and Qusim [20] and Rosen et al. [35].

The highest exchangeable K content in post-harvest soil was obtained in T_4 treatment (0.25 me/100 g soil) and the lowest value (0.14b me/100 g soil) was found in control (T_0) treatment (Table 3). All the treatments significantly increased K content in soil than the initial value (0.10 me/100 g soil). Results indicate that exchangeable K content was higher in soils treated with MSW slurry manures than treated with chemical fertilizer and control. Abdel and Hussain [31] showed that a significant effect of organic manures in improving the exchangeable K in soil. Similar results were also reported by Wong et al. [16], Mathew and Nair [32] and Rosen et al. [35].

Sulphur content in soil was significantly influenced by treatments after harvest of spinach (Table 3). The pre-plant sulphur was 19 ppm but after application of organic manures the highest sulphur content was recorded in treatment T_4 (30.38 ppm). The lowest value of sulphur (16.00 ppm) was in the control. The sulphur content in the post-harvest soils was higher in soils treated with MSW slurry compared to the soils treated with the chemical fertilizer. Smith and Hughes

[36] found that application of organic fertilizers increased available sulphur content in soil compared to that of inorganic fertilizer.

The MSW slurry did not affect soil temperature significantly (Table 3). The maximum soil temperature was measured in T_1 treatment where slurry was applied as 10 ton/ha and minimum in T_0 treatment. The decreasing order of soil temperature at 30 DAS was $T_1>T_2>T_3>T_4>T_0$. On the contrary, Smith and Hughes [36] measured highest temperature (69℃) in compost made from market and garden refuse whereas compost made purely from garden refuse the highest temperature was 53℃. Giri and Singh [37] observed soil temperature under straw mulch was lower at 2-4 pm.

3.2 Physical Quality of Spinach (Yield Component and Yield)

3.2.1 Plant height

Result from Table 4 indicated that the plant height significantly differed among treatments. The highest plant height (27.31 cm) was observed in treatment T_4 and the lowest (23.30 cm) was found in T_0 (control) treatment. This result was possibly due to higher rate of N levels in organic manures. Budhan et al. [38] also reported that application of cattle manure resulted in increase of plant height.

3.2.2 Leaf number

Result from Table 4 showed that leaf number varied among different treatments. The number of leaves ranged from 8.00 to 9.89. The highest number of leaves was obtained in treatment T_4 (20 ton/ha slurry + 50% of the recommended fertilizers) and lowest in the control. Possible reason behind this result is MSW slurry added higher phosphorus, potassium content in soil than conventional farming. Kant and Kumar [39] reported that organic manures resulted in increase of number of tiller of rice.

3.2.3 Rate of photosynthesis

The MSW slurry did not affect rate of photosynthesis significantly (Table 4). The maximum rate of photosynthesis was measured in T_0 treatment (36.12 μmol CO_2 m^{-2} s^{-1}) and minimum in T_4 treatment (31.09 μmol CO_2 $m^{-2}s^{-1}$). The decreasing order of rate of photosynthesis at 30 DAS was $T_0>T_1>T_3>T_2>T_4$.

3.2.4 Individual plant weight (g)

Result from Table 4 showed that individual plant weight varied among different treatments. The highest individual plant weight was obtained in treatment T_2 (54.44 g) and the lowest in the treatment T_3 (41.11 g). Similarly total dry matter of many field crops improved due to application of MSW [40]. Possible reasons behind this result is MSW slurry added higher phosphorus, potassium content in soil than conventional farming.

3.2.5 Total yield

There was significant effect on total yield (t/ha) of spinach (Table 4). Maximum yield (9.85 t/ha) was recorded in 10 ton/ha slurry + 50% of the recommended fertilizers treated treatment (T_3) and minimum (7.00 t/ha) was in T_1 treatment.

Many previous studies have showed potential of biogas slurry to improve soil quality and crop yield. Yu et al. [41] reported that biogas slurry is cheap source of plant nutrients and offer extra benefits to soil fertility and fruit quality of tomato as well as tomato cultivation. Islam et al. [42] reported that biogas slurry improved production and quality of maize fodder. Wu et al. [43] reported that application of biogas slurry to oil-seed rape increased protein, iron, manganese, copper and zinc content of the rapeseed and yield of the rape plant. LiNa et al. [44] reported that application of biogas slurry to rice field resulted similar or slightly higher yield compared to chemical fertilizers. TongGuo et al. [45] reported that application of biogas slurry can effectively promote the yield of green pepper, tomato and cucumber and change the qualities of vegetables. Treatment of seed with biogas slurry can significantly increase germination, root length and seedling vigour index of sesamum, sunflower and safflower crops [46]. Improvement of yield and crop quality of many other crop types with the application of biogas slurry has also been reported [2,47,48].

Table 3. Soil pH, moisture, organic matter, N, P, K and S content of post-harvest soil as influenced by different treatments

Treatments	Soil pH	Soil moisture (%)	Organic matter (%)	Total N (%) in soil	Available P (ppm) in soil	Exchangeable K (me /100 g) in soil	Available S (ppm) in soil	Soil temperature (℃)
T_0	6.37a	22.62bc	1.20	0.090b	15.42c	0.14b	16.00d	12.25
T_1	6.45a	22.37c	1.25	0.130ab	14.62c	0.16b	18.57c	12.59
T_2	6.20b	23.76b	1.27	0.156a	19.63a	0.21a	27.17b	12.46
T_3	6.25b	23.15bc	1.16	0.143ab	10.11d	0.21a	27.42b	12.45
T_4	6.16b	25.25a	1.19	0.169a	18.12b	0.25a	30.38a	12.36
LSD	0.165	0.762	0.119	0.0028	0.540	0.011	0.675	0.236
CV (%)	3.16	2.99	8.82	3.50	3.19	11.54	3.6	3.74
Pre plant data	6.42	23.32	0.68	0.11	12.00	0.10	19.00	-

Figure(s) in the column having common letter(s) do not differ significantly at 5% level or at 1% level; T_0 = No slurry, T_1 = 10 ton/ha slurry, T_2 = 20 ton/ha slurry, T_3 = 10 ton/ha slurry + 50% of the recommended fertilizers, T_4: = 20 ton/ha slurry + 50% of the recommended fertilizers

Table 4. Growth and yield component of spinach at 38 DAS as influenced by different treatments

Treatments	Plant height (cm)	No. of leaves/ plant	Rate of photosynthesis (μmol co_2 m^{-2} s^{-1})	Individual plant weight(gm)	Yield / plot (kg)	Total yield (t / ha)
T_0	23.30b	8.00c	36.12	45.56c	2.90c	7.25c
T_1	23.91b	8.45bc	35.89	47.78b	2.73c	7.00c
T_2	24.37b	9.50ab	35.24	54.44a	3.50b	8.75ab
T_3	26.86a	9.72a	35.44	41.11d	3.94a	9.85a
T_4	27.31a	9.89a	31.09	42.22d	3.70ab	8.31bc
LSD	0.623	0.687	2.803	0.901	0.197	0.862
CV(%)	3.28	6.93	7.42	3.79	5.38	9.63

Figure(s) in the column having common letter(s) do not differ significantly at 5% level or at 1% level; T_0 = No slurry, T_1 = 10 ton/ha slurry, T_2 = 20 ton/ha slurry, T_3 = 10 ton/ha slurry + 50% of the recommended fertilizers, T_4: = 20 ton/ha slurry + 50% of the recommended fertilizers

4. CONCLUSION

MSW slurry from biogas plant showed positive impact on growth and yield contributing characters of spinach and soil parameters and organic matter status as well as improving the fertility status of soil. Use of MSW as bio-fertilizer in agriculture will offer an alternative to manage these MSW which may pose health hazard and to convert them into value added products. Though further research is needed to make bio-fertilizer from MSW, it has the advantages of being less expensive, non-hazardous and eco-friendly and it can contribute to better agriculture production in Bangladesh.

COMPETING INTERESTS

Authors have declared that no competing interests exist.

REFERENCES

1. Economic Census. BBS (Bangladesh Bureau of Statistics), Statistics Division, Ministry of Planning, Dhaka, Government of Bangladesh; 2012.
2. Arthurson V. Closing the Global Energy and Nutrient Cycles through Application of Biogas residue to Agricultural Land–Potential Benefits and Drawbacks. Energies. 2009;2:226-242.
3. Halder PK, Paul N, Hoque ME, Hoque ASM, Parvez MS, Rahman MH, Ali M. Municipal solid waste and its management in Rajshahi City, Bangladesh: A Source of Energy. International Journal of Renewable Energy Research. 2014;4(1): 168-175.
4. Islam MR, Abedin MZ, Rahman MZ, Begum A. Use of selected wastes as sustainable agricultural inputs. Progressive Agriculture. 2009;20(1&2):201-206.
5. Drozd J. The risk and benefits associated with utilizing composts from municipal solid waste (MSW) in Agriculture. In Lynch JM, Schepers JS, Unver I. (eds.) Innovative soil-plant systems for sustainable agricultural practices. Organisation for Economic Co-operation and Development (OECD), Paris, France. 2003;211-226.
6. Ghaly AE, Alkoaik FN. Effect of Municipal Solid Waste Compost on the Growth and Production of Vegetable Crops. American Journal of Agriculture and Biological Sciences. 2010;5(3):274-281.
7. Ahmed SA, Ali M. Partnership for solid waste management in developing countries: Linking theories to realities. Habitat International. 2004;28:467-479.
8. Karim Z. Fertilizer recommendation guide. The Bangladesh Agricultural Research Council, New Airport Road, Dhaka. Soils Publication. No. 41; 1997.
9. Mian MJA, Eaqub M. Effect of intensive fertilization and manuring on some chemical properties of soils. Bangladesh Journal of Agricultural Sciences. 1980; 8(2):165-173.
10. Rahman MM, Sampa MSA. Combined effects of bradyrihizobial strains, municipal solid waste compost and fertilizers on nodulation, n content and uptake of soybean. Journal of Environmental Science and Natural Resources. 2012; 5(2):85-90.
11. Salakinkop SR, Shivaprasad P, Raguramulu Y, Naidu R. Composting of

coffee pulp. Journal of Coffee Research. 2003;31:153-160.
12. Hossain N, Islam M, Alamgir M, Kibria MG. Growth response of Indian spinach to biogas plant residues. IOSR Journal of Pharmacy and Biological Sciences. 2014; 9(2):1-6.
13. Cherif H, Ayaria F, Ouzaria H, Marzoratib M, Brusettib L, Hassena A et al. Effects of municipal solid waste compost, farmyard manure and chemical fertilizers on wheat growth, soil composition and soil bacterial characteristics under Tunisian arid climate. European Journal of Soil Biology. 2009; 45:138-145.
14. Garg RN, Pathak H, Das DK, Tomar RK. Use of flyash and biogas slurry for improving wheat yield and physical properties of soil. Environmental Monitoring and Assessment. 2005;107:1-9.
15. Dauda SN, Ajayi FA, Ndor E. Growth and yield of water melon (Citrulluslanatus) as affected by poultry manure application. Journal of Agriculture, Forestry and the Social Sciences. 2008;4:121-124.
16. Wong JWC, Ma KK, Fang KM, Cheung C. Utilization of manure compost for organic farming in Hong Kong. Bio-resource Technology. 1999;67:6-43.
17. Nehra AS, Hooda IS, Singh KP. Effect of integrated nutrient management on growth and yield of wheat (*Triticurn aestivum* L.). Indian Journal of Agronomy. 2001;45:112-17.
18. Suresh KD, Sneh G, Krishna KK, Mool CM. Microbial biomass carbon and microbial activities of soils receiving chemical fertilizers and organic amendments. Archives of Agronomy and Soil Science. 2004;50:7-641.
19. Furukawa Y, Hasegawa H. Response of spinach and komatsuna to biogas effluent made from source-separated kitchen garbage. Journal of Environmental Quality. 2006;35:1939-1947.
20. Iftikhar A, Qusim M. Influence of various potting media on growth and nutrient uptake efficiency of Scindapsus aureus. International Journal of Agriculture & Biology. 2003; 5: 594–597.
21. Okalebo JR, Gathua KW, Woomer PL. Laboratory methods of soil and plant analysis: A working manual. (2nd Ed.). TSBF-CIAT and Sacred Africa, Nairobi; 2002.
22. Black CA. Methods of soil analysis Part II and I. American Society of Agronomy Incorporation Publication, Madison, Wisconsin, USA; 1965.
23. Page AL, Miller RH, Keney DR. Methods of Soil Analysis. Part II, 2nd Edn. American Society of Agronomy Incorporation Madison. Wiscomin, USA; 1989.
24. Mahmoodabadi M, Amirabadi Z, Amini S, Khazaeipoul K. Fertilization of soybean plants with municipal solid waste compost under leaching and non-leaching conditions. American-Eurasian Journal of Agriculture & Environmental Sciences. 2010;8(1):55-59.
25. Olsen SR, Cole CV, Wataneble FS, Dean LA. Estimation of available phosphorus in soils by extraction with sodium biocarbonate. U.S. Department of Agriculture Circular. 1994;207-215.
26. Gomez KA, Gomez AK. Statistical procedures for Agricultural Research. 2nd edn. John Wiley and Sons. New York, 1984;207-215.
27. Yadav AC, Sharma SK, Batra BR. Effect of sodic water, FYM and gypsum on the soil, growth and yield of brinjal. Annals of Agriculture & Biological Research. 2002; 7(1):73-77.
28. More SD. Effect of farm waste and organic manure on soil properties, nutrient availability and yield of rice, wheat grain on sodicvertisol. Journal of Indian Society of Soil Science. 1994;42(2):253-256.
29. Sharma MP, Bali SV, Gupta DK. Crop yield and properties of Inceptisol as influenced by residue management under rice-wheat cropping sequence. Journal of Indian Society of Soil Science. 2000;48(3):506-509.
30. Malewar GU, Badole SB, Ismail S. Influence of fly ash with and without FYM and fertilizer on physico-chemical properties of sunflower and cotton growing soils. Annals of Agriculture Research. 2000;21(2):187-191.
31. Abdel NG, Hussein AHA. Effect of different manure sources on some soil properties and sunflower plant growth-Soil Physical and chemical properties. Alexandria Journal of Agricultural Research. 2010; 46(1):227-251.
32. Mathew T, Nair SS. Physical and chemical properties of rice soils as influenced by organic and inorganic sources of nutrients. Indian Journal of Agricultural Research. 1997;31(4):257-261.
33. Ahmad B, Hassan S, Bakhsh K. Factors Affecting Yield and Profitability of Carrot in

Two Districts of Punjab. International Journal of Agriculture and Biology. 2005; 7(5):794-798.
34. Salakinkop SR, Hunshal CS. Studies on production and characterization of enriched urban waste composts and their influence on crops productivity. Journal of Environmental Science & Engineering. 2008;50:83-88.
35. Rosen CJ, Bierman PM, Olson D. Swiss chard and alfalfa responses to soils amended with municipal solid waste incinerator ash: Growth and chemical elemental composition. Journal of Agricultural and Food Chemistry. 1994; 42(6):1361-1368.
36. Smith DC, Hughes JC. Changes in chemical properties and temperature during the degradation of organic wastes subjected to simple composting protocols suitable for small scale farming and quality of the mature compost. South African Journal of Plant and Soil. 2002;19(2):53-60.
37. Giri G, Singh RR. Influenced of straw mulch and transpiration suppressants on soil temperature, dry matter and total biomass production and nutrient uptake by dry land wheat. Indian Journal of Agricultural Sciences. 1985;55(4):256-261.
38. Budhan MN, Palaniappan SP, Rangassamy A. Effects of farm waste and green manure on low land rice. Indian Journal of Agronomy. 1991;36(2):251-252.
39. Kant S, Kumar R. A comparative study on the effect of four soil amendments on the uptake of Fe, Mn and Yield of rice in salt affected soil. Indian Journal of Agricultural Chemistry. 1994;27(283):59-70.
40. Salakinkop SR, Hunshal CS. Influence of long term utilization of domestic sewage on performance of wheat and health of farmers. Journal of Ecobiology. 2008;22: 45-51.
41. Yu FB, Luo XP, Song CF, Zhang MX, Shan SD. Concentrated biogas slurry enhanced soil fertility and tomato quality. Acta Agriculturae Scandinavica. 2010; 60(3):262-268.
42. Islam MR, Rahman SME, Rahman MM, OH DH, Ra CS. The effects of biogas slurry on the production and quality of maize fodder. Turkish Journal of Agriculture and Forestry. 2010;34:91-99.
43. Wu J, Yang Q, Yang G, Shen F, Zhang X-H, Zhang Y-Z. Effects of Biogas Slurry on Yield and Quality of Oil-Seed Rape. Journal of Plant Nutrition. 2013; 36(13):2084-2098.
44. LiNa J, Qiang W, DingJiang C, AiFen L, ShiMing Q, JianRong F, JiaMei W. Effects of paddy field disposal of biogas slurry on rice production, soil quality and environmental safety. Journal of Agro-Environment Science. 2011;30(7):1328-1336.
45. TongGuo G, Nan C, WeiQun L, BaoZhen L, HongLi Y. Effect of highly efficient nutrient solution of biogas slurry on yield and quality of vegetables. Journal of Agricultural Science & Technology-Hunan. 2011;12(4):567-570.
46. Shakuntala NM, Jagadeesh KS. Influence of Biogas spent slurry on seed quality parameters of sasamum, sunflower and safflower crops. Karnataka Journal of Agricultural Sciences. 1996;9(1):67-72.
47. Gurung JB. Review of Literature on Effect of Slurry Use on Crop Production. Biogas Support Programme. 102 p. BSP Lib Temp No. 20; 1997.
48. Ji-Dong W, Yun C, Zhi-zhou Z, Hong-Bo MA. Effects of combined application of biogas slurry with chemical fertilizers on fruit qualities of *Prunus persica* L. and soil nitrogen accumulation risk. Plant Nutrition and Fertilizer Science. 2013;19(2):379-386.

Permissions

All chapters in this book were first published in JAERI, by SCIENCE DOMAIN international; hereby published with permission under the Creative Commons Attribution License or equivalent. Every chapter published in this book has been scrutinized by our experts. Their significance has been extensively debated. The topics covered herein carry significant findings which will fuel the growth of the discipline. They may even be implemented as practical applications or may be referred to as a beginning point for another development.

The contributors of this book come from diverse backgrounds, making this book a truly international effort. This book will bring forth new frontiers with its revolutionizing research information and detailed analysis of the nascent developments around the world.

We would like to thank all the contributing authors for lending their expertise to make the book truly unique. They have played a crucial role in the development of this book. Without their invaluable contributions this book wouldn't have been possible. They have made vital efforts to compile up to date information on the varied aspects of this subject to make this book a valuable addition to the collection of many professionals and students.

This book was conceptualized with the vision of imparting up-to-date information and advanced data in this field. To ensure the same, a matchless editorial board was set up. Every individual on the board went through rigorous rounds of assessment to prove their worth. After which they invested a large part of their time researching and compiling the most relevant data for our readers.

The editorial board has been involved in producing this book since its inception. They have spent rigorous hours researching and exploring the diverse topics which have resulted in the successful publishing of this book. They have passed on their knowledge of decades through this book. To expedite this challenging task, the publisher supported the team at every step. A small team of assistant editors was also appointed to further simplify the editing procedure and attain best results for the readers.

Apart from the editorial board, the designing team has also invested a significant amount of their time in understanding the subject and creating the most relevant covers. They scrutinized every image to scout for the most suitable representation of the subject and create an appropriate cover for the book.

The publishing team has been an ardent support to the editorial, designing and production team. Their endless efforts to recruit the best for this project, has resulted in the accomplishment of this book. They are a veteran in the field of academics and their pool of knowledge is as vast as their experience in printing. Their expertise and guidance has proved useful at every step. Their uncompromising quality standards have made this book an exceptional effort. Their encouragement from time to time has been an inspiration for everyone.

The publisher and the editorial board hope that this book will prove to be a valuable piece of knowledge for researchers, students, practitioners and scholars across the globe.

List of Contributors

N. Mahmoud
National Center for Research, Khartoum, Sudan

M. S. Elhakeem, A. H. Abdallah and M. Kordofani
Department of Botany, Faculty of Science, Khartoum University, Sudan

A. M. M. Al-Naggar and R. Shabana
Department of Agronomy, Faculty of Agriculture, Cairo University, Giza, Egypt

M. M. Abd El-Aleem and Zainab El-Rashidy
Wheat Research Department, FCRI, Agricultural Research Centre (ARC), Giza, Egypt

Karima Bouhafa, Amal Hadiddou, Khadija Bouchoufi and Abdelkader Sdouq
Regional Agricultural Research Center, P.O.Box 578, Meknes, Morocco

Karima Bouhafa and Lhoussaine Moughli
Department of Soil Science, Hassan II Institute of Agronomy and Veterinary Medicine, P.O.Box 6202, Rabats-Instituts, Rabat, Morocco

M. A. Abul-Soud, M. S. A. Emam, A. M. H. Hawash, M. H. Mohammed and Z. Y. Maharik
Central Laboratory for Agricultural Climate, Agricultural Research Centre, Giza, Egypt

Olufunke O. Olayode and Theodora N. Osuji
Department of Forestry, Wildlife and Fisheries Management, Ekiti State University, Ado-Ekiti, Nigeria

E. M. Muindi
Department of Crop Science, Pwani University, Kenya

E. Semu, J. P. Mrema and P. W. Mtakwa
Department of Soil Science, Sokoine University of Agriculture, Tanzania

C. K. Gachene
Department of Land Resource Management, University of Nairobi, Kenya

Fidèle Tonalta Ngaryo
Department of Biology, University Adam Barka of Abéché, Faculty of Sciences and Technology, B.P 1173, Abéché, Tchad

Ampa-Kande Badiatte and Léonard-Élie Akpo
Department of Crops Sciences, University Cheikh Anta Diop, Faculty of Sciences and Technology, B.P 5005, Dakar, Sénégal

Venceslas Goudiaby
Département des Sciences Biologiques, Université du Québec à Montréal, Case Postale 8888, succursale Centre-ville, Montréal (Québec) H3C 3P8, Canada
Wageningen University, Forest Ecology and Forest Management Group, P.O. Box 47 6700AA Wageningen, The Netherlands

Azza Ibrahim Ebid
Department of Biology, Faculty of Science, Princess Nora bint Abdulrahman University, Saudi Arabia

A. O. Omonona
Department of Wildlife and Ecotourism Management, Faculty of Agriculture and Forestry, University of Ibadan, Nigeria

T. A. Jarikre
Department of Veterinary Pathology, Faculty of Veterinary Medicine, University of Ibadan, Nigeria

J. C. Nwite and S. O. Nwafor
Department of Crop Production Technology, Federal College of Agriculture, Ishiagu, Ebonyi State, Nigeria

B. E. Eneruvie
Department of Agricultural Technology, Akanu-Ibiam Federal Polytechnic, Unwana, Afikpo, Ebonyi State, Nigeria

G. M. B. Bohm
Departamento de Gestão Ambiental, Instituto Federal Sul-rio-grandense- IFSul, Praça 20 de setembro 455, Pelotas, 96015-360, RS, Brazil

L. Amarante
Departamento de Bioquímica, Faculdade de Agronomia Eliseu Maciel, Universidade Federal de Pelotas - UFPel, Campus Universitário, Capão do Leão, 96010-900, RS, Brazil

E. M. Bohm
Gestão Ambiental, Instituto Federal Sul-rio-grandense-IFSul, Praça 20 de Setembro 455, Pelotas, 96015-360, RS, Brazil

C. V. Rombaldi
Departamento de Ciências e Tecnologia Agroindustrial, Faculdade de Agronomia Eliseu Maciel, Universidade Federal de Pelotas - UFPel, Campus Universitário, Capão do Leão, 96010-900, RS, Brazil

M. I. Genovese
Departamento de Alimentos e Nutrição Experimental, Faculdade de Ciências Farmacêuticas da USP, São Paulo, Brazil

Vincent K. Payne, Ndah Germaine, Fusi Ngwa, Cedric Yamssi, Leonelle Megwi and Mpoame Mbida
Department of Animal Biology, Faculty of Science, University of Dschang, P.O.Box 067 Dschang, Cameroon

Fonteh A. Florence
Department of Animal Production, Faculty of Agriculture, University of Dschang, P.O.Box 222 Dschang, Cameroon

Bilong Bilong C. Felix
Department of Animal Biology and Physiology, Faculty of Science, University of Yaoundé 1, P.O.Box 812, Yaoundé, Cameroon

Ngonda Saasa and Ethel M'kandawire
Department of Disease Control, School of Veterinary Medicine, University of Zambia, P.O.Box 32379, Lusaka 10101, Zambia

King Shimumbo Nalubamba and Joyce Siwila
Department of Clinical Studies, School of Veterinary Medicine, University of Zambia, P.O.Box 32379, Lusaka 10101, Zambia

Walter Muleya
Department of Biomedical Sciences, University of Zambia, School of Veterinary Medicine, P.O.Box 32379, Lusaka 10101, Zambia

Milton Mwenda
Department of Veterinary Services, Ministry of Agriculture and Livestock, Mwandi 10101, Zambia

Youki Itoh
Laboratory of Wildlife Ecology, Department of Environmental Symbiotic Science, College of Agriculture, Food and Environmental Sciences, Rakuno Gakuen University, Japan

O. S. Olabode and A. O. Sangodele
Department of Crop Production and Soil Science, Ladoke Akintola University of Technology, Ogbomoso, Oyo State, Nigeria

F. A. Akinpelu
Department of Environmental and Crop Protection, Ladoke Akintola University of Technology, Ogbomoso, Oyo State, Nigeria

M. A. Al-Sulaiman and A. M. Aboukarima
Community College, Huraimla, Shaqra University, P.O.Box 300, Huraimla 11962, Saudi Arabia

A. M. Aboukarima
Agricultural Engineering Research Institute, Agricultural Research Centre, Egypt

M. I. Rumainul and F. M. Aminuzzaman
Department of Plant Pathology, Faculty of Agriculture, Sher-e-Bangla Agricultural University, Sher-e-Bangla Nagar, Dhaka-1207, Bangladesh

H. Bello and A. O. Musa
Department of Biological Sciences, Ahmadu Bello University, Zaria, Nigeria

Md. Sariful Islam, Bikash Chandra Sarker and Najim Uddin
Department of Agricultural Chemistry, Hajee Mohammad Danesh Science and Technology University, Dinajpur-5200, Bangladesh

K. M. Manirul Islam and Shah Md. Shakil Ahmed
Faculty of Agriculture, Hajee Mohammad Danesh Science and Technology University, Dinajpur-5200, Bangladesh

M. A. Abua
Department of Geography and Environmental Science, University of Calabar, Calabar, Nigeria

S. W. Ashua
Department of Geography, Cross River State College of Education, Akamkpa, P.M.B. 1171, Calabar, Nigeria

N. M. Rolli and T. P. Giraddi
BLDEA's Degree College, Jamkhandi, 587301, Karnataka, India

S. B. Gadi
Department of Botany, JSS College, Dharwad, 580004, Karnataka, India

E. D. Oruonye, B. B. Ezekiel, H. G. Atiku, E. Baba and N. I. Musa
Department of Geography, Taraba State University, P.M.B. 1167, Jalingo, Taraba State, Nigeria

M. Rahman and M. M. Mian
School of Environmental Science, Bangladesh Agricultural University, Mymensingh, 2202, Bangladesh

M. M. Mian
School of Urban Construction and Environmental Engineering, Chongqing University, 174 Shazhengjie, Shapingba, Chongqing, 400044, China

V. E. Nethaji Mariappan, A. Hari Nivas and S. Parthiban
Centre for Remote Sensing and Geoinformatics, Sathyabama University, Rajiv Gandhi Road, Jeppiaar Nagar, Chennai - 600 119, India

T. Kanmani
M. E. Environmental Engineering, Sathyabama University, Rajiv Gandhi Road, Jeppiaar Nagar, Chennai - 600 119, India

Muhammad Rusdy
Department of Forage Crops and Grassland Management, Faculty of Animal Science, Hasanuddin University, Makassar, Indonesia

Damoun Razmjoue
Department of Range and Watershed Management, Faculty of Natural Resources, Behbahan Khatam Alanbia University of Technology, Behbahan, Iran

Zahra Zarei
The Range management, Behbahan Khatam Alanbia University of Technology, Behbahan, Iran

Md. Rabiul Islam and Md. Abdul Baten
Department of Environmental Science, Bangladesh Agricultural University, Mymensingh-2202, Bangladesh

Md. Nur Hossain and Md. Fakruddin
Institute of Food Science and Technology (IFST), BCSIR, Dhaka-1205, Bangladesh

Khusnade Rabbi
Ulipur Government College, Ulipur, Kurigram, Bangladesh